ADVANCES IN CATALYSIS
AND RELATED SUBJECTS

VOLUME II

ADVANCES IN CATALYSIS

AND RELATED SUBJECTS

VOLUME II

EDITED BY

W. G. FRANKENBURG
Lancaster, Pa.

V. I. KOMAREWSKY
Chicago, Ill.

E. K. RIDEAL
London, England

1950

ACADEMIC PRESS INC., PUBLISHERS
NEW YORK, N. Y.

ACADEMIC PRESS INC.
125 East 23rd Street
NEW YORK 10, N. Y.

PRINTED IN THE UNITED STATES OF AMERICA

CONTRIBUTORS TO VOLUME II

OTTO BEECK, *Shell Development Company, Emeryville, California*

V. N. IPATIEFF, *Universal Oil Products Company, Riverside, Illinois*

T. H. JAMES, *Research Laboratories, Eastman Kodak Co., Rochester, New York*

CHARLES KEMBALL, *Department of Physical Chemistry, The University, Cambridge, England*

ALWIN MITTASCH, *Heidelberg, Baden, Germany*

LOUIS SCHMERLING, *Universal Oil Products Company, Riverside, Illinois*

GEORGE-MARIA SCHWAB, *Department of Inorganic, Physical and Catalytic Chemistry, Institute Nicolaos Canellopoulos, Piraeus, Greece*

FREDERICK SEITZ, *Carnegie Institute of Technology, Pittsburgh, Pennsylvania*

J. H. SIMONS, *Fluorine Laboratories, The Pennsylvania State College, State College, Pennsylvania*

PREFACE

With this second volume of *Advances in Catalysis,* the editors have continued their efforts to present the many facets of the catalytic process. A number of highly qualified men have contributed to this volume. From the theoretical treatments of elementary processes between molecules reacting at solid surfaces to the technically important actions of fluoride catalysts, and to catalytic polymerizations of olefins, the reader will become acquainted with manifold ideas and with some typical experimental results relating to catalytic phenomena. Our lack of a complete understanding of catalytic action, and our corresponding inability to "predict" the best way of achieving a desired catalytic reaction, make it indispensable for everyone working in this direction to familiarize himself with the experience of others, even if such experience was gathered in remote sectors of this vast field.

W. G. Frankenburg
V. I. Komarewsky
E. K. Rideal

December, 1949

CONTENTS

Catalytic Phenomena Related to Photographic Development

By T. H. James, *Research Laboratories, Eastman Kodak Co., Rochester, New York*

Catalysis and the Adsorption of Hydrogen on Metal Catalysts

By Otto Beeck, *Shell Development Company, Emeryville, California*

Hydrogen Fluoride Catalysis

By J. H. Simons, *Fluorine Laboratories, The Pennsylvania State College, State College, Pennsylvania*

Entropy of Adsorption

By Charles Kemball, *Department of Physical Chemistry, The University, Cambridge, England*

The Fundamental Principles of Catalytic Activity

FREDERICK SEITZ

Carnegie Institute of Technology, Pittsburgh, Pennsylvania

CONTENTS

INTRODUCTION

One of the great triumphs of the development of atomic theory is the framework it has given us with which to visualize chemical reactions in terms of moving atoms. This framework, which is based upon the results of combining quantum mechanics and statistical mechanics, provides us with quantitative answers only in the simplest cases; that is, for problems which usually are only of academic interest. In most cases we must be content, at least for the present, with qualitative pictures. This situation may be altered in the course of time if suitable computing techniques are developed, for the great obstacle to quantitative application of the fundamental theory arises from the difficulty encountered in solving the equations with ordinary mathematical methods, even when the simplest molecules are involved. In spite of this obstacle, the approach to the understanding of chemical phenomena which is based on those concepts which may be gleaned from the fundamental theory represents the only satisfactory pathway to follow at the present time. For it is believed that the equations of quantum mechanics and statistical mechanics contain within them the "truth" of chemistry in the sense that these equations would tell us exactly the way in which nature behaves, were we clever enough to solve them. If, in the future, the improvement of computational techniques permits us to handle these equations as accurately as we desire, the new treatment will probably not result in any radical changes of the concepts which have been gleaned from the theory thus far by the relatively crude methods available at the present time. Instead, it is most probable that we will be given more quantitative data in places where we have empirical information or none at all at the present time. The qualitative framework will probably remain almost unchanged.

It will be convenient in the following discussion to focus our attention on the heterogeneous solid catalyst. Systems of this type exhibit most of the characteristics that typify catalysts and hence serve as useful reference materials.

The field of catalysis represents one of the most intricate branches of chemistry. The typical catalyst is useful not merely because of the average or bulk properties of the material in it, but rather because of the combination of the bulk properties and the very specific form in which the particular catalytic specimen appears. The ultimate development of catalytic theory in quantitative form will require an understanding not only of the average valence characteristics of the atoms in it, but of the unusual arrangements and electronic structures which are possible; for example, of the unusual arrangements of atoms that are possible near the surface in a crystalline catalyst. In view of this underlying intricacy we must expect that the detailed quantitative understanding of catalytic activity will take place slowly, and will be among the last branches of chemistry to be fully clarified.

The following discussion of catalytic activity can be viewed only as a preliminary attempt to describe the important features of the field in terms of the language which has arisen out of the development of atomic theory, and which it is felt must eventually become the *lingua franca* of all of chemistry if the tradition which desires to describe chemical phenomena in terms of atomic behavior is continued. This language has arisen out of the pioneer work of Wigner, Pelzer, Eyring, Polanyi (1), and their many associates and coworkers during the very fruitful period of activity in the vicinity of 1930 when the evolution of atomic mechanics first made it possible to discuss chemical phenomena, particularly rate processes, from a unified and selfconsistent viewpoint.

II. The Energy Surface

Each atom consists of a central, positively charged nucleus which contains practically all of the mass of the atom, and which is surrounded by a cloud of electrons. From the standpoint of chemistry, the internal energy of the atom may be measured in terms of the energy of the electrons.

The energy of the electrons may be broken into component parts. Those which raise the total energy of the atomic molecular system, and hence raise the energy content of one gram atom or one mol of the substance, will be termed positive, whereas those which lower the energy content will be termed negative. The energy of the electron in the atoms is composed of three terms:

(1) The kinetic energy of the electrons, which is always a positive quantity.

(2) The energy of the coulomb attraction between the negatively charged electrons and the positively charged nucleus. If the electrons are regarded as having zero energy when at an infinite distance from the nucleus, this attractive energy between the electrons and nucleus is always a negative quantity.

(3) The energy of the repulsive coulomb interaction between the similarly charged electrons. This energy is always positive if chosen to be zero for two infinitely separated electrons.

If the atom is not at rest, the kinetic energy of motion of the atom as a whole must be included in the total energy. For the moment we shall assume that it is at rest.

Isolated atoms or ions are stable only because the sum of these three energy terms is negative, that is, only because the attraction of the nucleus for the electron cloud overbalances the kinetic energy and mutual repulsive forces of the electrons. A singly charged negative halogen ion, such as Cl^- can be stable even in a vacuum because the attraction of the nucleus for the extra electron is great enough to compensate for the additional positive energy terms. On the other hand, this attraction is not large enough to permit the addition of a second extra electron.

One of the most fundamental concepts of atomic theory is that the observed, or permitted, energy states of the electrons in atoms are discrete, that is, the total energy of the electrons cannot take arbitrary continuous values but is restricted to fixed discrete numbers, characteristic of the atom. This feature of atomic systems is automatically contained in quantum mechanics and cannot be understood on the basis of classical or Newtonian mechanics. If a group of atoms which are isolated from one another are placed in a region of sufficiently low temperature, they will eventually, in the equilibrium state, find themselves in the lowest energy level, regardless of the level they were in initially. In the simplest case the excess energy will be radiated away. From the standpoint of thermodynamics, this attainment of the lowest energy state at equilibrium can be expressed in terms of the general principle derived from thermodynamics, that the stable state of a system confined to a fixed volume is that for which the free energy is a minimum.

$$A = E - TS \tag{1}$$

Here, E is the energy of the system; T is the temperature, which may be expressed in degrees Kelvin; and S is the entropy. If the system is at very low temperatures, the term TS in equation (1) may be neglected so that the stable state is that of lowest energy. If the temperature T is not zero, the system will not usually be in the state of lowest energy, but will be excited because of the influence of thermal agitation. Under these conditions the TS term in equation (1) has an important influence in determining the state of the system. Since S increases with the disorder of the system, the free energy in equation (1) is minimized, as the temperature is elevated, by passing to states of higher and higher disorder. In the case of the system consisting of atoms which are isolated from one another but are at thermal equilibrium with their surroundings, not all atoms will be excited in the same way at the same time when T is greater than zero; in fact, the energy of any given atom will fluctuate from time to time. The condition that (1) be a minimum describes the average state of the system, independent of these *thermal fluctuations.* The entropy S for any atomic system may be expressed in terms of the atomic characteristics by means of the equation

$$S = k \log N(E) \tag{2}$$

Here, k is Boltzmann's constant, namely $1.39 \cdot 10^{-16}$ ergs/deg.; and $N(E)$ is the number of independent ways in which the system of atoms may have the total energy E. The subject of quantum statistics is concerned with the detailed determination of the quantity $N(E)$.

When the atoms in a system become close enough to interact with one another, that is, when the system is no longer simply a group of isolated atoms, the total energy is comprised of five terms instead of three. Three of these are identical with the three described above; namely, the kinetic energy of the electrons, the energy of coulomb attraction between the electrons and the oppositely charged nuclei, and the energy of coulomb repulsion of the electrons. The fourth energy term is the energy of coulomb repulsion between the nuclei of the interacting atoms and the fifth is the kinetic energy of motion of the relatively massive nuclei relative to one another. The fourth term is negligible as long as the atoms are widely separated from one another, but becomes appreciable as soon as the atoms are close enough to interact. The fifth energy term is, for a bound system, the counterpart of the kinetic energy of translational motion of isolated atoms. In the simplest cases this kinetic energy is associated with vibrational motion of atoms relative to one another as, for example, the vibrational motion of the atoms in a crystal lattice.

The general condition for the stability of a cluster of atoms is again that the free energy (1) be a minimum. At very low temperatures this condition, as we have seen, is synonymous with the condition that the energy be a minimum. In the absolute sense, the condition of minimum energy is usually satisfied only if all the atoms in the system form one or two condensed phases, usually crystalline, for all atoms have an attraction for one another if not pressed too closely together. However, if the temperature is not actually at the absolute zero, the system may exist as a gas of molecules in which the interatomic bonding forces are large compared with the forces between molecules. This is the situation for oxygen or nitrogen above 80°K. at atmospheric pressure. In such cases the gain in entropy associated with the transition to the vapor phase more than compensates for the loss in binding energy that accompanies the separation of molecules in the formation of the gas phase as soon as the temperature is sufficiently large that the TS term in (1) becomes important.

Consider a single molecule composed of a group of n atoms, and let us suppose for the purpose of discussion that the nuclei can be held fixed at arbitrary distances relative to one another. In the simplest case the molecule will possess two atoms, and we may restrict our attention to such a case for the moment. For each value of the interatomic distance the electrons may take on a set of discrete energy values, the counterpart of the discrete levels of isolated atoms. We shall call these *electronic energy levels*, even though the energy of coulomb repulsion of the nuclei is included. For each of these levels the electron cloud possesses a definite distribution; the energy is determined by summing the various kinetic and potential energy terms. Since the nuclei are held fixed their kinetic energy of motion will not be included in this sum. We shall focus our attention on the lowest electronic energy state because it usually represents the state of interest for chemical problems. The electronic energy of the molecule will be different for each value of the interatomic separation. Both experiment and theory show that this energy varies in the manner shown in Fig. 1 as the interatomic distance is varied. For separations so large that two atoms no longer interact, the electronic energy approaches a constant equal to the sum of the energy of the isolated atoms. For very close separations, the energy rises very rapidly because of the large repulsive force between nuclei. In general, the electronic energy is a minimum for a certain interatomic spacing, designated by r_m in the diagram.

If the two nuclei are not held fixed, but are allowed to move freely, and if there is no external source of excitation, so that the molecule can be regarded as if at the absolute zero of temperature, the separation of the atoms will eventually take up values very close to r_m. Classically, the equilibrium separation would actually be r_m, however, quantum effects cause the atoms to vibrate relative to one another about r_m even when at the absolute zero. The permitted vibrational states of the molecule are discrete, just as are the allowed energy states of the electrons in an isolated atom. In the states of higher vibrational energy the two atoms vibrate over a larger range of r than in the lower states. If the vibrational energy becomes sufficiently large, the atoms may actually fly apart.

Figure 1 shows the energy of the lowest electronic state as the interatomic distance is varied. The energy of the next lowest electronic

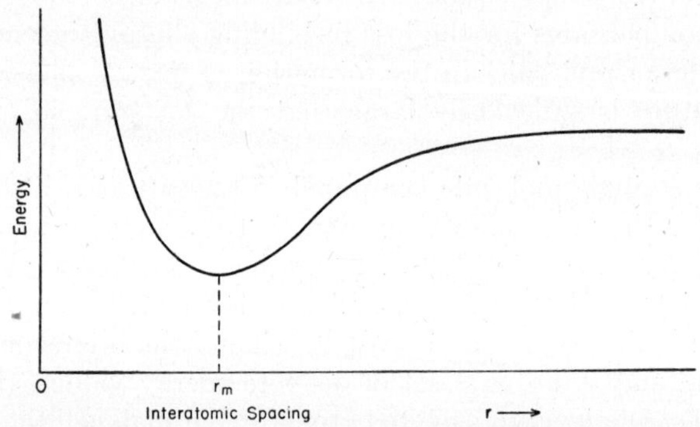

FIG. 1. The electronic energy of the lowest state of a diatomic molecule. The horizontal axis represents the interatomic distance. This energy includes the coulomb repulsion of the two nuclei as well as the kinetic and potential energy terms involving the electrons. The curve rises steeply for small interatomic distances in accordance with the coulomb repulsion of the nuclei. For large interatomic spacing the curve approaches a constant value, corresponding to the energies of the isolated constituent atoms.

state, analogous to the first excited state of a single atom, usually varies in the manner shown in Fig. 2. In most diatomic molecules the excited state is of interest only at very high temperatures, or when the molecule is irradiated with light of sufficient energy to cause a transition from the ground state to the excited state.

It should be noted that, in addition to possessing electronic and vibrational energy, a diatomic molecule may possess kinetic energy associated with the rotation and translation of the molecule as a whole. These two additional terms will be of little concern to us in most of this discussion and will be taken into account only when special need arises.

Consider now a molecule which has more than two atoms in it. The electrons will have a definite set of energy states for each of the positions the nuclei may take, just as in the diatomic case. These electronic energy levels will vary continuously as the nuclei are moved. If the molecule contains n atoms, the possible positions of all the nuclei will be described by giving $3n$ coordinates since each atom possesses three coordinates in space. Only one of these coordinates is important for determining the electronic energy in the diatomic case since the electron

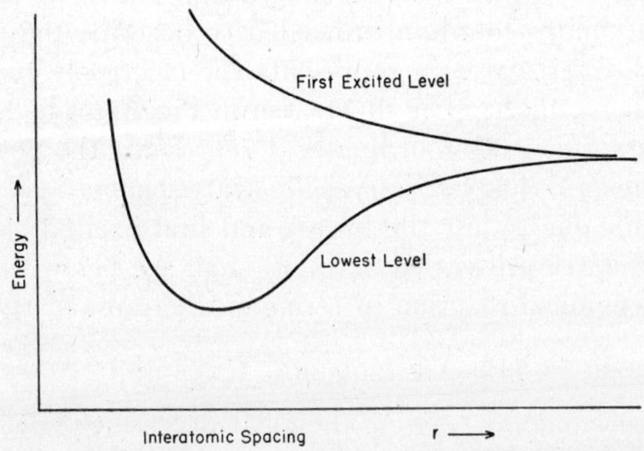

FIG. 2. This diagram is similar to Fig. 1 with the exception that two energy curves are shown. The upper curve represents the first excited electronic state of the molecule.

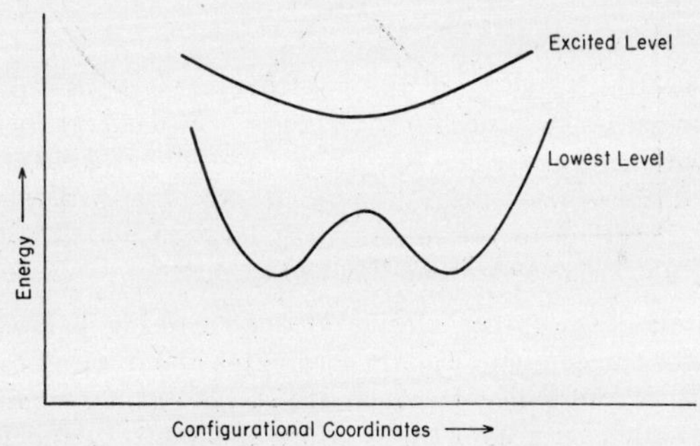

FIG. 3. A schematic diagram showing a hypothetical section through the energy surface for a polyatomic molecule. The horizontal axis represents the positional coordinates of the nuclei. Since the total number of these is $3n$, if there are n atoms in the molecule, it is clear that a multidimensional diagram would be required to represent the dependence of energy on these variables in complete detail. Two electronic levels are shown.

configuration does not change if the interatomic spacing is kept fixed. For this reason the dependence of the electronic energy upon the nuclear coordinates can be represented by a simple two-dimensional diagram of the type shown in Figs. 1 or 2. When there are more than two atoms an accurate diagram of this type can be represented only in a space of many dimensions. The exact number required is $3n - 5$ for a nonlinear molecule. The coordinates which do not need to be considered are those having to do with rotation and translation of the molecule as a whole, since these will have a relatively small influence upon the electrons. It is possible in a schematic way to preserve the simplicity of Figs. 1 and 2 by using a two-dimensional diagram in which the horizontal axis symbolizes the many coordinates needed to describe the position of the nuclei, and the vertical axis represents the electronic energy. A schematic diagram of this type is illustrated in Fig. 3. The horizontal axis, labelled "configurational coordinates," represents the positional coordinates of the nuclei. The two curves illustrate schematically the variation of the electronic energy for the lowest and first excited electronic states as these coordinates are varied. As we shall see below, it is possible to describe any chemical reaction in terms of diagrams of this type.

III. An Isomeric Reaction

The simplest general type of chemical process to consider is an isomeric reaction in which a molecule undergoes an internal rearrangement

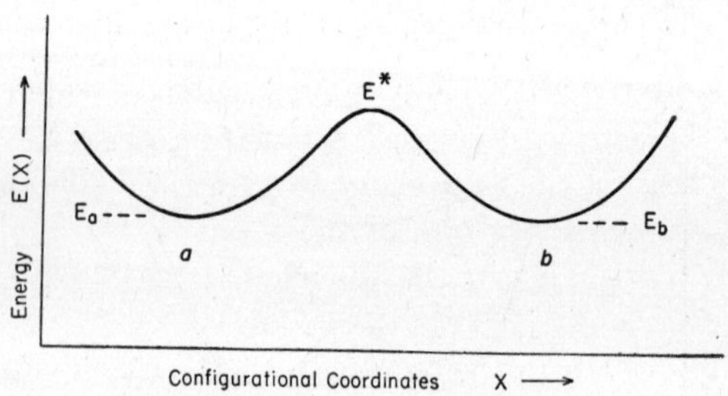

Fig. 4. A section of the energy surface, analogous to Fig. 3, for an isomeric transition. The values of the configurational coordinates about a and b correspond to the two isomeric forms of the molecule which are stable against small atomic displacements. The electronic energies for the two forms, namely E_a and E_b, are assumed to be the same, although this need not necessarily be the case. This section of the energy surface is assumed to pass through the saddle point in the potential range separating the two minima a and b. The energy of the saddle point is E^*.

from one structure to another as, for example, when a group on a hydrocarbon chain shifts its position relative to the chain. We shall assume that the two isomeric forms have essentially the same energy, although the discussion given below would be valid even if they do not. In this case, the electronic energy curve pertinent for the reaction can be represented in the form shown in Fig. 4. This shows the energy for the lowest electronic state as the nuclei are rearranged from the positions going with one of the isomeric forms to the other. The electronic energy has minima for the two coordinate arrangements (a and b in the diagram) which correspond to the two isomeric forms. In each form the nuclear coordinates undergo oscillations about the values corresponding to the minima. The higher the temperature, the larger the amplitude of this oscillation will be. It is to be noted that the two isomeric forms have distinct chemical existence because the two arrangements of nuclei are stable against small displacements of the atoms from the equilibrium positions; that is, the two arrangements correspond to relatively stable structures. At least in principle, each could be obtained and studied experimentally at sufficiently low temperatures.

Let us now imagine the change in electronic energy of the molecule as we move the nuclei from one equilibrium position to the other. We shall imagine this being done so slowly that the electrons maintain their equilibrium at each stage of the process, as is assumed in the construction of diagrams such as Figs. 1 to 4. The energy will vary in different ways for each route we take in passing between the two minima. Figuratively speaking, the variation in energy is similar to the variation in height as one climbs over a mountain range separating two valleys. In any such journey the paths which have particular practical interest are those which pass over the passes or saddles in the mountain range. There will be corresponding paths in our energy surface which pass over the lowest point in the energy "range" separating the two minima. The energy curve connecting the two minima in Fig. 4 is assumed to represent the variation of energy along one of these paths, the highest energy on the path, designated by E^* in the diagram, is the height of the saddle point.

The importance of the paths passing over the saddle point in chemical phenomena is based upon the fact that reactions which proceed spontaneously as a result of thermal stimulation, that is, all chemical reactions which proceed spontaneously and reversibly at a finite temperature, follow such paths. This principle rests upon the following circumstances: Consider a molecule (or more generally a system) whose configurational coordinates are initially at the position of a minimum on the surface. The probability that this system will be found at a point on the surrounding energy surface that is at a height ΔE above the minimum, decreases as ΔE increases.

In other words, it is much more probable that the system will pass from one energy minimum to another over a saddle point than along any other route, which passes through a point of higher energy, even though alternate routes are not strictly forbidden. According to statistical mechanics, the relative probability that the molecule will find itself at a given point X on the energy surface is given by the "Boltzmann factor"

$$P(X) = e^{-A(X)/RT} \tag{3}$$

where $A(X)$ is the free energy for a mol of molecules having the configurational coordinates X, R is the gas constant and T is the absolute temperature. $A(X)$ usually increases as X rises on the energy surface near a minimum point so that the relative probability (3) decreases. The ratio of the probability that the system will be at the saddle point having energy E^* and entropy S^* and the probability that it will be at the energy minimum a having energy E_a and entropy S_a is, according to (3)

$$P^*/P_a = e^{-(E^*-TS^*)/RT}e^{-(E_a-S_aT/RT} = e^{-[(E^*-E_a)-T(S^*-S_a)]/RT} \quad (4)$$

In most practical cases the quantity $T(S^* - S_a)$ is sufficiently small compared with $E - E_a$ that the magnitude of this ratio is determined by the factor

$$e^{-(E^*-E_a)/RT} \quad (5)$$

The quantity

$$Q_a = E^* - E_a \quad (6)$$

is called the *activation energy* for the transition over the saddle point from the configuration a. In the case under consideration this transition will lead to the final state b. By analogy the quantity

$$Q_b = E^* - E_b \quad (7)$$

is the activation energy for the transition from b to a over the saddle point.

It is conceivable that the quantity $T(S^* - S_a)$ in (4) will have a more important effect than Q_a in determining the likelihood that the system will make the transition from a to b. This will be the case, for example, if the activation energy is very small. Experience shows, however, that Q_a is the most important quantity in the great majority of practical cases.

Whenever the system finds itself at the saddle point it will pass into one of the valleys on either side in a short time. The rate at which the system moves through the saddle point can be determined in a relatively simple way because the energy of the system is nearly constant for a short distance in the direction of the path through the saddle point. The frequency factor which determines the rate of flow of systems which have reached the barrier is kT/h, where k is Boltzmann's constant, T is the absolute temperature and h is Planck's constant.

In our example of an isomeric transition then, we may conclude that the speed with which the reaction takes place in either direction is determined by the activation energies Q_a and Q_b. The larger these quantities are the slower the reaction in either direction will be. In the case shown in Fig. 4, Q_a and Q_b are assumed to be equal since E_a and E_b are taken as equal. This need not be the case and usually will not, unless the two isomeric states happen to be symmetrically identical as, for example, when a group jumps between two identical positions on a benzene ring.

It is interesting to note that the relative probability of finding the system with the configuration a and b in the equilibrium state is, according to (3)

$$P_a/P_b = e^{-(A_a-A_b)/RT} \tag{8}$$

This shows, in accordance with thermodynamics, that the state having lowest free energy is preferred. For example, (8) is large compared with unity if A_a is small compared with A_b.

It is a necessary condition for equilibrium that the number of molecules which make a transition in one direction over the saddle point in unit time just equal the number making the transition in the opposite direction. There will be a preponderance of transitions in a given direction only if the initial system starts with a nonequilibrium distribution, that is, with a distribution in which the ratio of numbers of molecules in the states a and b is different from that given by (8). If the ratio of a to b molecules is larger than that given by the ratio (8), there will be a preponderance of transitions from a to b until the equilibrium ratio is established, and vice versa. If the molecules are sufficiently separated that they do not interact with one another when at the average spacing, the likelihood that any given molecule in, say, state a will make the transition to state b is independent of the total number in state a or state b. The attainment of equilibrium will come about purely through the action of the principle of mass action.

It is to be emphasized that since the saddle point does not represent a stable configuration, because there are configurations arbitrarily close to it that have lower energy, only a negligible fraction of the molecules in an assembly will have configurations near the saddle point at any given time. Molecules whose configuration coordinates are near the saddle point are said to be in the "activated state." The relative number in this state will be small compared with the number near points such as a and b where the energy surface has minima.

If the isomeric transition is taking place under conditions where the molecules cannot interact with one another or with the walls of the vessel in which they are contained other than in such a way as to transfer energy or momentum, the activation energy Q_a and Q_b will be determined entirely by the internal characteristics of the molecule and will be what might be termed "natural constants" of the system, analogous to the heat of formation. Thus under these conditions the reaction rates for the isomeric transitions will be fixed by the internal constitution of the molecules. At a given temperature the reaction rate can be changed only by interposing some external agent which *has the effect of altering the activation energy for the reaction.* This is the essential role of a catalyst in any chemical reaction. In accordance with conventional terminology it is also expected that the catalyst will not be consumed in the reaction.

In the simplest case the catalyst will react with the activated molecule in such a way that the energy surface is distorted, as shown in Fig. 5. In this case the molecules which are in the state *a* or *b* are practically unaffected by the catalyst, as is indicated by the fact that the energy surface near the two minima is undistorted. On the other hand, the

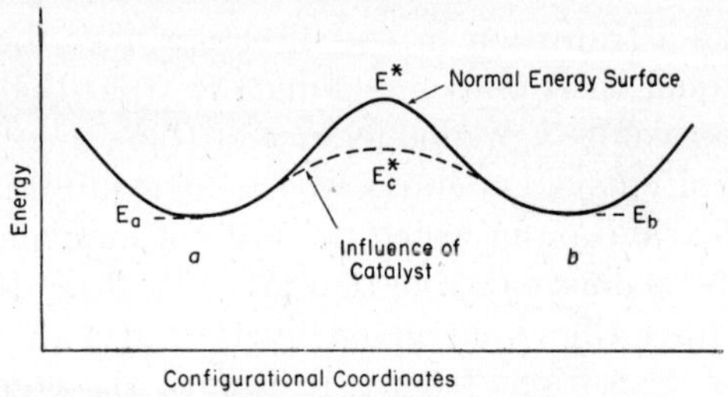

FIG. 5. A schematic representation of the influence of a catalyst on the energy surface in the ideal case. The full curve represents the normal energy curve, whereas the dotted curve represents the energy curve in the presence of the catalyst. Both curves are assumed to be sections through the saddle points. In this example the portions of the energy surface corresponding to the isomers *a* and *b* are practically unaffected; however, the activated molecules are strongly attracted so that the activation energy for the isomeric transition is lowered. The activated molecule is still less stable than the isomers.

energy E^* is depressed because the activated molecules are made relatively less unstable in the presence of the catalyst. In this case the isomeric transition would be speeded by the presence of the catalyst.

The ideal situation described in the last paragraph is met whenever one is dealing with a solid catalyst on the surface of which the normal isomeric molecules are weakly adsorbed. In such a case the molecules near configurations *a* and *b* are influenced relatively little by the presence of the solid. However, since the solid has a catalytic effect, the activated molecules must interact relatively strongly so that the activated state is attained more easily in its presence. It is conceivable that a given solid would interact so strongly with the activated complex that this configuration actually would be stabilized relative to the isomers *a* and *b* (see Fig. 6). In this case the solid surface would become covered with a stable layer of molecules having the configuration associated with the activated state, and it would not act as a catalyst.

It is also conceivable that a solid would act as an effective catalyst even though it interacts fairly strongly with the normal isomers (Fig. 7). In this case the normal molecules would be strongly adsorbed on the surface; however, the activation energy would still be lower on the surface than in the gas phase if the situation shown in Fig. 7 prevails,

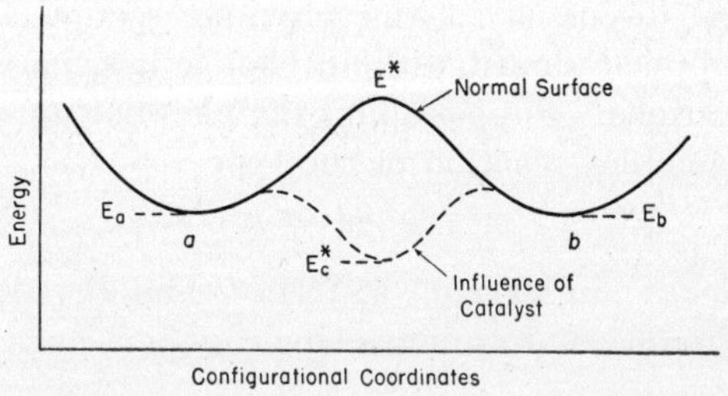

FIG. 6. This diagram is similar to Fig. 5, with the exception that the activated molecule reacts so strongly with the catalyst that the energy E_c* corresponding to the activated molecule is lower than E_a and E_b. In this case the activated configuration is stabilized on the surface of the catalyst. The molecules may become attached so strongly that the surface is poisoned.

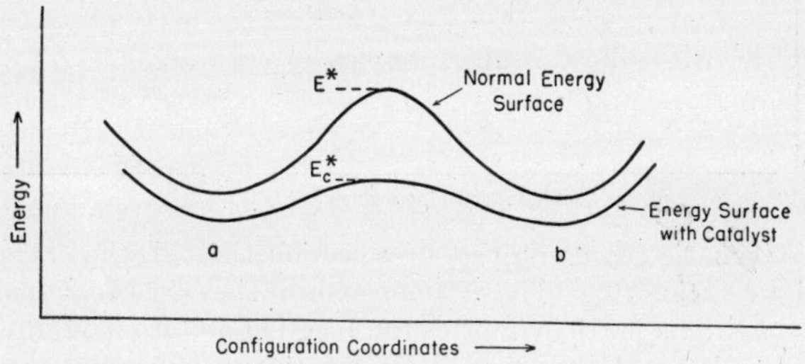

FIG. 7. This diagram resembles Fig. 5. However, the energy surface is lowered for the isomeric configurations *a* and *b* as well as for the activated molecule in the presence of the catalyst. Hence the normal molecules will be more strongly adsorbed than in the case shown in Fig. 5. The activation energy is lowered by a larger amount than the energy of the normal molecules, so that the surface still has a conventional catalytic action.

and the adsorbed molecules would undergo the isomeric transition more rapidly than isolated molecules in the gas phase. A catalyst which has the effect shown in Fig. 7 would evidently have disadvantages because the surface would have a tendency to become covered with tightly adsorbed molecules which would prevent others from enjoying the benefits of being near the surface.

It is clear that the catalytic agent need not be a solid, for any additional agent, such as an admixture of foreign gas, which will interact more strongly with the activated configuration than with the normal configurations, could lower the activation energy and hence speed up the reaction, provided it does not interact so strongly that the molecules attached to it become permanently bound and hence "poison" its action.

IV. More Complex Reaction

The principles discussed in the previous section for an isomeric reaction can be applied almost without change in general conception to more complex reactions. Suppose, for example, that we are considering a somewhat more complex reaction of the type

$$A + B \rightleftarrows C + D \tag{9}$$

in which molecules A and B react to form C and D. The molecules A and B can be regarded as forming a single system even when they are

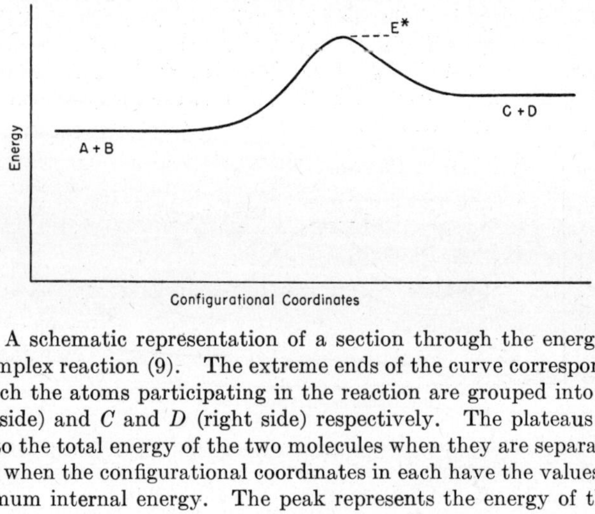

FIG. 8. A schematic representation of a section through the energy surface for the more complex reaction (9). The extreme ends of the curve correspond to the two cases in which the atoms participating in the reaction are grouped into molecules A and B (left side) and C and D (right side) respectively. The plateaus in each case correspond to the total energy of the two molecules when they are separated from one another and when the configurational coordinates in each have the values corresponding to minimum internal energy. The peak represents the energy of the system of atoms along a path passing through the saddle point for the change from $A + B$ to $C + D$.

not in intimate contact, and a single energy surface can be used to represent the energy of the pair. This surface will possess a broad plateau for those values of the coordinates of the nuclei for which the molecules are not sufficiently close to interact. However, it will have all the characteristics of the surface for a single molecule in regions where the pair do interact.

The corresponding energy surface of the pair of molecules C and D will join continuously to that for A and B (Fig. 8) since the same atoms enter into both pairs of molecules. Whether these atoms are joined together to form $A + B$ or $C + D$ depends upon the region of configuration space occupied by the nuclear coordinates. In general, the two regions will be separated by an energy "mountain range" over which the system must pass if a reaction is to take place. The plateau on one side of this range corresponds to the state in which one has molecules A and B; the plateau on the other side corresponds to the combination C and D. One of the plateaus corresponds to configuration a in the case of the polymeric arrangement described in the preceding paragraph and the other corresponds to the configuration b. In making the transition from one plateau to the other, the system will usually pass over the lowest saddle in the range. The height of this saddle relative to the plateaus determines the activation energies for the forward and backward processes in the reaction (9). The speeds with which these reactions occur will depend primarily on these activation energies in the normal case. As in the example of the isomeric reaction the rate may be speeded by introducing an agent, such as a solid surface, which lowers the height of the saddle point relative to the plateaus.

It is evident that the same qualitative reasoning can be applied to any reaction which can be written in the form of a conventional chemical equation: The reactants on both sides correspond to regions of configuration space which are separated by an energy barrier. The differences between various reactions lie in the atomic coordinates involved and the intricacy of the potential surface.

There is one important difference between the case of an isomeric reaction and a more general one of the type (9). In any nonisomeric reaction there is an essential chemical difference between the reactants and the products of chemical reaction. It is possible that one of the reactants or one of the products will be strongly adsorbed on the surface. Since they differ chemically it is not necessary that both reactants and products be adsorbed equally, as in the case of an isomeric reaction. Should one of the reaction products be adsorbed strongly, the reaction may be blocked because the active regions become sheathed with an obstructing layer. On the other hand, the reaction need not be blocked if one of the reactants is strongly adsorbed since the reaction itself will tend to remove this obstruction. This topic has been made the subject of detailed discussion by Frankenburg (2).

V. Solid Catalysts

It is interesting to speculate on the conditions which make a given solid a good or poor catalyst. According to the discussion of the previous sections the characteristics of a good catalyst when viewed from the standpoint of fundamentals of chemical binding, are as follows:

a) The catalyst should lower the activation energy associated with the reaction of interest. On the other hand, it should not lower it so completely that the activated state becomes more stable than the normal state. In the latter case the surface may become blocked.

b) The normal molecules should not be so strongly adsorbed on the surface that the initial layer of molecules which reach the surface stay and prevent others from reacting there. On the other hand, they should be weakly absorbed to permit exchange between unreacted species.

Solids may be classified into four types which form a suitable basis for our discussion, namely:

1) Valence crystals
2) Metals
3) Ionic crystals
4) Molecular crystals

Valence crystals, of which diamond is the prototype, are characterized by strong directional bonding. The atoms in the solid prefer to take up very definite positions relative to one another; the energy of the lattice is very sensitive to the relative angular orinetation of the atoms. The bonding forces in these materials are, in fact, very similar to those between the atoms in ideal organic molecules, which usually lie in the short rows of the periodic chart. The surface layer of atoms in a valence crystal which terminates abruptly at one atomic plane would be highly unsaturated because the adjoining layer of atoms to which the atoms in the surface layer would normally be bound is missing. A surface layer of this type, if it could be obtained, would be highly reactive and probably would form strong bonds with a molecule that came near the surface, particularly if the internal forces in the molecule were not as strong as those within the crystal. In fact, a freshly cleaved surface of diamond would probably react strongly with molecules that arrive there from the surrounding atmosphere and would become saturated rapidly because of the addition of a tightly bound layer of foreign material. For this reason it probably would be very difficult to obtain an ideally clean surface of a valence crystal for catalytic investigations. It follows, however, that a surface of this type would probably be a poor catalyst because a layer of the reacting molecules and reaction products would become tightly attached to it.

There are two general classes of metals. First, there are those which may be called simple metals in which the *d* and *f* shells are either completely filled or completely empty. The alkali and alkaline earth metals and metals such as copper, silver, and gold are of this type. The second group are the transition metals which have partly filled *d* or *f* shells, and the metals of the iron group and the rare earths belong in this family. The bonding forces in the first class are most completely understood. In these cases the forces between atoms are almost completely nondirectional—the energy of the lattice is more sensitive to the average interatomic spacing, that is, the average atomic volume, than to the detailed structure. In the second class in which there are incompletely filled inner shells, the forces between atoms are more strongly directional because of the bonding characteristics of the inner shell electrons; however, the forces are not as strongly directional as in ideal valence compounds.

A clean surface of a simple metal will have unsaturated forces and hence be capable of interacting with molecules. However, since the forces are not highly directional the bonding energy is probably almost independent of the configurational coordinates within the molecule. Thus, in the ideal case the activated configuration does not interact appreciably more strongly than a normal configuration, and the surface has relatively little catalytic effect. The bonding energy between the surface and a molecule may be large, particularly if the metal is highly electropositive and the molecule either is electronegative or has a dipole group which may be strongly attracted to the surface. In this case the molecule will become almost permanently attached to the surface.

On the other hand a transition metal, particularly if it is not strongly electropositive, may react more strongly with the activated complex than with the normal structure of a molecule. This will be the case if the interatomic spacing of the activated arrangement is just such as to provide the maximum binding energy with the partly directional forces of the atoms in the metal. It is clear that this behavior can be expected only when the spacing of the metal atoms and of the atoms in the activated molecule satisfy the proper relation. In other words, the catalytic action of the metal will depend in a highly specific manner both on the metal surface and on the molecules involved in the reaction. If one of the prominent crystal faces of the metal is effective in catalyzing a given reaction, the metal will normally be highly efficient per unit exposed area. On the other hand, if the most effective arrangement of atoms in the metal is that occurring on an unusual crystal facet or at a region of the surface where an irregularity occurs, the metal may normally be inefficient. In the second case it may prove possible to raise the catalytic efficiency by preparing the material in such a way that the unusual crystal facet or the irregularities occupy a large fraction of the exposed surface area.

It is clear from the preceding discussion that the transition metals have the ideal characteristics for catalytic activity, both because of their pronounced but relatively moderate directional binding, and because they are not strongly electropositive. It is worth noting that metals such as copper and zinc, which have nearly filled d-shells, may act like transition metals in cases in which their d-shell electrons may participate in bonding with the activated molecule.

The ideal salts, of which the alkali halides are prototypes, are composed of ions which behave in most chemical reactions like rigid, charged spheres. The completely formed stable layers on the surfaces of crystals of ideal salts usually are very nearly saturated and hence should not be effective catalytically since they would interact in a relatively mild manner with both the normal and the activated state of molecules. The surfaces of salts may have effective catalytic action in two cases, however. First, the regions of discontinuity on an imperfectly formed surface layer are much less saturated than the completed surface; there is an opportunity for catalytic activity at such irregular regions. Second, salts containing ions of the transition metals, such as iron and nickel, may be able to form semihomopolar bonds with adsorbed molecules when the atomic spacing of the surface layers and the spacing of atoms within the molecules are proper. Hence, as in the case of the transition metals, the normal surface layers of salts of this type may exert a very specific catalytic action in individual cases.

The type of catalytic activity that is associated with irregularities on the surface layer can be enhanced by increasing the curvature of the surface of the particles, which stimulates the formation of imperfect layers. Since the best means of increasing the surface curvature is to decrease particle size, this end can be achieved by the procedure which is commonly used to increase surface area.

Molecular crystals are loosely bound aggregates of molecules which possess a large amount of internal stability, such as crystals of highly stable organic molecules. Since the constituent molecules are almost completely saturated, crystals of this type usually exert little catalytic activity; they do not interact strongly with adsorbed molecules.

It is evident that a very broad range of solid catalytic systems becomes possible if we combine the various factors in solids which contribute to catalytic activity in different possible ways. In the case of metals for example, we may vary the composition of alloy systems, the pattern of crystallographic phases which occur, the size of the grains which occur and the size of the polycrystalline aggregates in which the grains are bound. A virtually infinite number of possible combinations of these and other variables can be effected. Clearly this is not a field in which detailed theoretical understanding can be achieved rapidly through application of first principles. Rather it is one in which empirical methods and sound theoretical judgment must be combined with the hope of achieving results through patient effort.

References

1. The basic papers of the field are as follows: Eyring, H., and Polanyi, M., *Z. physik. Chem.* **12B,** 279 (1931); Pelzer, H., Wigner, E., *Z. physik. Chem.* **15B,** 445 (1932); Wigner, E., *Z. physik. Chem.* **19B,** 203 (1932); Eyring, H., *J. Chem. Phys.* **3,** 107 (1935).
   The following publications present surveys of the field: Wigner, E., *Trans. Faraday Soc.* **34,** 29 (1938); Glasstone, S., Laidler, K. J., and Eyring, H., *Rate Processes in Chemical Reactions.* McGraw-Hill, New York, 1941.
2. Frankenburg, W., *Z. Elektrochem.* **39,** 269 (1939).

The Mechanism of the Polymerization of Alkenes

LOUIS SCHMERLING AND V. N. IPATIEFF

Universal Oil Products Company, Riverside, Illinois

CONTENTS

I. INTRODUCTION

1. Scope

The chemical literature contains very many papers on the polymerization of olefins, written both from the theoretical and practical standpoint. A discussion of the large variety of olefins, catalysts and conditions which have been investigated is beyond the scope of this article. Extensive summaries of experimental results may be found in a number of books (Baroni, 1; Burk *et al.*, 2; Egloff, 3; Ellis, 4; Ipatieff, 5; Thomas, 6). It is the primary purpose of this chapter to discuss the mechanism of the reaction, and particularly those mechanisms which have been proposed during the past fifteen years.

The discussion is limited to the reaction of alkenes. The mechanism of the polymerization of such substances as styrene, butadiene, vinyl esters, and acrylic esters is not included.

Furthermore this chapter deals chiefly with polymerizations which are catalyzed by acid-acting catalysts. A comprehensive discussion of not only the thermal but even the photochemical and free radical-initiated polymerizations is outside its scope. The free radical-initiated reactions include those which are induced by metal alkylies, peroxides, oxygen and certain other substances. They depend on free radical initiation of a chain reaction; whether or not these free radicals should be considered to be catalysts has been questioned because the radicals enter into the reaction chain and are part of the reaction product.

2. Preliminary Survey

Depending on the reaction conditions, alkenes may undergo either of two types of catalytic polymerization. The products of the first type, which may be termed *true polymerization*, consist of alkenes having molecular weights which are integral multiples of the monomer alkene. The second type, *conjunct polymerization*, yields a complex mixture of alkanes, alkenes, alkadienes, cycloalkanes, cycloalkenes, cycloalkadienes, and, in some cases, aromatic hydrocarbons; the products do not necessarily have a number of carbon atoms corresponding to an integral multiple of the monomer.

Interpolymerization of two different alkenes is termed cross polymerization or, more usually, copolymerization.

Of the gaseous olefins, isobutylene is the most readily polymerized catalytically and ethylene the least. On the other hand, ethylene is the most easily thermally polymerized alkene.

In general, the catalysts may be classified as acids and metal halides. As will be explained below, both types of catalysts are "acid-acting" catalysts in the modern sense of the term. Some metals (*e.g.*, sodium, copper, and iron) are catalysts for the polymerization of alkenes, especially ethylene. They are active probably because they can combine with one of the pi electrons of the alkene and form a free radical which can then initiate a chain reaction (p. 25).

The acid catalysts include sulfuric acid, phosphoric acid, hydrogen fluoride, dihydroxyfluorboric acid, and alkanesulfonic acids. Some heavy metal sulfates, phosphates, sulfides and other salts are also catalysts; it seems probable that they undergo partial reduction to acidic compounds during the polymerization reactions.

Certain oxides, particularly clays and synthetic silica-alumina composites, are very active polymerization catalysts. They probably owe their activity to the presence of acidic hydrogen.

The metal halide catalysts include aluminum chloride, aluminum bromide, ferric chloride, zinc chloride, stannic chloride, titanium tetrachloride and other halides of the group known as the Friedel-Crafts catalysts. Boron fluoride, a nonmetal halide, has an activity similar to that of aluminum chloride.

The outstanding features of a few of the more important catalysts will be outlined briefly here.

Sulfuric acid finds commercial application in the polymerization of isobutylene to diisobutylene (a mixture of the two 2,4,4-trimethylpentenes) and the copolymerization of isobutylene with the *n*-butylenes to obtain a more complex mixture of octylenes. Dilute (65–70%) acid is used at 20–35° for the polymerization of only the isobutylene. The copolymerization occurs in the so-called "hot acid" polymerization process in which the dilute sulfuric acid is employed at about 80–90°. If more concentrated sulfuric acid is used, particularly acid of concentration above about 90%, conjunct polymerization occurs even at −35°.

Propene undergoes little polymerization when treated with 96% sulfuric acid, the chief product being isopropyl hydrogen sulfate which yields isopropyl alcohol on hydrolysis. When 98% sulfuric acid is used, propylene is converted to conjunct polymer. Ethylene cannot be polymerized by sulfuric acid because the stable ethyl hydrogen sulfate and ethyl sulfate are formed; attempts to obtain the polymerization by increasing the reaction temperature are unsuccessful because oxidation occurs.

Phosphoric acid may be used for the polymerization of all the gaseous olefins. Ethylene is converted to ethyl phosphoric acid at temperatures below 250°. At higher temperatures, the ester decomposes to yield conjunct polymer including isobutane. Propylene undergoes either conjunct or true polymerization depending on whether the reaction temperature is above or below 300°. The butylenes undergo true polymerization chiefly.

Orthophosphoric acid and pyrophosphoric acid are preferred catalysts. Phosphorus pentoxide is catalytically active but no conclusive evidence has been described to show whether or not its activity depends on the presence of traces of water as promoter. Copper pyrophosphate and acid phosphates of cadmium are also good catalysts; that the former probably owes its activity to partial conversion to acid or acidic salt under the polymerization conditions seems to be shown by the fact that there is an induction period. A composite prepared by calcining kieselguhr impregnated with orthophosphoric acid (the so-called "solid phosphoric acid") has found wide commercial use.

Isobutylene polymerizes readily even at −80° in the presence of floridin (an aluminum hydrosilicate) which has been activated by ignition at about 300°. The product is true polymer consisting largely of a mixture of the two 2,4,4-trimethylpentenes. Polymerization at room temperature yields triisobutylene as the major product.

Aluminum chloride, boron fluoride and certain other Friedel-Crafts catalysts catalyze the polymerization of isobutylene, at temperatures below about −70°; recent work has indicated that the presence of a promoter such as water is usually necessary (see Section V). A rubber-like polymer is obtained.

Because the acid catalysts are usually not suitable for the polymerization of ethylene to a very high-boiling product, the action of the Friedel-Crafts catalysts on ethylene has been extensively investigated as a means of preparing lubricating oils. The polymerization of ethylene occurs at room temperature under superatmospheric pressure in the presence of aluminum chloride or boron fluoride, the respective hydrogen halides serving as promoters. Conjunct polymerization occurs yielding a water-white upper layer consisting chiefly of paraffins and cycloparaffins and a viscous red or red-brown lower-layer consisting of an addition complex of highly unsaturated aliphatic and cyclic hydrocarbons with the catalyst.

True polymerization of alkenes takes place only under special conditions in the presence of aluminum chloride and boron fluoride. Polymer containing a minimum of paraffinic material is obtained, for example, when ethylene is treated with a mixture of aluminum chloride and aluminum or of boron fluoride and nickel.

With the other metal halides and the other alkenes either true or conjunct polymerization may predominate depending largely on the particular catalyst and conditions.

II. Free Radical-initiated Polymerization

As has already been indicated, a thorough discussion of the mechanism of metal, metal alkyl-, oxygen- or peroxide-induced polymerizations is beyond the scope of this article. Chiefly for purposes of comparison, it seems worthwhile to present a brief outline of the free radical type of mechanism which best explains these reactions and which probably is valid also for the photochemical and thermal polymerizations, particularly in the lower temperature range.

The peroxide-induced polymerization of ethylene under 200–300 atmospheres pressure in the presence of a solvent yields a high molecular weight polymer (Staff Report, 7). Thus, a high quality wax of 2000–3000 molecular weight, melting at 105–110° and containing 0.7–1.3% oxygen is obtained in the presence of methanol and benzoyl peroxide at 110–120°. Polymerization at 1000–2000 atmospheres and

180–220° in the presence of 0.05–0.1% oxygen yields a harder polymer of 15,000–20,000 molecular weight and melting at about 110°. Its extremely good electrical properties account for what is probably its major use, namely, as an electrical insulating material, for example in coaxial cables. It has good chemical resistance and is apparently insoluble in all solvents at room temperature.

The first step in the peroxide-induced reaction is the decomposition of the peroxide to form a free radical. The oxygen-induced reaction may involve the intermediate formation of a peroxide or a free radical olefin-oxygen addition product. (In the case of thermal and photochemical reactions, the free radical may be formed by the opening up of the double bond or, more probably, by dissociation of a carbon-hydrogen bond; in metal alkyl-induced reactions, decomposition of the metal alkyl yields alkyl radicals.)

The mode of formation of the free radicals is rather unimportant for the present purpose since they may be considered to serve only to initiate the reaction. The free radical, R′—, reacts with the alkene (e.g., $CH_2{=}CHR$, where R may be hydrogen or alkyl) in the following manner (Staudinger, 8; Flory, 9; Price, 10):

Chain propagation:

$$R'{-} + CH_2{=}CHR \rightarrow R'{-}CH_2{-}\underset{|}{C}HR \quad (1)$$

$$R'{-}CH_2{-}\underset{|}{C}HR + (m+1)CH_2{=}CHR \rightarrow R'{-}CH_2{-}\underset{R}{\underset{|}{C}}H(CH_2\underset{R}{\underset{|}{C}}H)_mCH_2{-}\underset{|}{C}HR \quad (2)$$

Chain transfer:

The polymeric free radical may become saturated by removal of a hydrogen atom from a molecule $R''CH_2R'''$ which may be a monomer, a polymer, or a solvent molecule

$$R'{-}CH_2{-}\underset{R}{\underset{|}{C}}H(CH_2{-}\underset{R}{\underset{|}{C}}H)_mCH_2{-}\underset{|}{C}HR + R''CH_2R''' \rightarrow R'{-}CH_2{-}\underset{R}{\underset{|}{C}}H(CH_2{-}\underset{R}{\underset{|}{C}}H)_mCH_2{-}CH_2R + R''\underset{|}{C}HR''' \quad (3)$$

Alternately, the polymeric free radical may transfer hydrogen to the monomer.

$$R'{-}CH_2{-}\underset{R}{\underset{|}{C}}H(CH_2{-}\underset{R}{\underset{|}{C}}H)_mCH_2{-}\underset{|}{C}HR + CH_2{=}CHR \rightarrow R'CH_2{-}\underset{R}{\underset{|}{C}}H(CH_2{-}\underset{R}{\underset{|}{C}}H)_mCH{=}CHR + CH_3\underset{|}{C}HR \quad (4)$$

The free radicals, $R''\underset{|}{C}HR'''$ and $CH_3\underset{|}{C}HR$, formed in the reactions of equations 3 and 4 start new chains as in equations 1 and 2.

Chain termination:

Two radicals (not necessarily alike) may couple (eq. 5) or disproportionate (eq. 6):

$$2R'(CH_2—CHR)_nCH_2—\underset{|}{C}HR \rightarrow R'(CH_2—CHR)_nCH_2—CHR—CHR—CH_2(CHR—CH_2)_nR' \quad (5)$$

$$2R'(CH_2—CHR)_nCH_2—\underset{|}{C}HR \rightarrow R'(CH_2—CHR)_nCH{=}CHR + R'(CH_2—CHR)_nCH_2—CH_2R \quad (6)$$

The thermal polymerization of isobutylene (at 370–460° and 540–5350 p.s.i.) is of particular interest because it yields 1,1,3-trimethylcyclopentane rather than 2,4,4-trimethyl-1- and -2-pentene (McKinley et al., 11). This cyclic dimer amounted to as much as 45.9% of the total liquid product when the reaction was carried out at 400° and 540 p.s.i. It was suggested that its formation might involve one of three mechanisms:

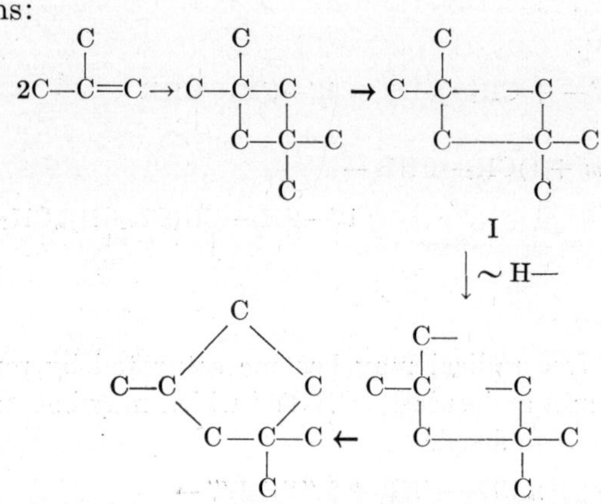

III II (7)

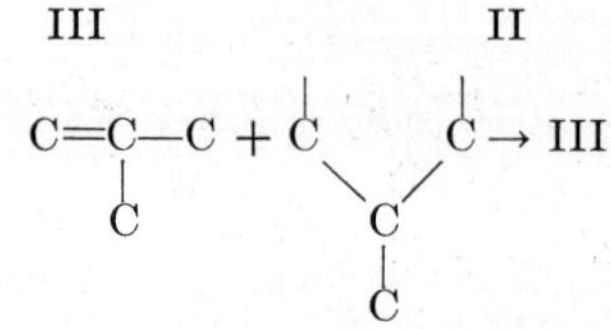 (8)

IV

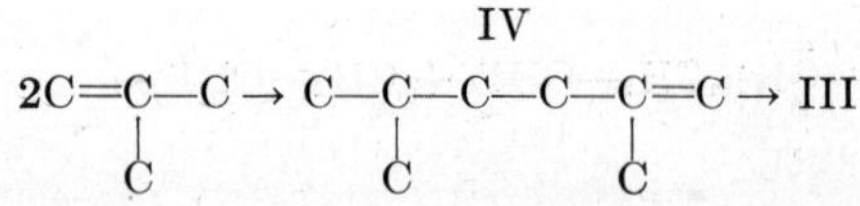 (9)

V

It is the opinion of the present authors that isomerization of a tertiary alkyl radical to a primary radical as in the formation of II from I is improbable. The formation of IV is similarly unlikely. The cyclization of V by intramolecular alkylation seems quite plausible; however, equation 9 does not explain either the formation of V or its subsequent cyclization. The following mechanism has the advantages that, like the generally accepted free radical-initiated mechanisms, it postulates a chain reaction and that the intramolecular alkylation step is directly analogous to that proposed for thermal alkylation, namely addition of an alkyl radical to the double bond of the alkene (Frey and Hepp, 12). The method of formation of the chain initiator, R′—, again is not critical since R′—, merely starts the first cycle of the chain reaction; it may be formed by decomposition of the isobutylene.

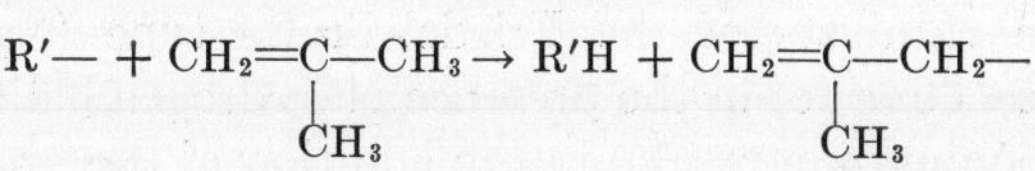 (10)

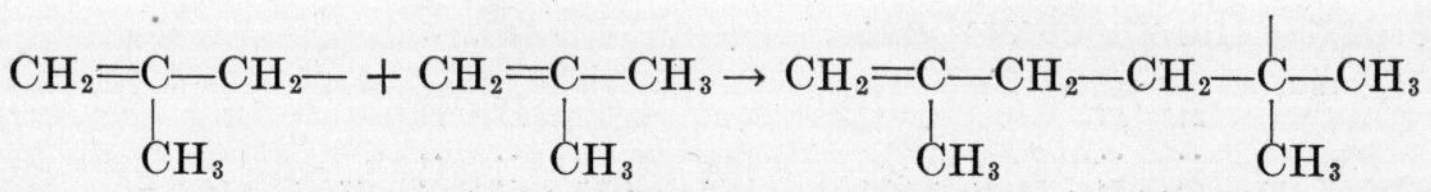 (11)

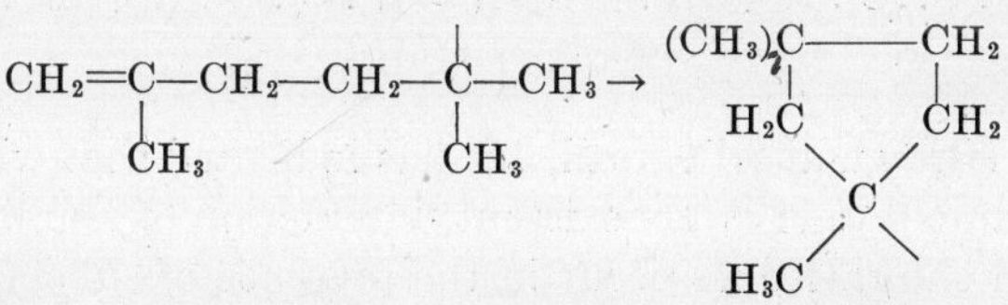 (12)

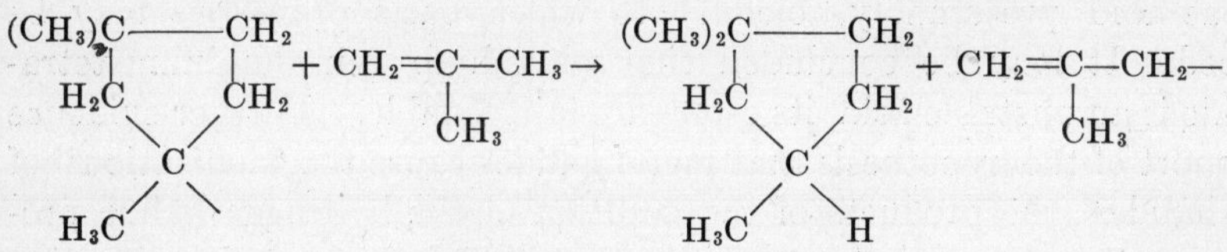 (13)

III. TRUE POLYMERIZATION

1. Carbonium Ion Mechanism

The most widely accepted mechanism (Whitmore, 13) for the polymerization of olefins involves the so-called carbonium ions. In accordance with this mechanism a carbonium ion (usually a tertiary ion) adds to the olefin to form a higher molecular weight carbonium ion which then yields the olefin polymer by elimination of, usually, a proton. With acid catalysts (e.g., sulfuric acid) the initial carbonium ion is formed by addition of the hydrogen ion from the acid to the extra electron pair in the double bond (the pi electrons):

$$H_3C : \underset{\displaystyle CH_3}{\underset{\cdot\cdot}{C}}::CH_2 + H^+ + OSO_3H^- \rightleftharpoons H_3C : \underset{\displaystyle CH_3}{\underset{\cdot\cdot}{\overset{+}{C}}} : CH_3 + OSO_3H^- \qquad (14)$$

Calculation of the proton affinities of the carbon atoms of the doubly-bonded pairs in propylene and isobutylene has shown that the proton affinity of the end carbon atom is greater in each case (Evans and Polanyi, 14). This means that the proton will add to the double bond at the CH_2 more readily than at the $CHCH_3$ or $C(CH_3)_2$. Hence, if the addition of HX to a double bond proceeds by way of initial addition of a proton, the hydrogen atom will become attached to the carbon atom holding the greater number of hydrogen atoms. Markownikoff's rule has thus been interpreted in terms of proton affinities which in turn are calculated from bond strengths and ionization potentials.

With halide catalysts of the Friedel-Crafts type (e.g., aluminum chloride or boron fluoride) in the presence of hydrogen halide the formation of the carbonium ion results in the addition of the proton from the promoter to the pi electrons:

$$H : \overset{\displaystyle H}{\overset{\cdot\cdot}{C}}::\overset{\displaystyle H}{\overset{\cdot\cdot}{C}} : H + HCl + AlCl_3 \rightleftharpoons H : \underset{+}{\overset{\displaystyle H}{\overset{\cdot\cdot}{C}}} : \underset{\displaystyle H}{\underset{\cdot\cdot}{\overset{\displaystyle H}{\overset{\cdot\cdot}{C}}}} : H + AlCl_4^- \qquad (14a)$$

Recent work (Brown and Pearsall, 15) has indicated that while hydrogen aluminum tetrachloride is nonexistent, interaction of aluminum chloride and hydrogen chloride does occur in the presence of substances (such as benzene and presumably, olefins) to which basic properties may be ascribed. It may be concluded that while hydrogen aluminum tetrachloride is an unstable acid, its esters are fairly stable. Further evidence in support of the hypothesis that metal halides cause the "ionization" of alkyl halides (the products of the addition of the hydrogen halide promoters to the olefins) is found in the fact that exchange of radioactive chlorine atoms for ordinary chlorine atoms occurs when *tert*-butyl chloride is treated with aluminum chloride containing radioactive chlorine atoms; the hydrogen chloride which is evolved is radioactive (Fairbrother, 16).

The halide catalysts are electron acceptors, and, in the absence of hydrogen halide promoter, the active complex is presumably formed by the addition of the catalyst to the olefin (Hunter and Yohe, 17; cf. Whitmore, 18):

$$\begin{array}{ccccccccccccccc} & & \overset{\cdot\cdot}{:Cl:} & & H & & H & & & \overset{\cdot\cdot}{:Cl} & : & H & & H & \\ & & \cdot\cdot & & \cdot\cdot & & \cdot\cdot & & \cdot\cdot & \cdot\cdot & & \cdot\cdot & & \cdot\cdot & \\ :Cl & : & Al & + & C & :: & C & \rightarrow & :Cl & :Al & : & C & : & C^+ & \qquad (15) \\ \cdot\cdot & & \cdot\cdot & & \cdot\cdot & & \cdot\cdot & & \cdot\cdot & \cdot\cdot & & \cdot\cdot & & \cdot\cdot & \\ & & \underset{\cdot\cdot}{:Cl:} & & H & & H & & & \underset{\cdot\cdot}{:Cl} & : & H & & H & \end{array}$$

The metal halide thus functions in similar manner to the proton and may be considered to be an acidic catalyst (cf. Luder and Zuffanti, 19). The catalyst-olefin complex differs in one significant respect from the product formed by the addition of the proton (or the corresponding acid) to the olefin: the halide catalyst is a neutral but electronically deficient molecule and combines with the pi electrons of the double bond to form a coordinate bond between the carbon atom and the aluminum or boron. On the other hand, the addition of the positive proton to the double bond results in the formation of a true (covalent) link between carbon and hydrogen. In other words, the complex, while it contains an electron-deficient (hence, positive) carbon atom, is in itself electronically neutral; the product of the addition of a proton to the alkene contains a similar carbon atom but is itself electrically positive. It has been suggested (Whitmore and Meunier, 20) that this difference is related to the fact that metal halide catalysts tend to yield much higher polymers than do the acid (proton) catalysts.

Silica-alumina catalysts are acidic and presumably can furnish protons for the formation of carbonium ions (Kazanskii and Rozengart, 21).

The carbonium ion can undergo certain characteristic reactions (Whitmore, 13). These include (1) union with a negative ion, X, having a complete octet of electrons, the net result being the simple addition of HX to the double bond; (2) elimination of the same or a different proton to give the same or a new olefin; (3) rearrangement of the carbon skeleton followed by the loss of the proton to give a new olefin; and (4) polymerization, which merely involves the addition of an olefin by way of its extra electron pair to the carbon atom having the sextet of electrons:

$$(H_3C : \overset{H_3C}{\underset{H_3C}{\ddot{\underset{..}{C}}}})^+OSO_3H^- + H_2C::\underset{CH_3}{\underset{..}{C}} : CH_3 \rightarrow (H_3C : \overset{H_3C}{\underset{H_3C}{\ddot{\underset{..}{C}}}} : \overset{H}{\underset{H}{\ddot{\underset{..}{C}}}} : \underset{CH_3}{\underset{..}{C}} : CH_3)^+ + OSO_3H^- \qquad (16)$$

The product is a larger carbonium ion which can undergo any of the changes indicated above, including further polymerization, which may follow one of two courses: the higher molecular weight carbonium ion may either (1) add to a second molecule of the olefin or (2) lose a proton to give the olefin polymer to which a carbonium ion may then add.

a. Methylpropene (*Isobutylene*). The dimerization of isobutylene in the presence of the various acid catalysts, including clays, and of metal halides promoted by hydrogen halides may be illustrated by the following equations. For reasons of convenience, the anions are not indicated but it must constantly be remembered that the negative portion of the ionic pair plays an essential, though lesser, role in the reaction.

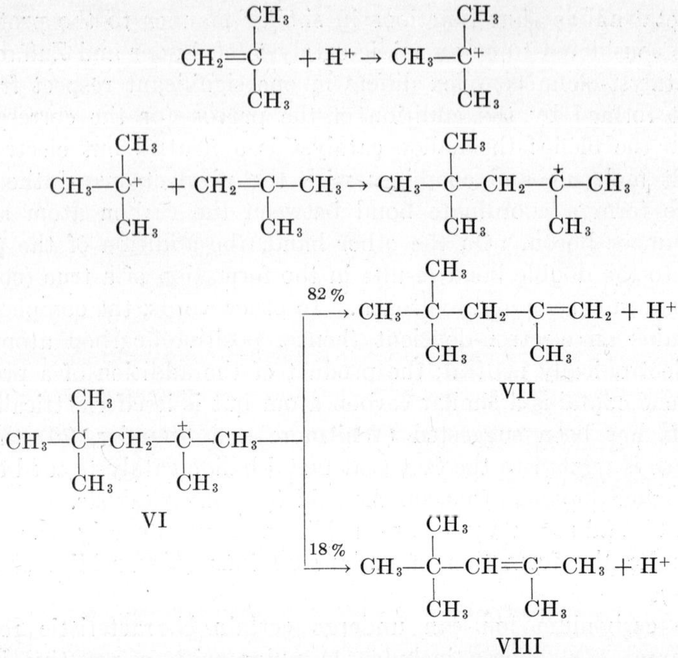

(17) (18) (19) (20)

The two diisobutylenes, namely 2,4,4-trimethyl-1-pentene (VII) and 2,4,4-trimethyl-2-pentene (VIII), are produced in the ratio of about four to one in the presence of dilute sulfuric acid (McCubbin and Adkins, 22; Whitmore and Church, 23). The former is formed in larger amount because elimination of a proton from the methyl groups adjacent to the electron deficient carbon atom in VI takes place more readily than from the neopentyl group. Furthermore, there are six hydrogens attached to the methyl carbons and only two protons in the methylene group of the neopentyl system. The proton eliminated in the formation of the olefin dimer is, of course, available for further addition to the isoolefin to form a tertiary carbonium ion which can continue the reaction chain.

Data illustrating the relative difficulty with which a proton is released from the neopentyl group have been obtained by investigation of the dehydration of a number of tertiary alcohols. Dehydration of dimethylneopentylcarbinol by different procedures yielded, in every case, an olefin mixture consisting of about 80% of 2,4,4-trimethyl-1-pentene (i.e., dehydration from a methyl group) and 20% of 2,4,4-trimethyl-2-pentene (dehydration from the neopentyl group) (Whitmore *et al.*, 24). Similarly, methylethylneopentylcarbinol yielded largely 3,5,5-trimethyl-2-hexene (dehydration from the ethyl group), less than 5% of 2,2,4-trimethyl-3-hexene (dehydration from the neopentyl group), and only traces of 4,4-dimethyl-2-ethyl-1-pentene (dehydration from the methyl group) (Whitmore and Laughlin, 25). Further, diethylneopentylcarbinol gave about 90% of 5,5-dimethyl-3-ethyl-2-hexene (dehydration from an ethyl group) and less than 10% of 2,2-dimethyl-4-ethyl-3-hexene (dehydration from the neopentyl group) (Whitmore and Rohrmann, 26). Methyl-*n*-butylneopentylcarbinol yielded over 80% of 2,2,4-trimethyl-4-octene (dehydration from the *n*-butyl group), about 10% of 4,4-dimethyl-2-*n*-butyl-1-pentene (dehydration from the methyl group), and a trace of 2,2,4-trimethyl-3-octene (dehydration from the neopentyl group) (Whitmore and Rohrmann, 26).

As has already been mentioned, carbonium ions are usually written without indicating the presence of negative ions, even though such anions are essential and the environment often affects the type of product obtained. Also, it has been suggested that the acid-carbonium ion system can act as a reservoir for carbonium ions (Whitmore, 18):

$$R^+ + H_2SO_4 \rightleftarrows R : \overset{\overset{\text{H}}{\cdot\cdot}}{\underset{\cdot\cdot}{O}} : SO_3H \rightleftarrows R : \overset{\cdot\cdot}{\underset{\cdot\cdot}{O}} : SO_3H + H^+ \tag{21}$$

It was further stated, "Esters are often assumed as intermediates in various processes of the types in which we are interested. It should be noted, however, that any use of a pure neutral ester is naturally ineffective since the ester by itself probably has no more tendency to give ions than does water by itself or hydrogen chloride by itself. On the other hand, the ester in the presence of a suitable donor of protons can form carbonium ions."

An investigation (Whitmore *et al.*, 27; cf. McCubbin, 28) of the isomers present in triisobutylene has shown that it consists of approximately 55% of 4-methylene-2,2,6,6-tetramethylheptane (IX), 35% of 2,2,4,6,6-pentamethyl-3-heptene (X), and about 5% each of 2,4,4,6,6-pentamethyl-1-and -2-heptene (XI and XII). The formation of the first two involve the addition of a *tert*-butyl carbonium ion to 2,4,4-trimethyl-1-pentene (eq. 22). Formation of the last two involves the addition of the carbonium ion (VI) formed by the addition of the *tert*-butyl carbonium ion to isobutylene (or of a proton to diisobutylene) to a second molecule of isobutylene (eq. 25). These two additions occur in the ratio of about 9 to 1.

Contrary to the reports of earlier workers (Lebedev and Kobliansky, 29) 2,4,4-trimethyl-3-*tert*-butyl-2-pentene does not seem to be present in the triisobutylene produced in the presence of sulfuric acid (Whitmore *et al.*, 27). Hence, condensation of *tert*-butyl carbonium ion with 2,4,4-trimethyl-2-pentene apparently does not occur, presumably because of

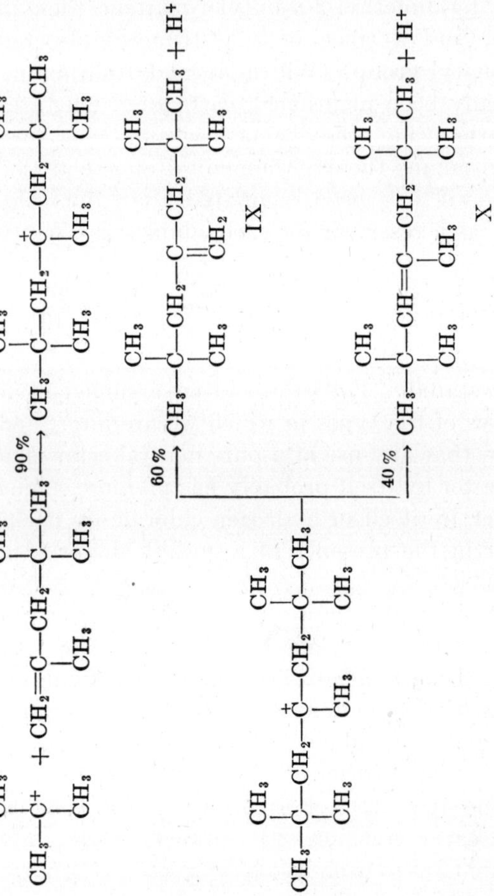

(22)

(23)

(24)

(25)

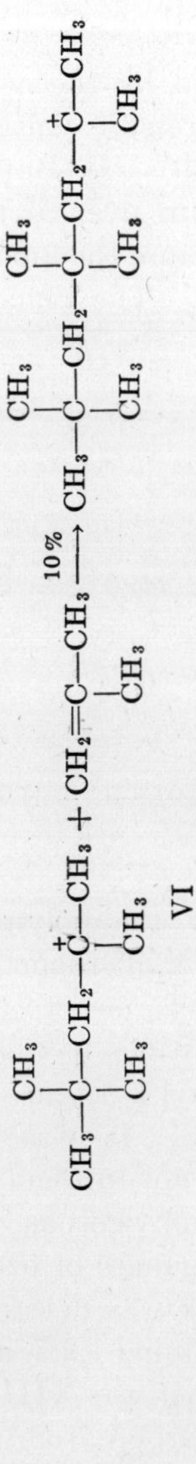

(26)

(27)

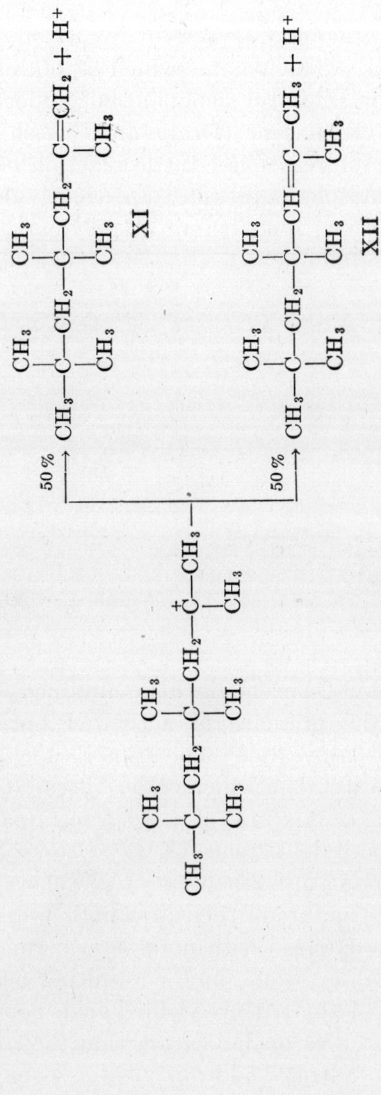

steric reasons: such a reaction would put two tertiary butyl groups on one carbon atom (eq. 28). Furthermore, 2,4,4-trimethyl-2-pentene isomerizes readily (by way of the carbonium ion VI) to 2,4,4-trimethyl-1-pentene, to which the *tert*-butyl carbonium ion does add.

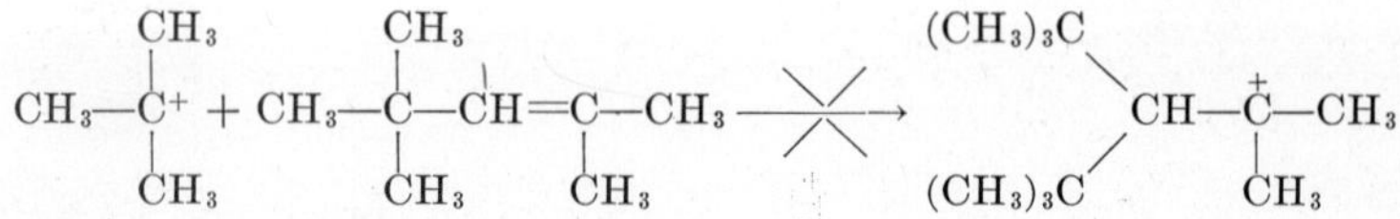 (28)

b. 2-Butene. The octene which is obtained by the action of 75% sulfuric acid on *sec*-butyl alcohol at 80° under pressure consists primarily of 3,4-dimethyl-2-hexene (Drake and Veitch, 30). *sec*-Butyl ether is a byproduct of the reaction. Little reaction occurs even at temperatures of 100° at atmospheric pressure, apparently because the 2-butene escapes from the reaction zone before it polymerizes. The reaction may be formulated as follows:

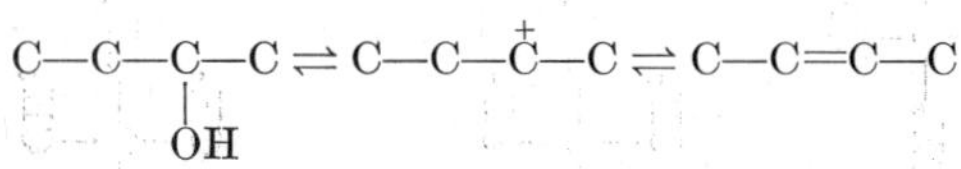 (29)

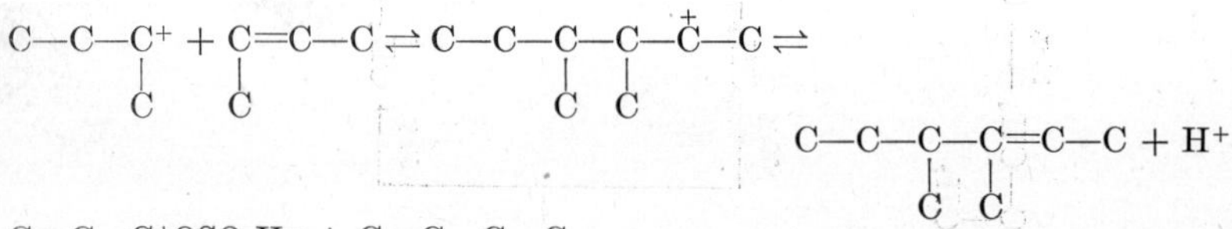 (30)

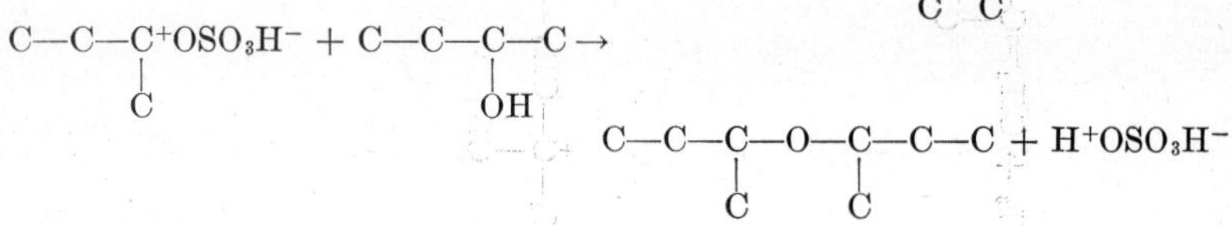 (31)

3,4-Dimethyl-2-hexene is also obtained by the polymerization of 2-butene in the presence of activated floridin at room temperature (Lebedev and Orlov, 31). The trimer which is formed to the extent of 15-18% of the total polymer under these conditions has been reported to consist of a mixture of 3,4,5,6-tetramethyl-2-octene (XIII) and 3,4,5,6-tetramethyl-4-octene (XV) (Orlov, 32). It was stated that 3,4,5-trimethyl-4-ethyl-2-heptene (XIV) was not present. However, the proofs of structure of the compounds were not rigorous. The trimer fractions were oxidized with potassium permanganate or ozone, and the number of carbon atoms in the resulting acids was determined. The fact that one of the trimers yielded acids containing 1,2,7, and 9 carbon atoms was taken as an indication that the trimer was XIII rather than XIV since the latter could not yield a seven carbon atom acid readily. The actual structure of the acids was not proved in any case. The car-

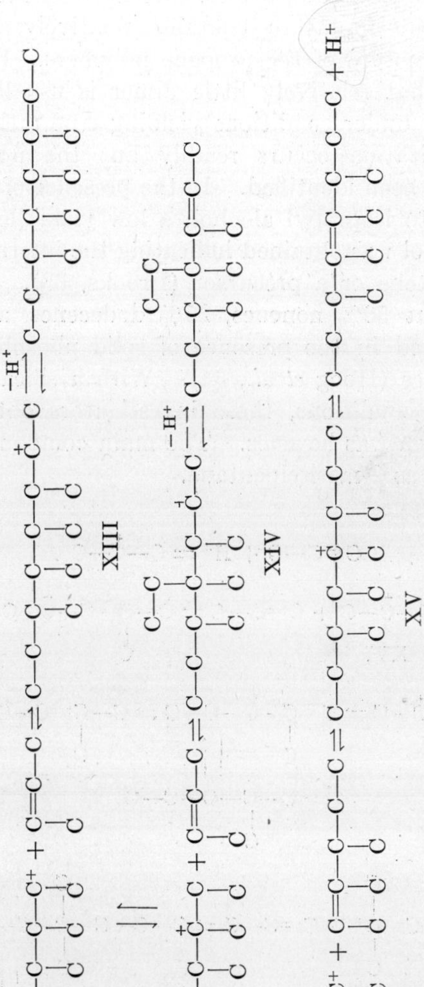

(32) XIII

(33) XIV

(34) XV

bonium ion mechanism explains the formation of the trimers as indicated in equations 32, 33 and 34; it will be noted that formation of XV involves the addition of a butyl carbonium ion to an octene, which presumably is the predominant reaction in the trimerization of isobutylene.

c. Propene. The literature contains relatively little information concerning the structure of the propene polymers. Principally, this is due to the fact that relatively little dimer is usually obtained when propene is treated under conditions yielding true rather than conjunct polymer. Trimerization occurs readily but the nonenes which are obtained have not been identified. In the presence of 90—92% sulfuric acid, in addition to isopropyl alcohol, a low yield (less than 0.2%) of 2-methyl-2-pentanol was obtained indicating the intermediate formation of 2-methyl-2-pentene or a precursor (Brooks, 33). A liquid product consisting of about 50% nonenes, 25% dodecenes, and less than 5% hexenes was formed in the presence of solid phosphoric acid at 160° and 10 atmospheres (Hoog *et al.*, 34). With a silica-alumina catalyst under unspecified conditions, these investigators obtained a polymer about 50% of which was hexenes. The main component of the hydrogenated product was 2-methylpentane.

$$\mathrm{C{=}C{-}C + H^+ \rightleftharpoons C{-}\overset{+}{C}{-}C} \quad (35)$$

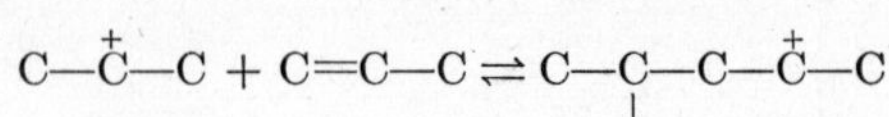

XVI

(36)

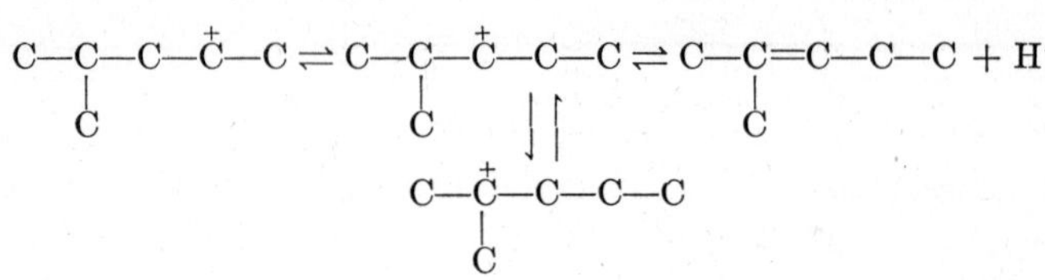

(37)

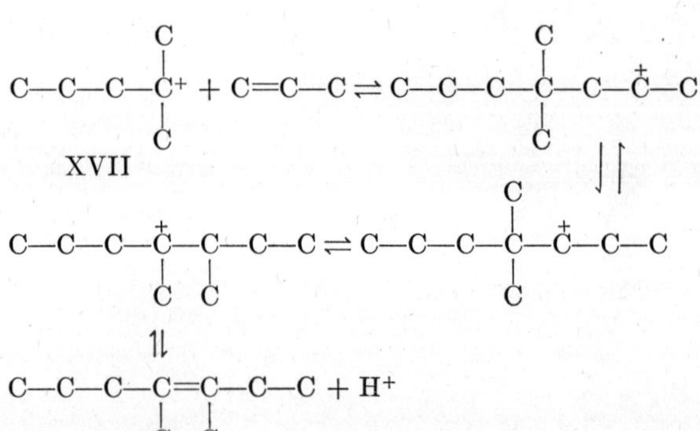

XVII

(38)

The reason for the low yield of hexene and high yield of nonene apparently lies in the relative reactivity of the 2-methyl-2-pentene and the original propene. In other words, the tertiary hexyl carbonium ion (XVII) adds to the olefinic double bond more rapidly than does the secondary propyl ion (XVI).

d. Methylbutenes. The action of 75% sulfuric acid on methylisopropylcarbinol at 76-80° resulted in the following yields of product: trimethylethylene, 1%; 3-methyl-2-pentene, 3%; methyl isopropyl ketone, 1%; diisobutylenes, 1%; 2,3,4,4-tetramethyl-1-pentene, 2%; other nonenes, 1%; 3,4,5,5-tetramethyl-2-hexene, 45%; 3,5,5-trimethyl-2-heptene, 35%; and higher polymers, 5% (Whitmore and Mosher, 35; cf. Drake *et al.*, 36). The formation of the minor products is readily explained. Trimethylethylene is a normal product of the dehydration of methylisopropylcarbinol (eq. 39). 3-Methyl-2-pentene (XXII) and isobutylene are formed by the β-scission of the carbonium ion XXI (Whitmore and Mosher, 37); the isobutylene polymerizes to yield the 2,4,4-trimethylpentenes and copolymerizes with trimethylethylene and other pentenes to yield, respectively, 2,3,4,4-tetramethyl-1-pentene and the other nonenes. β-Scission takes place relatively readily in carbonium ions containing the grouping $R_3C—C—C^+$ (Whitmore and Stahly, 38; Whitmore and Mosher, 37); it is the reverse of polymerization. The methyl isopropyl ketone was formed by the oxidation of the carbinol; sulfur dioxide was always observed during the polymerization reaction.

The formation of the decenes is indicated by equations 39 to 41.

3,5,5-Trimethyl-2-heptene (XIX) is formed by loss of a proton from the ethyl group in XVIII. Neither 3,5,5-trimethyl-3-heptene formed by a correspomding loss of a proton from the methylneopentyl group or 2-ethyl-4,4-dimethyl-1-hexene formed by loss of a proton from the methyl group were isolated "in spite of painstaking searches." This is in line with the above-mentioned evidence which shows that loss of a proton from a methylene group alpha to a gem-dimethyl group is more difficult than is the loss of a proton from a methylene group which is not so inactivated. Also, it is to be expected that a proton on a secondary carbon atom will be removed more readily than one in a methyl group. Direct evidence to show that 3,5,5-trimethyl-2-heptene is the most stable trimethylheptene isomer under the conditions used was obtained by investigating the dehydration of 3,5,5-trimethyl-3-heptanol. In the presence of copper sulphate at the reflux temperature of the mixture, 3,5,5-trimethyl-2-heptene and 3,5,5-trimethyl-3-heptene were obtained in the ratio of 6 to 1. On the other hand, dehydration with 75% sulfuric acid at 80° (the conditions under which the methyliso-

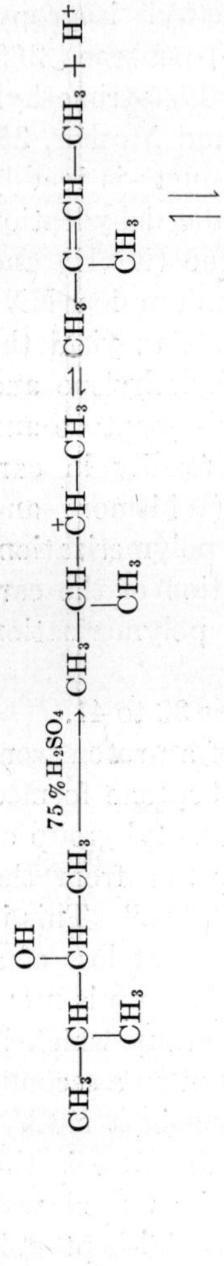

(39)

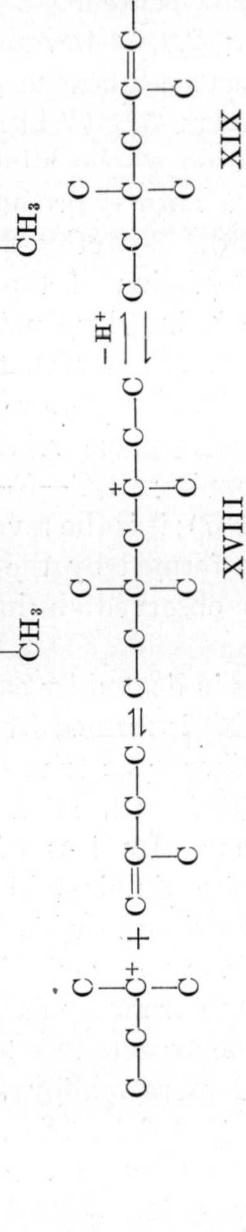

(40)

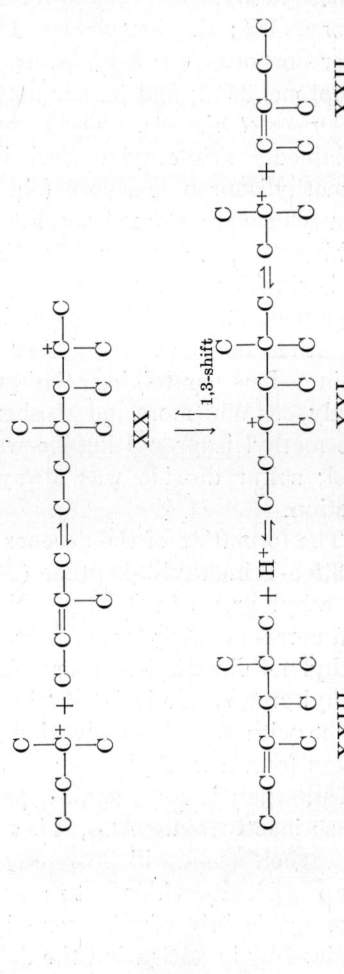

(41)

propylcarbinol was treated) yielded these compounds in the ratio of 20 to 1; i.e., the secondary carbon atom in the ethyl group lost a proton twenty times as readily as did the one in the neopentyl system. In neither dehydration was the methylene isomer, namely 2-ethyl-4,4-dimethyl-1-hexene detected; i.e., dehydration from the methyl group occurred least readily.

Similarly, the fact that 3,4,5,5-tetramethyl-2-hexene (XXIII) is formed from XXI while 3,4,5,5-tetramethyl-3-hexene (i.e., 2,2,3,4-tetramethyl-3-hexene) is not, again illustrates the difficulty of removing a proton from a neopentyl carbon atom, even when it is a tertiary carbon atom.

Formation of XXI involves a 1,3-shift of a methyl group in XX. Absence of a decene having the 2,3,4,4-tetramethylhexene structure may be explained by assuming that the intermediate carbonium ion XX undergoes the 1,3-shift of a methyl group as fast as it is formed, yielding the more stable carbonium ion XXI. Furthermore, it is postulated that XX is less stable and undergoes depolymerization more readily, than XXI; a *tert*-pentyl group is lost by a carbonium ion much more readily than is a *tert*-butyl group (Whitmore and Mosher, 37).

It has been suggested (Kline and Drake, 60) that the occurrence of a 1,3-shift of a methyl group is unlikely and that a rearrangement such as that of XX to XXI actually involves successive 1,2-shifts. On the other hand, the failure to isolate products corresponding to the intermediates formed by the 1,2-shifts as well as steric considerations indicate that 1,3 shifts are quite probable (Whitmore and Mosher, 37). The intermediate formation of XXIV with its two quarternary carbon atoms seems less probable than does the formation of XXI by a direct 1,3-shift (see equation 42 on page 40).

e. 2,3-Dimethylbutenes. In the presence of boron trifluoride, tetramethylethylene was converted to 2,2,3,5,6-pentamethyl-3-heptene (XXVII) and a trimethylheptylethylene (Brunner and Farmer, 39). Similar results were obtained with 80% sulfuric acid (Whitmore and Meunier, 20); the dimer fraction consisted of about 50% of 2,2,3,5,6-pentamethyl-3-heptene, about 25% of 2,2,4,6,6-pentamethyl-3-heptene (XXXIII), about 10% of 2,3,4,6,6-pentamethyl-2-heptene (XXXV), and about 0.2% of 1,1-dineopentylethylene (XXXIV, 2-neopentyl-4,4-dimethyl-1-pentene). However, none of these products corresponds to the simple addition of a carbonium ion formed from the tetramethylethylene to another molecule of the tetramethylethylene. In each case rearrangement apparently took place both before and after the initial addition of the carbonium ion to the olefin.

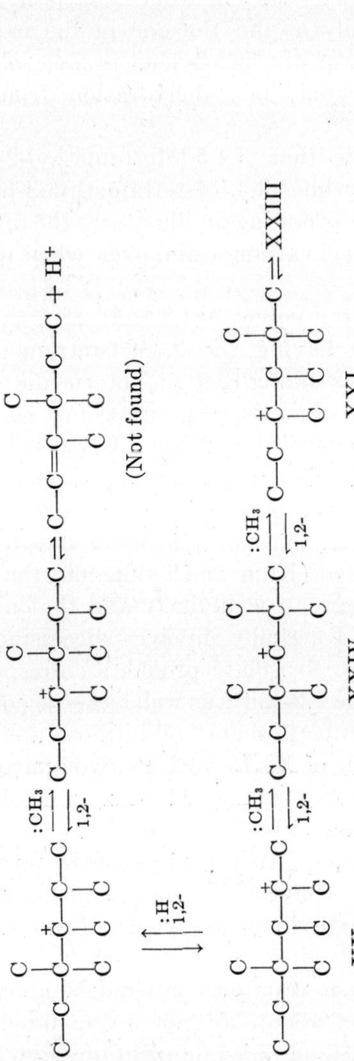

(42)

In the presence of acid catalysts, tetramethylethylene exists as an equilibrium mixture of its related isomers (Laughlin *et al.*, 40):

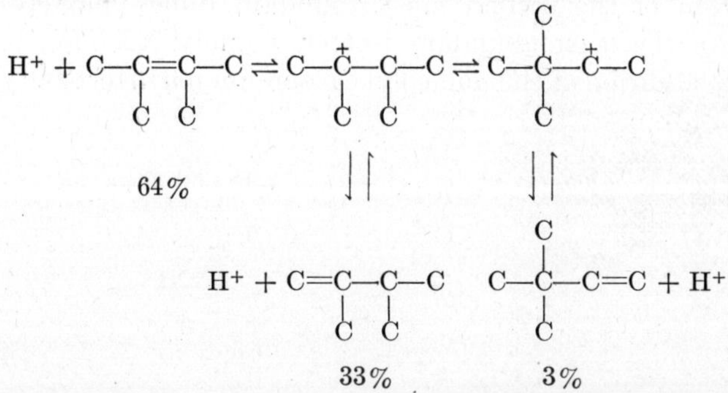

(43)

The following mechanism was proposed (Whitmore and Meunier, 20) to explain the formation of the principal dimer, XXVII:

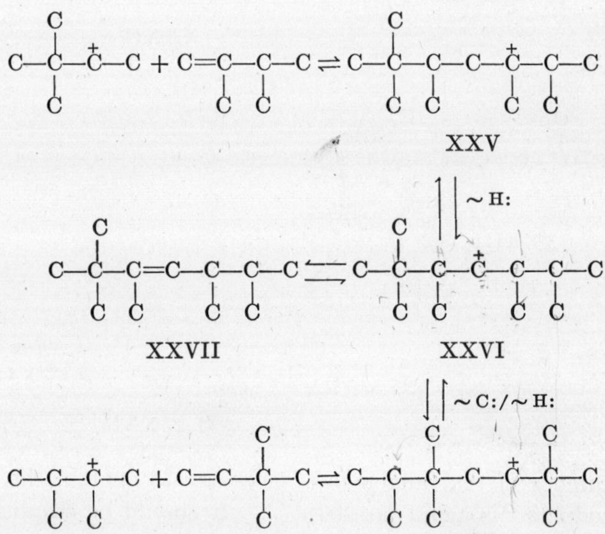

(44)

Formation of XXVII by loss of a proton from the tertiary carbon atom in the neopentyl group of XXVI rather than from the tertiary carbon atom in the secondary isopentyl group is, however, hardly to be expected. Furthermore, the addition of a tertiary olefin to a secondary carbonium ion is also unexpected (compare the results on the copolymerization of *n*-butylene with isobutylene, page 46). A somewhat more likely combination consists of the addition of the tertiary carbonium ion XXVIII to *tert*-butylethylene (XXIX) followed by a 1,3-shift of a methyl group and a subsequent hydrogen shift to yield XXVI; however, formation of XXVII again involves the unlikely proton elimination.

Formation of the second most abundant dimer, namely XXXIII, as well as of the least abundant isomer, namely XXXIV, presumably involves the addition of the pinacolyl carbonium ion to tert-butylethylene:

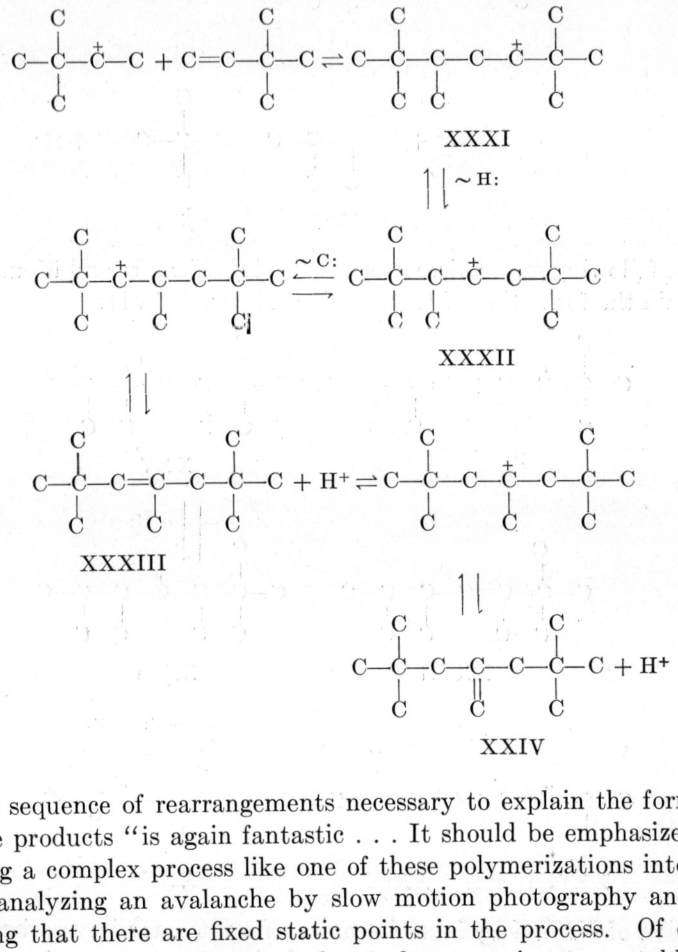

XXXI, XXXII, XXXIII, XXIV (45)

The sequence of rearrangements necessary to explain the formation of these products "is again fantastic . . . It should be emphasized that breaking a complex process like one of these polymerizations into steps is like analyzing an avalanche by slow motion photography and then assuming that there are fixed static points in the process. Of course, both the slide and the chemical change keep moving to a stable end-point" (Whitmore and Meunier, 20).

While there was apparently no olefin corresponding to the carbonium ion XXXII present in the dimer formed in the presence of 80% sulfuric acid, it has been reported that such a compound, namely, 2,2,3,6,6-pentamethyl-3-heptene, is the principal product when 84% acid is used (Whitmore, 18).

Formation of the third most abundant product (XXXV) also apparently involves the intermediate formation of carbonium ion (XXX) by way of the addition of the tertiary carbonium ion to *tert*-butyl ethylene. The ion apparently undergoes the shift of a hydrogen and then of a methyl with final loss of a proton to yield the observed dimer:

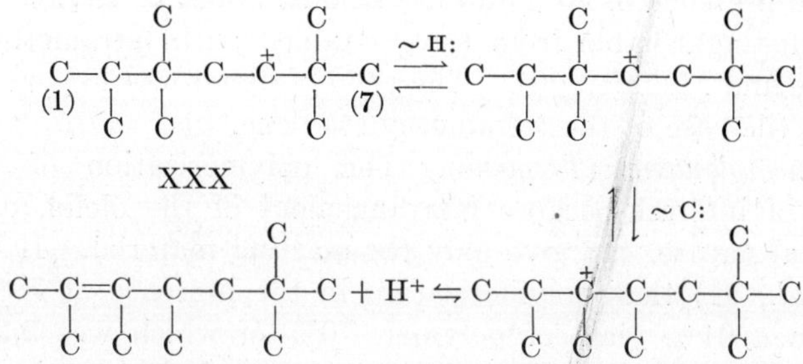

XXX

XXXV

(46)

The shift of the hydrogen in a neopentyl system (i.e., the hydrogen attached to carbon atom (4) in XXX) in preference to that of the methyl groups of the neopentyl system (i.e., one of the two methyl groups attached to carbon atom (6)) would probably not have been predicted. Such a migration of a methyl group would yield 2,3,5,5,6-pentamethyl-2-heptene, a product which was not found.

There was no evidence of the presence of the products to be expected from the intermediate condensation of either carbonium ion with the tetramethylethylene or of the tertiary carbonium ion with 2,3-dimethyl-1-butene. It was pointed out that this was not strange because the additions would involve, respectively, (1) the union of the tetra-substituted olefin and a tertiary carbonium ion or a sterically hindered secondary carbonium ion (pinacolyl) and (2) the addition of a tertiary carbonium ion with a disubstituted olefin. This failure of a sterically hindered ion to add to sterically hindered double bonds is similar to the failure of *tert*-butyl carbonium ion to add to 2,4,4-trimethyl-2-pentene (cf. eq. 28). It is not, however clear to the present authors why the combination of the tertiary carbonium ion with the 2,3-dimethyl-1-butene should not be expected, particularly in view of the postulated combination of this ion with *tert*-butylethylene (eq. 44), of *tert*-butyl carbonium ion with 2,4,4-trimethyl-1-pentene (eq. 22) and of the 2,4,4-trimethyl-2-pentyl carbonium ion (VI) with isobutylene (eq. 25).

No olefins corresponding to the nonrearranged primary addition ions XXV, XXX and XXXI were found "in spite of years of careful search" (Whitmore and Meunier, 20).

It seems necessary to conclude that more data are necessary before the mechanism of the dimerization of tetramethylethylene will be completely understood and before it will be possible to predict accurately which isomers will be produced by the dimerization of higher molecular weight olefins.

Polymerization of 2,3-dimethyl-1-butene (1-methyl-1-isopropylethylene) in the presence of 80% sulfuric acid at about 0° gave a mixture of dimers indistinguishable from that obtained with tetramethylethylene (Whitmore and Meunier, 20). The yield of dimer was 43% as compared to 62% in the case of the tetramethylethylene.

f. Trimethylbutene (Triptene). The polymerization of trimethylbutene is of interest because rearrangement of the olefin, unless of a very radical nature, can give only the starting material. It was found (Cook *et al.*, 41) that polymerization in the presence of 75% sulfuric resulted in a 91% yield of polymer, 70% of which was 2,2,3,5,5,6,6-heptamethyl-3-heptene. The minor products of the reaction consisted of 3.1% of unreacted triptene, 0.9% of 8- to 10-carbon atom olefins, 3.0% of 10-carbon atom olefins, 9.0% of 11- to 14-carbon atom olefins and 12.0% of residue. The formation of the heptamethylheptene is to be expected on the basis of the carbonium ion mechanism:

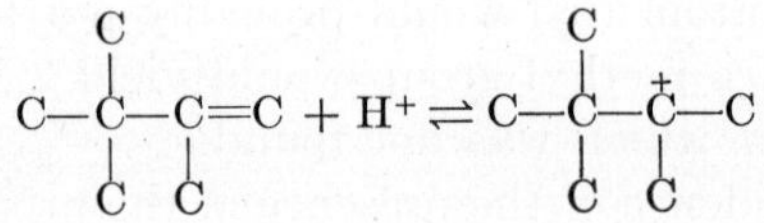

(47)

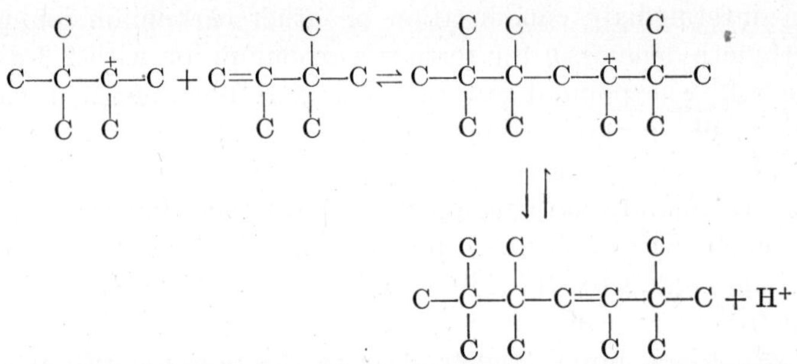

(48)

g. Methylpropene plus Propene. The phosphoric acid catalyzed copolymerization of isobutylene and propene yields a copolymer consisting of a mixture of 2,2- and 2,3-dimethyl-χ-pentene as shown by the fact that hydrogenation yields 2,2- and 2,3-dimethylpentane. The polymer obtained in the presence of "solid" phosphoric acid at 135° and 38 atmospheres from an approximately equimolecular mixture of the two alkenes contained 40-45% heptenes which yielded a heptane fraction consisting largely of 2,3-dimethylpentane together with a lesser amount of 2,2- and a trace of 2,4-dimethylpentane (Ipatieff and Schaad,

42). At 180° and 40 atmospheres pressure, other investigators obtained a 65% yield of polymer hydrogenation of which yielded a liquid containing 54% of heptane fraction consisting of approximately 20% of 2,2-dimethylpentane and 80% of 2,3-dimethylpentane (Hoog *et al.*, 34).

Similarly, hydrogenation of the copolymer of isobutylene and propene formed in the presence of dihydroxyfluorboric acid at 15–20° yielded a liquid product, 67% of which was heptane (Brooks, 43). At least 95% of the heptane was 2,3-dimethylpentane. In the absence of isobutylene, propene does not undergo polymerization when treated with hydroxyfluoboric acid at 0–40°.

The formation of the heptenes may be indicated by:

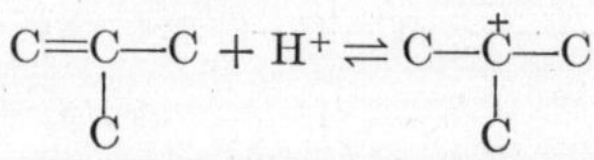
(49)

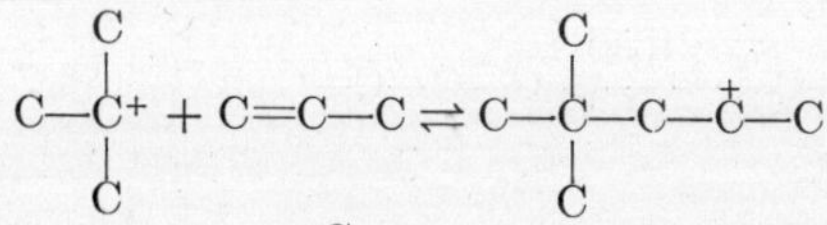
(50)

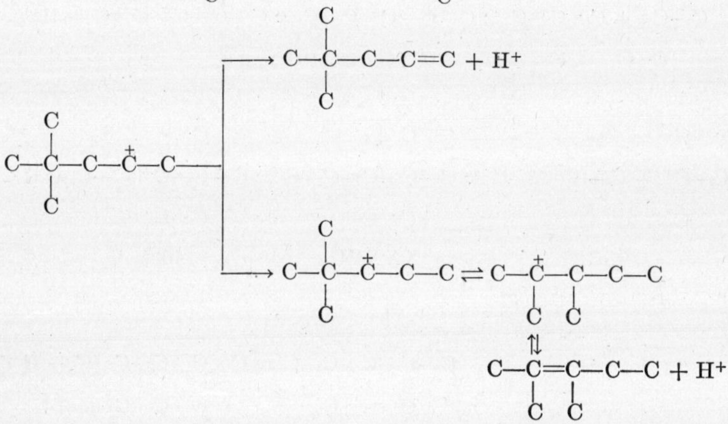
(51)
(52)

It is significant to note that dehydration of 4,4-dimethyl-2-pentanol (methylneopentylcarbinol) at reflux temperature by means of about 5% by weight of 100% sulfuric acid yielded 4.5 parts of 4,4-dimethyl-2-pentene and 1.0 parts of 4,4-dimethyl-1-pentene (Whitmore and Homeyer, 44). This is contrary to expectations since it would be predicted that loss of hydrogen from the methyl group rather than from the methylene group of the neopentyl radical would predominate. Also from the structure of the heptane produced in the copolymerization experiments, rearrangement to 2,3-dimethyl-2-pentene would be expected. If the above mechanism (eqs. 49–52) is correct, such rearrangement will occur when the methylneopentylcarbinol is treated with phosphoric acid or dihydroxyfluorboric acid under the conditions of the copolymerizations.

h. Methylpropene plus 2-Butene. The copolymerization of isobutylene with the *n*-butylenes was investigated by adding *tert*-butyl alcohol to a mixture of *sec*-butyl alcohol and 75% sulfuric acid at 64° (Whitmore *et al.*, 45). Equimolecular quantities of the alcohols were used. The octenes, which were obtained in 78% yield, consisted of diisobutylenes (25%), 3,4,4-trimethyl-2-pentene (40%), and 2,3,4-trimethyl-2-pentene (35%). The formation of all these products may be explained on the basis of the addition of a *tert*-butyl ion to an olefin:

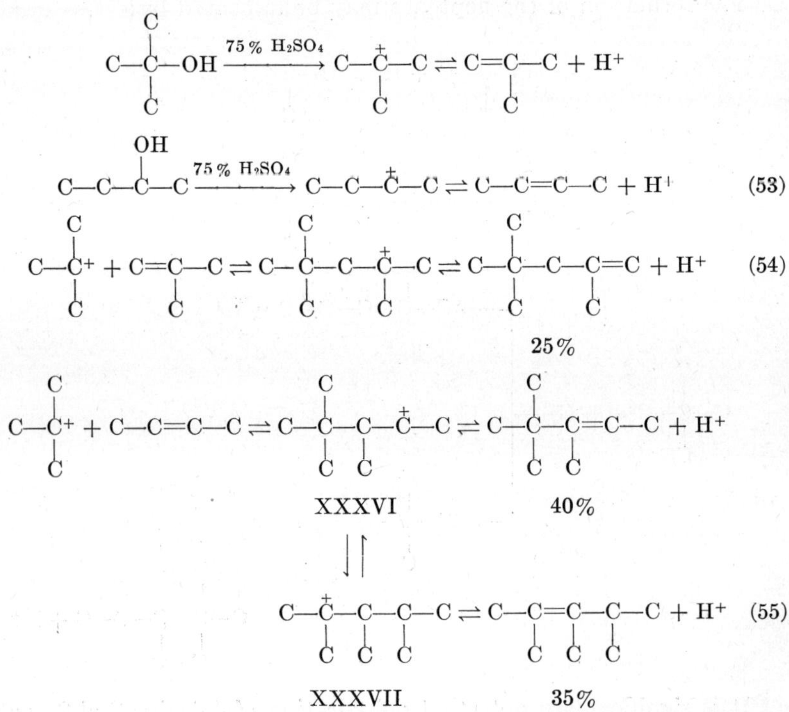

Formation of XXXVII apparently again involves a 1,3-shift of a methyl group (cf. eq. 41).

The octenes appear to be formed by the addition of the *tert*-butyl ion to methylpropene and 2-butene to the extent of 25 and 75%, respectively, about half of the product in the latter case being formed by rearrangement. None of the products which would be formed by the addition of the *sec*-butyl carbonium ion to the 2-butene or to isobutylene were obtained. The rate of addition of the *sec*-butyl ion to the olefinic double bond is apparently too slow; this finds analogy in the comparative ease of addition of *tert*-butyl chloride and *sec*-butyl chloride to ethylene in the presence of Friedel-Crafts type catalysts (Schmerling, 46). 1-Butene is not formed as an intermediate because loss of a proton occurs more readily from the methylene group than from the methyl group.

Copolymerization of isobutylene with 2-butene in the presence of hydroxyfluorboric acid at 20–25° followed by hydrogenation of the polymer yielded a liquid product 36% of which was an octane fraction consisting (by infrared analysis) of 4% 2,2,4-trimethylpentane, 27% 2,2,3-trimethylpentane and 69% 2,3,4-trimethylpentane (Brooks, 43). Rearrangement of carbonium ion XXXVI to carbonium ion XXXVII evidently occurs to a larger extent with this catalyst than with sulfuric acid.

i. Methylpropene plus 1-Butene. Infrared analysis of the hydrogenated octenes (about 43% of the polymer) obtained by the copolymerization of isobutylene and 1-butene in the presence of dihydroxyfluoboric acid at 20° showed that the octane product consisted of 5.3% 2,2-dimethylhexane, 28.4% 2,3-dimethylhexane, 10.0% 2,2,4-trimethylpentane, 19.3% 2,2,3-trimethylpentane, and 37.0% 2,3,4-trimethylpentane (Brooks, 43). The trimethylpentanes are formed via the selfpolymerization of the isobutylene and the copolymerization of the isobutylene and 2-butene formed by intermediate isomerization of the 1-butene:

$$\mathrm{C{=}C{-}C{-}C + H^+ \rightleftharpoons C{-}\overset{+}{C}{-}C{-}C \rightleftharpoons C{-}C{=}C{-}C + H^+} \quad (56)$$

The dimethylhexanes are formed by copolymerization of the isobutylene with 1-butene:

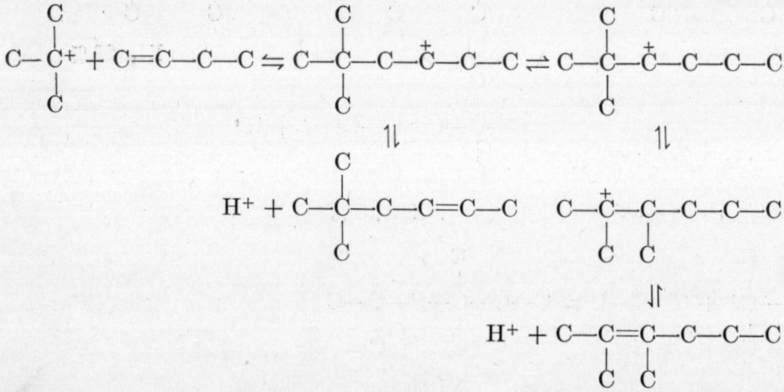

(57)

j. Methylpropene Plus Methylbutenes. Less definite results were obtained in the copolymerization of the isobutylene and isoamylenes. The reaction of equimolecular amounts of *tert*-butyl alcohol and *tert*-amyl alcohol with 65% sulfuric acid at 80° yielded isobutylene, 0.5%; isoamylenes, 30%; diisobutylenes, 22%; nonenes, 17%; diamylenes, 6%; triisobutylenes, 6% and higher polymers, 1.5% (Whitmore and

Mixon, 47). Ozonolysis of the nonene mixture (which could not be separated by fractionation) indicated that it consisted of the following nonenes: 2,3,4,4-tetramethyl-1- and -2-pentene, 50 and 10% respectively; 3,5,5-trimethyl-2 and -2-hexene, 23 and 5%, respectively; and 2,4,4-trimethyl-2-hexene, 10%. As is shown in the equations given below, the yields indicate that 88% of the nonenes were formed by the addition of a *tert*-butyl ion to an amylene as compared with 12% formed by the addition of a *tert*-pentyl ion to isobutylene. The comparatively slow rate of reaction of the *tert*-pentyl ion is also indicated by the fact that only 0.5% of isobutylene was obtained as compared with 30% of isoamylenes; also, only 6% of diamylenes were formed as compared to 22% of diisobutylenes.

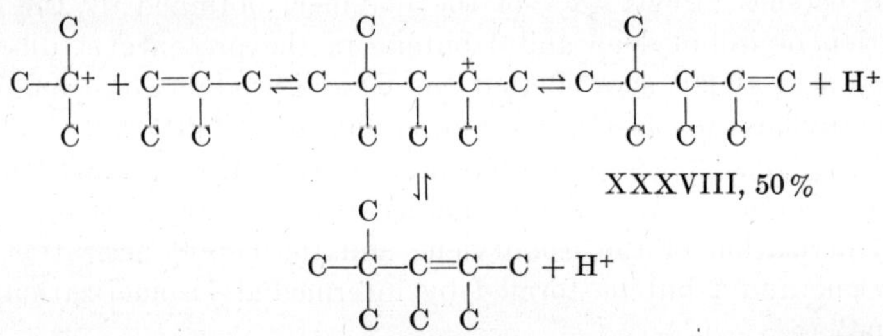

XXXVIII, 50%

XXXIX, 10% (58)

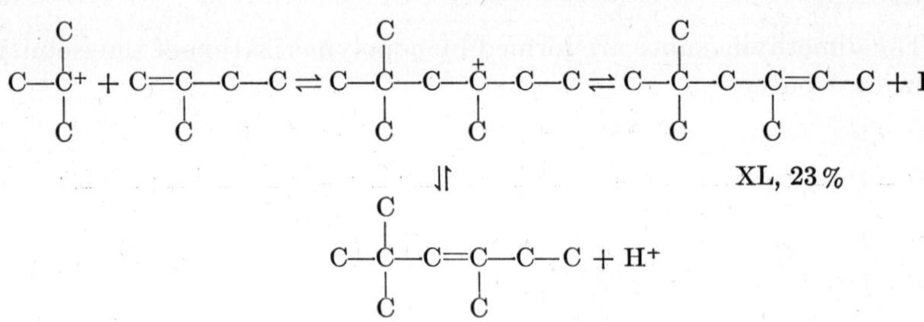

XL, 23%

XLI, 5% (59)

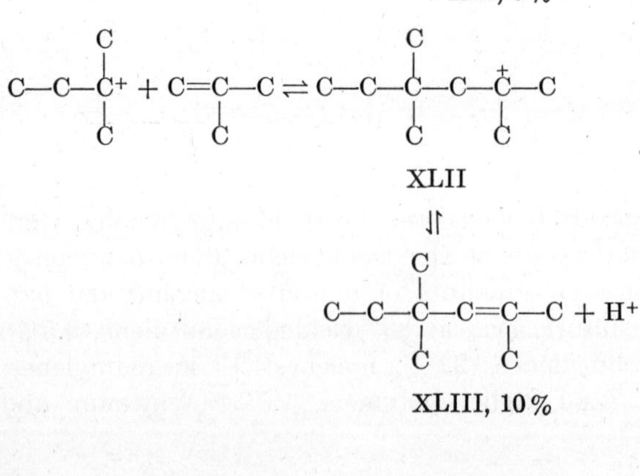

XLII

XLIII, 10% (60)

The fact that about five times as much of 2,3,4,4-tetramethyl-1-pentene (XXXVIII) was obtained as its 2-isomer (XXXIX) indicates that the loss of a proton from either of the two methyl groups takes place about five times as easily as do the loss of the proton on the tertiary carbon atom that is part of the neopentyl system. Similarly, the relative amounts of 3,5,5-trimethyl-2-hexene and its 3-isomer (XL and XLI) indicates that the loss of a proton from the ethyl group occurs about five times as readily as from the neopentyl group; no loss of a proton from the methyl group appears to have occurred. By analogy with the formation of the two isomeric diisobutylenes from the carbonium ion VI it would be expected that the carbonium ion XLII which leads to the formation of 2,4,4-trimethyl-2-hexene (XLIII) would yield the 1-isomer in about four to five times the amount of the 2-isomer. The failure to find any of the 1-isomer was "little less than startling" (Whitmore and Mixon, 47).

k. 1- and 2- Butene plus Propene. The reaction of a mixture of "*n*-butylene" and propene over a solid phosphoric acid at 260° and 40 atmospheres pressure resulted in a 53% yield of liquid product, 38% of the *n*-butylene and 64% of the propene having reacted (Hoog *et al.*, 34). The hydrogenated product contained only 30% of heptane, the chief component of which was 3-methylhexane:

$$\mathrm{C{=}C{-}C{-}C + H^+ \rightleftharpoons C{-}\overset{+}{C}{-}C{-}C \rightleftharpoons C{-}C{=}C{-}C + H^+} \qquad (61)$$

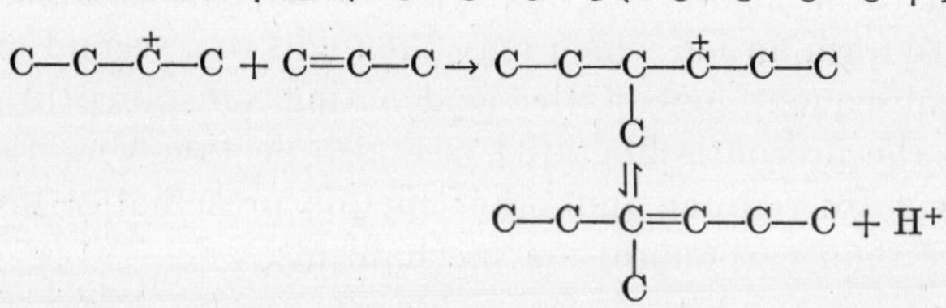

(62)

l. Methylbutenes Plus Propene. A mixture of methylbutenes consisting of 87.5% 2-methyl-2-butene and 12.5% 2-methyl-1-butene was copolymerized with propene in the presence of dihydroxyfluorboric acid at 20° (Brooks, 43). The polymer was hydrogenated and distilled. Infrared analysis of the octane fraction (66.4% of the total product) gave 54.8% 2-methyl-3-ethylpentane, 42.5% 3,4-dimethylhexane and 2.7% unidentified. The formation of the octenes is indicated as follows

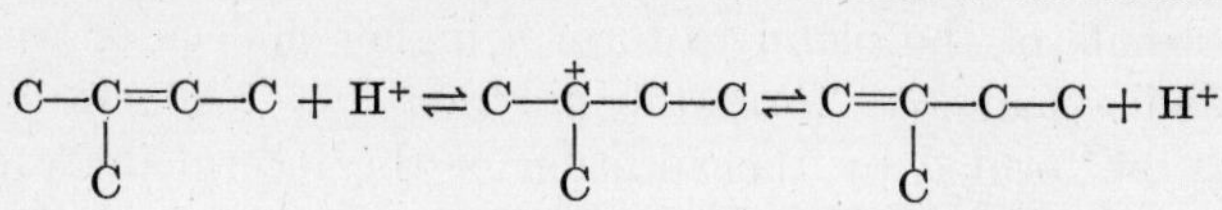

(63)

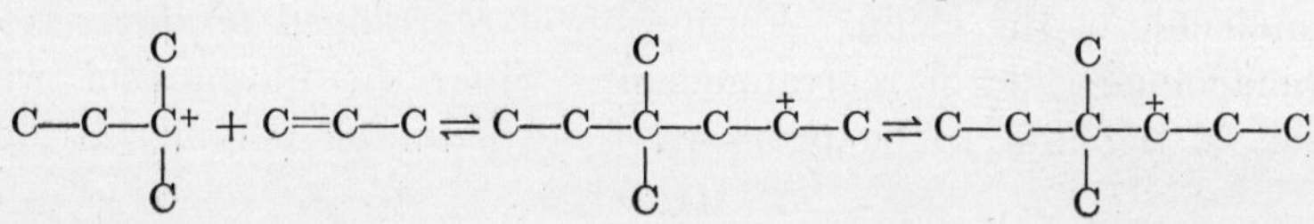

(64)

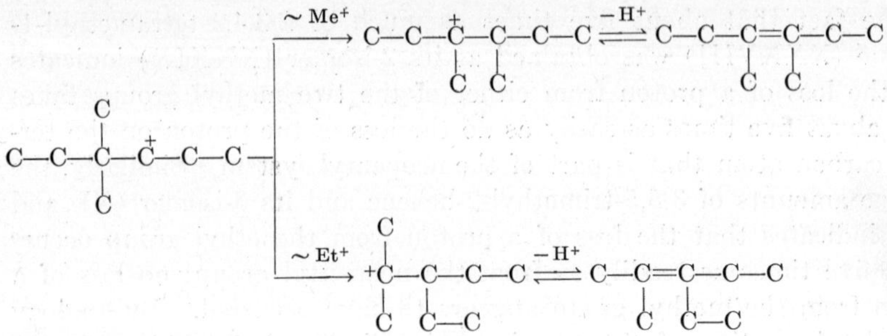

(65)

(66)

m. Conclusions. An important advantage of the carbonium ion mechanism is that it gives a common basis for a large number of hydrocarbon reactions including the polymerization of alkenes, the alkylation of alkanes (Bartlett *et al.*, 80; Schmerling, 48) and of aromatic hydrocarbons (Price, 49) and the isomerization of alkanes (Bloch *et al.*, 50). Its chief disadvantage is that it is not possible to predict which of several possible isomers will actually be obtained. The objection has also been made (Bergmann, 51) that not all polymerizations are catalyzed by acids; it was pointed out that dimerization may take place in the absence of a proton; for example when diphenylene is polymerized by stannic chloride or α-methylstyrene by sodium. However, as was shown above, such substances as boron trifluoride and stannic chloride, may be considered to be acid-acting materials which are capable of adding to the olefin to form an ion which may then add to a second molecule of olefin with subsequent loss of the acid acting substance in much the same way as the proton is liberated (see also Section V). Polymerizations catalyzed by sodium and other metals presumably involve free radical rather than carbonium ion mechanisms.

2. Ester Mechanisms

a. Ester plus Olefin. A mechanism which is closely related to the carbonium ion mechanism assumes that the condensation takes place by way of addition of an alkyl ester to the olefin (Lwow, 52; cf. Kondakow, 53, and Katsuno, 54). The first step of the mechanism consists in ester formation by the reaction of the olefin with the acid, the addition occurring in accordance with Markowinkoff's rule; the ester adds to another molecule of the olefin to form a higher molecular weight ester which then dissociates to form the polymer and regenerate the acid. Trimer may be formed by the addition of the intermediate ester to a third molecule of the olefin. Formation of rearranged products is due to a pinacolone type of rearrangement. Since this mechanism, which was one of the first to be proposed to explain the polymerization of olefins in the presence of sulfuric acid, is in a sense a molecular expression of the carbonium ion mechanism, it will obviously explain the formation of the same products as does the carbonium ion mechanism. While it inherently describes the same sequence of steps as does the carbonium ion mechanism, it is somewhat less satisfactory in that it does not give as good a picture of the driving force of the reaction. The difference between the two mechanisms may perhaps be considered to be one of language rather than of kind, with the carbonium ions giving an insight into the inner mechanism. It should be remembered, of course, that in the carbonium ion theory the word "ion" has a different connotation than it does in inorganic chemistry; the degree to which the ester is dissociated may actually lie somewhere between the undissociated ester as in the ester mechanism and the free ions as usually written (although actually believed to be very short-lived) in the carbonium ion mechanism. Nevertheless, the postulation that the true catalyst in the catalytic polymerization of olefins is a hydrogen ion or other electron acceptor (hence, an acid) appears to merit general acceptance.

The interaction of an ester and an olefin was also the basis of another early polymerization mechanism (Berthelot, 55). It was postulated that the acid portion of the ester and a hydrogen atom from the olefin are eliminated in the following manner:

$$CH_3-\underset{CH_3}{\overset{CH_3}{\underset{|}{\overset{|}{C}}}}-OSO_3H + HCH{=}\underset{CH_3}{\underset{|}{C}}-CH_3 \rightarrow CH_3-\underset{CH_3}{\overset{CH_3}{\underset{|}{\overset{|}{C}}}}-CH{=}\underset{CH_3}{\underset{|}{C}}-CH_3 + H_2SO_4 \quad (67)$$

Presumably, elimination of allylic hydrogen may also occur:

$$CH_3-\underset{CH_3}{\overset{CH_3}{\underset{|}{\overset{|}{C}}}}-OSO_3H + HCH_2-\underset{CH_3}{\underset{|}{C}}{=}CH_2 \rightarrow CH_3-\underset{CH_3}{\overset{CH_3}{\underset{|}{\overset{|}{C}}}}-CH_2-\underset{CH_3}{\underset{|}{C}}{=}CH_2 + H_2SO_4 \quad (68)$$

This mechanism, however, does not take into consideration rearrangements which result in the formation of products having a different carbon skeleton than the primary product.

b. Two Esters. An investigation of the polymerization of olefins in the presence of phosphoric acid catalyst led to the postulation that the polymerization involved the interaction of two molecules of phosphoric acid ester (Ipatieff, 56):

$$2CH_3-\underset{CH_3}{\underset{|}{C}}{=}CH_2 + 2H_3PO_4 \rightarrow 2CH_3-\underset{CH_3}{\overset{CH_3}{\underset{|}{\overset{|}{C}}}}-OPO(OH)_2 \quad (69)$$

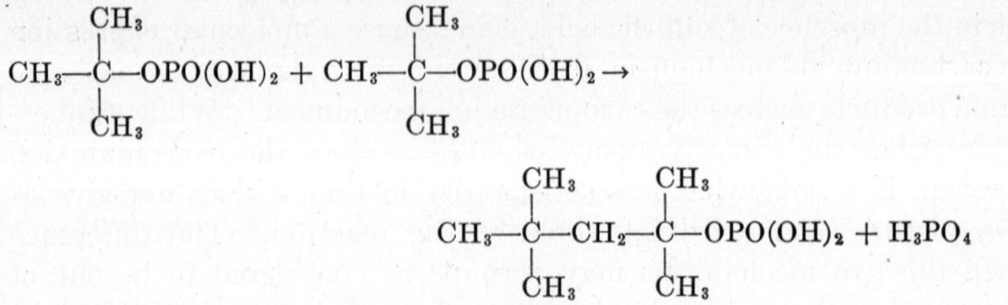

(70)

XLIV

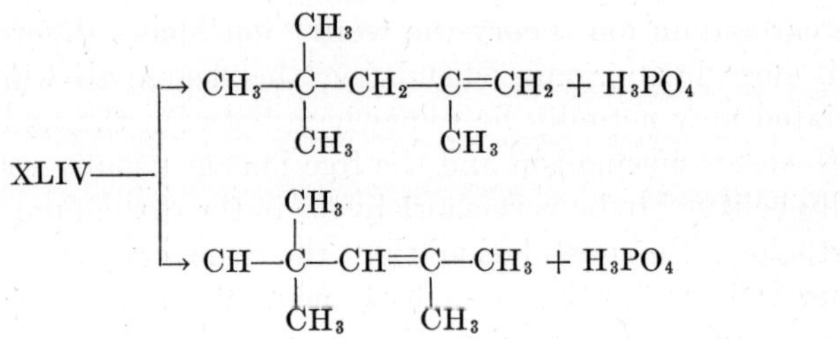

(71)

(72)

Formation of products of rearrangement may be looked upon as occurring by way of loss of hydrogen from a carbon atom which is not adjacent to the carbon atom holding the phosphate radical. This results in the transitory formation of a cyclopropane or cyclobutane ring which then opens to yield the rearranged olefin. Thus, in the copolymerization of isobutylene with 2-butene, the intermediate ester may react in the following ways:

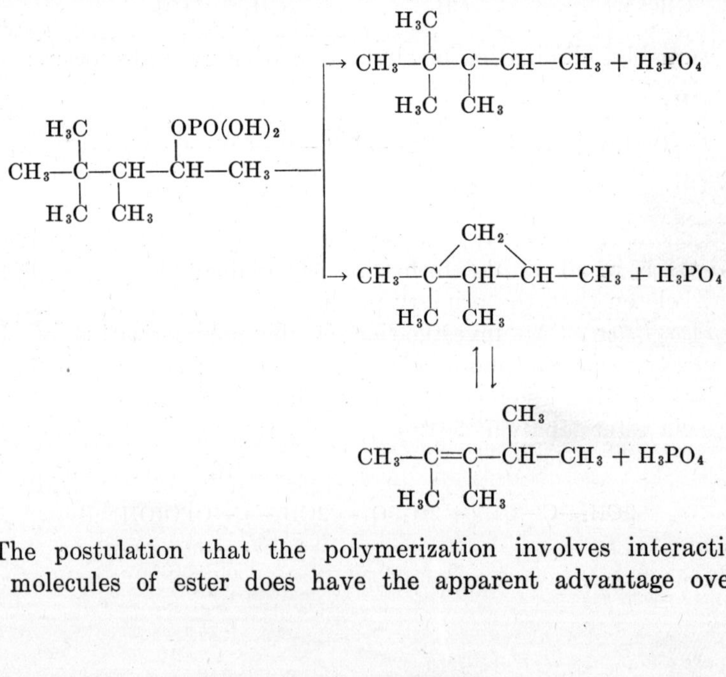

(73)

(74)

(75)

The postulation that the polymerization involves interaction of two molecules of ester does have the apparent advantage over the postulation that it involves the reaction of an ester or a carbonium ion with an olefin in that in many cases the olefin first dissolves in the catalyst and then separates as a polymer; this is particularly true in conjunct polymerization with strong sulfuric acid. It has been stated (Norris and Joubert, 57) that "The accepted explanation of polymerization, namely that is consists of the condensation of the alkyl sulfuric acid and the unsaturated hydrocarbon, does not appear to be correct in the case of at least four of the five [pentene] isomers [isopropylethylene being the exception] because these hydrocarbons can be made to pass completely into solution as the alcohol or acid ester before polymerization begins."

Similarly, the reaction of propene with 90% orthophosphoric acid at 125° at an initial pressure of 10 atmospheres yielded a homogeneous liquid product which presumably contained isopropyl phosphate (Ipatieff, 56). When this liquid was heated at 150° two layers were formed. The upper layer consisted of propylene polymer and the lower layer was phosphoric acid which was capable of polymerizing additional propene.

On the other hand, a study of the kinetics of the polymerization of propene in the presence of *dilute* phosphoric acid (10–50% by weight) at 260–350° and 170–410 atmospheres has shown that the rate of polymerization is proportional to the square of the gas-phase propene concentration and the first power of the acid concentration, indicating that the polymerization involves addition of an ester to an olefin rather than interaction of two molecules of ester (Monroe and Gilliland, 58). It was pointed out, however, that both mechanisms may occur side by side and that under certain circumstances (as in dilute acid) one of these mechanisms may predominate over the other.

c. Complex Esters. A modification of the "two ester" mechanism was proposed as the result of an investigation of the polymerization of olefins in the presence of trideuterophosphoric acid (Farkas and Farkas, 59). It was found that when ethylene, propylene, and isobutylene were polymerized in the presence of trideuterophosphoric acid, both the polymer and the recovered unreacted olefin contained up to 48% deuterium depending largely on the reaction time and temperature. Under similar conditions, the polymerization and the exchange reactions were far slower with *n*-butylene than with isobutylene; the reactions with ethylene and propylene were still slower. It was stated that since the various proposed polymerization mechanisms predict different degrees of exchange, the experimental results permit choosing between them. In calculating the amount of deuterium to be expected in the polymer and in the recovered olefin it was assumed that not only is the rate of polymerization constant at the beginning of the experiment but also the rate of exchange. (The validity of this assumption is, however, questionable. There seems to be no means for determining to what extent isobutylene, for example, will undergo deuterium exchange, by successive additions of deuterium ions and elimination of hydrogen ions before addition of the *tert*-butyl ion to isobutylene occurs.) It was concluded that the catalyst and the olefin molecule can combine in two different forms, one of which is produced from the ordinary ester by way of transfer of hydrogen atoms via the intermediate compound (XLV).

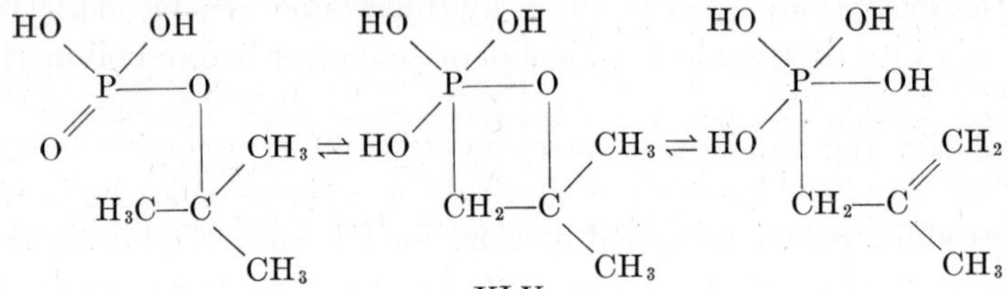

XLV

(76)

Interaction of the two forms, either on the same phosphoric acid molecule or on two adjacent molecules, yields the polymer:

e.g.,

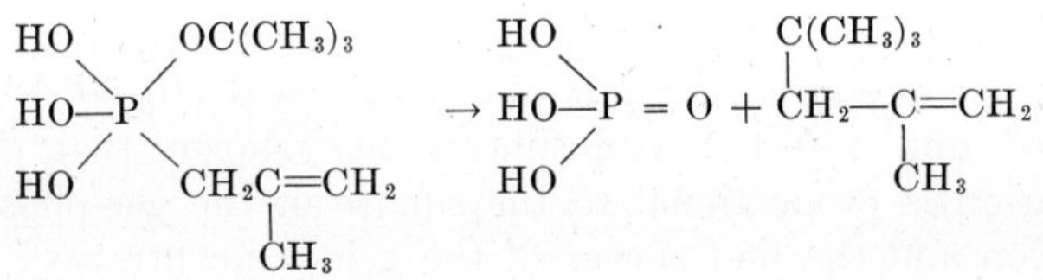

(77)

e.g.,

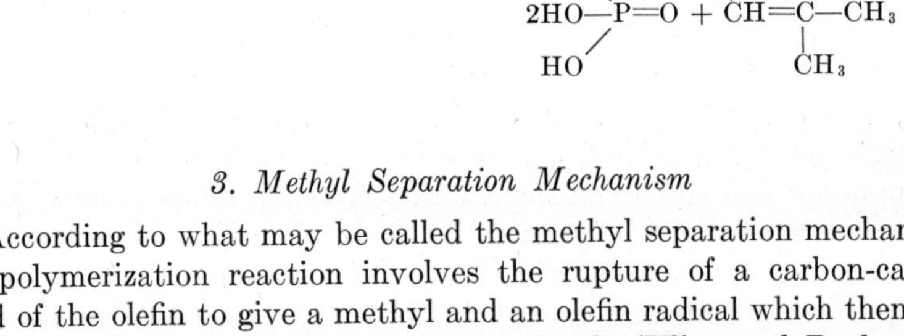

(78)

3. Methyl Separation Mechanism

According to what may be called the methyl separation mechanism, the polymerization reaction involves the rupture of a carbon-carbon bond of the olefin to give a methyl and an olefin radical which then add to the double bond of another olefin molecule (Kline and Drake, 60). The molecule of isobutylene, for example, behaves as though it were activated in the following manner: $CH_3{}^{-+}C(CH_3){=}CH_2$. The positive and negative signs indicate the relative electronegativities (Kharasch, 61) and are not intended to indicate ionization. The addition of these two fragments to the double bond of a nonactivated molecule of isobutylene then takes place in accordance with the relative electronegativities as follows:

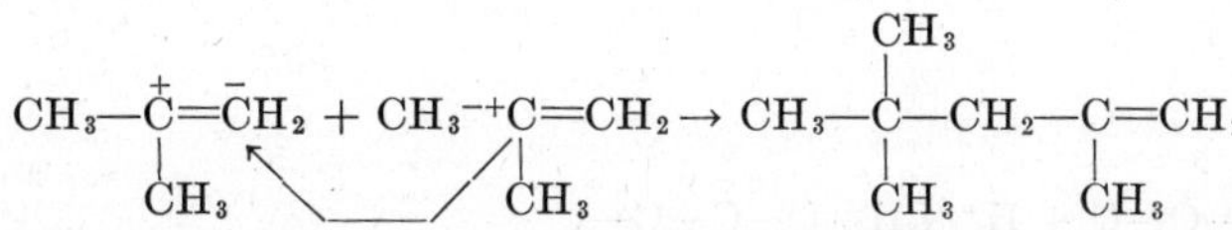

(79)

The formation of the other diisobutylene isomer is explained by assuming that isobutylene may also behave as $H^{+-}CH{=}C(CH_3)_2$:

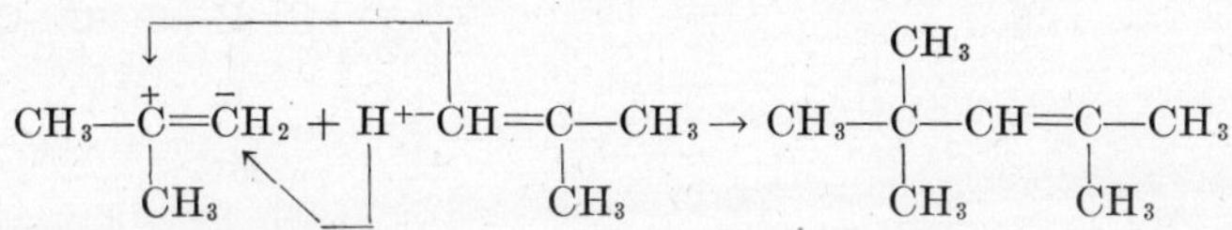

(80)

Since 2,4,4-trimethyl-1-pentene is obtained in about four times the amount as is 2,4,4-trimethyl-2-pentene it may be assumed that the "activation" takes place 80% by methyl separation and 20% by hydrogen separation.

In order to predict the formation of the isobutylene trimers, three alkenes may be considered to be activated, the original isobutylene and the two diisobutylenes. The various possibilities are shown in the following equations:

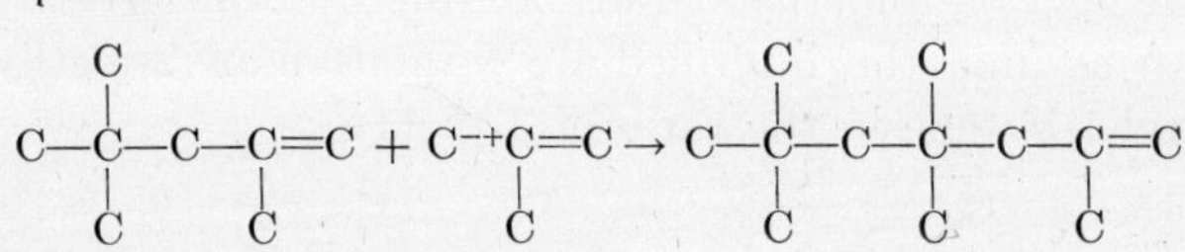

(81)

5%

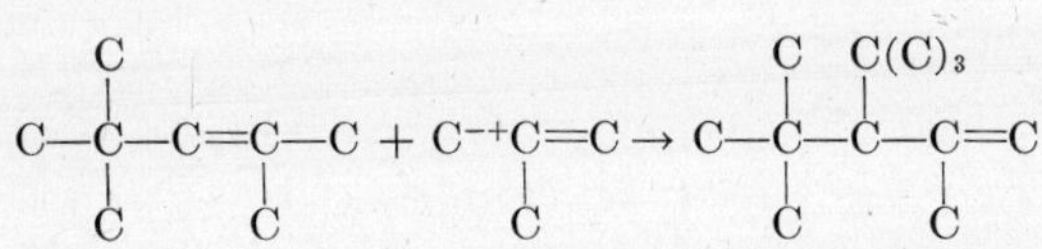

(82)

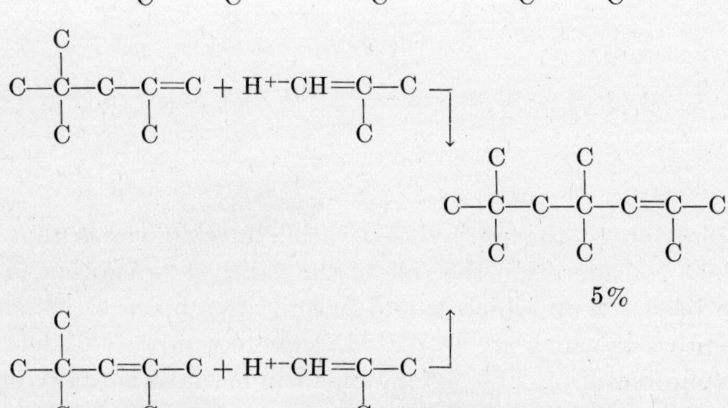

(83)

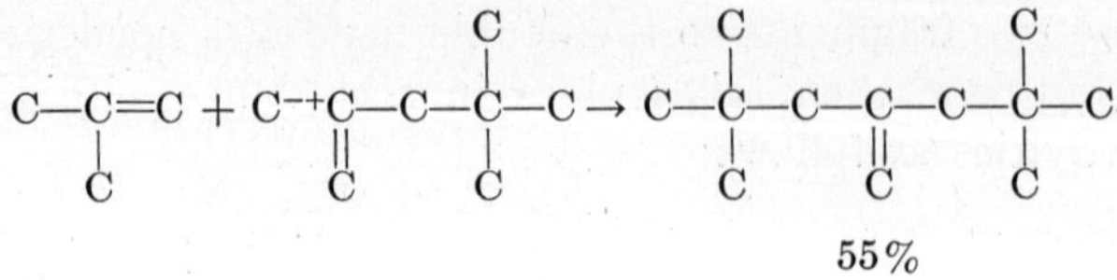

(84)

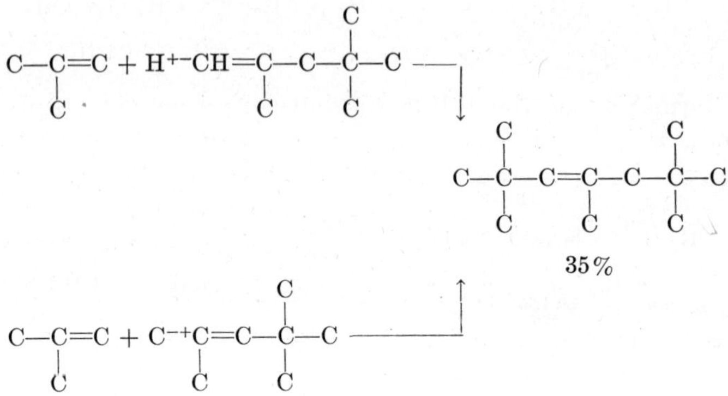

(85)

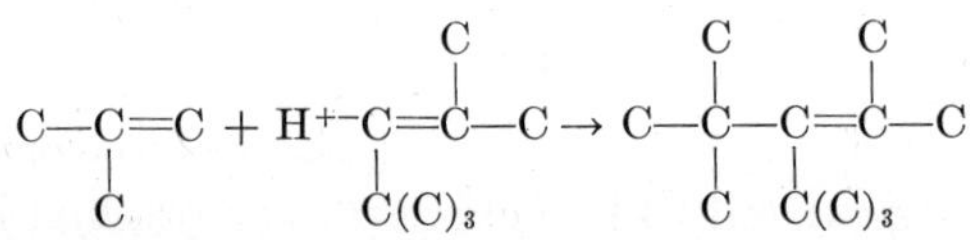

(86)

In order to explain the structures of the decenes by the action of 75% sulfuric acid on 3-methyl-2-butanol (methylisopropylcarbinol), it is necessary to assume the intermediate formation of 2-methyl-2-butene and 2-methyl-1-butene in the ratio of 3 to 1:

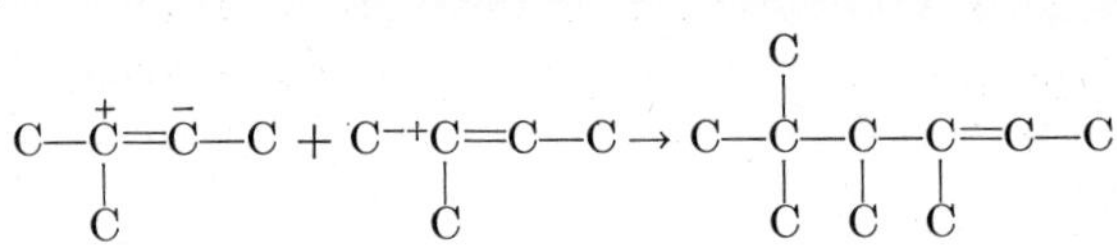

(87)

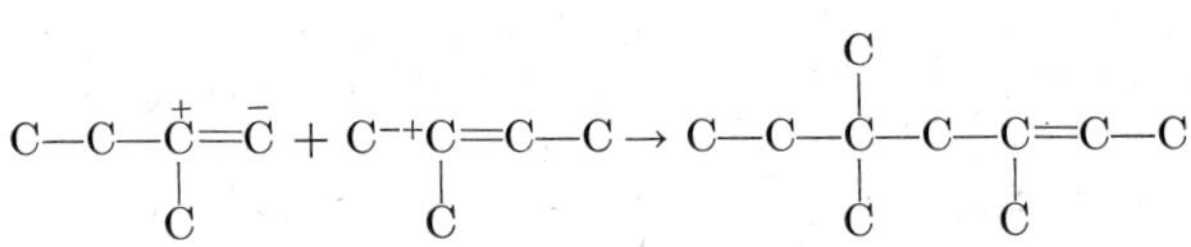

(88)

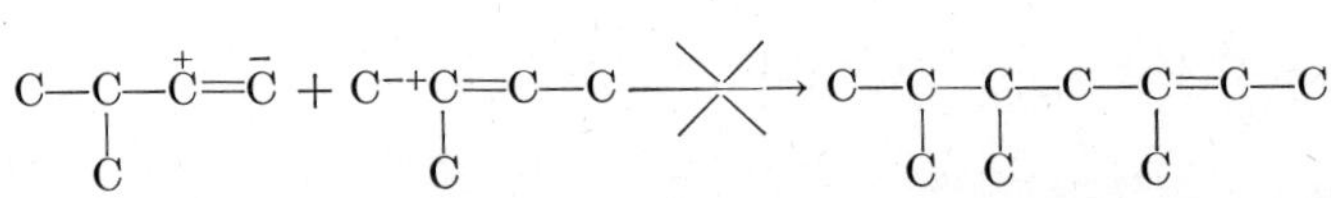

(89)

(Not found)

One objection to the methyl separation mechanism is that it does not take into consideration the role of the catalyst. Another is that no rules have been set up which would permit the choice of which olefins act as acceptors or which are activated. There is no way of determining which of a number of possible compounds will be those actually obtained;

it predicts that 4,5-dimethyl-2-hexene will be the major product of the polymerization of 2-butene together with a lesser amount of 3,4-dimethyl-2-hexene.

$$\mathrm{C{-}C{=}C{-}C + C^{-+}C{=}C{-}C \rightarrow C{-}\underset{\underset{C}{|}}{C}{-}\underset{\underset{C}{|}}{C}{-}C{=}C{-}C} \qquad (90)$$

(Not found)

$$\mathrm{C{-}C{=}C{-}C + H^{+-}\underset{\underset{C}{|}}{C}{=}C{-}C \rightarrow C{-}C{-}\underset{\underset{C}{|}}{C}{-}\underset{\underset{C}{|}}{C}{=}C{-}C} \qquad (91)$$

As was shown above, only the latter is obtained when *sec*-butyl alcohol is treated with 75% sulfuric acid at 80° under pressure. It is significant to note that this shortcoming of the methyl separation mechanism as compared to the carbonium ion mechanism was pointed out in a later paper (Drake and Veitch, Jr., 30) by one of the original proponents of the methyl separation mechanism.

4. Hydrogen Separation Mechanism

A mechanism which attempts to overcome the disadvantages of the methyl separation mechanism assumes that hydrogen, rather than methyl, separation occurs and that the subsequent addition to the second alkene molecule takes place in accordance with certain rules; rearrangement of the original olefin before, and of the primary polymer after, polymerization yields isomeric polymers (Wachter, 62; cf. Sparks *et al.*, 63). A carbon-carbon rupture might seem more reasonable than a carbon-hydrogen rupture in view of the lower energy of dissociation of the carbon-carbon bond (82.5 kcal.) as compared to that of the carbon-hydrogen bond (100 kcal.); however, the apparent activation energy required to break the carbon-carbon bond in ethylene and propene on the catalyst is higher than that required to break the carbon-hydrogen bond, and the presence of the double bond in the olefin markedly weakens the carbon-hydrogen bonds (Morikawa *et al.*, 64).

The general equation for the hydrogen separation mechanism may be formulated as follows (Wachter, 62):

$$\underset{\text{Acceptor}}{\mathrm{{-}\overset{|}{C}{=}\overset{|}{C}{-}}} + \underset{\text{Donor}}{\mathrm{H{-}\underset{|}{C}{=}\underset{|}{C}{-}}} \rightarrow \underset{\text{Primary product}}{\mathrm{H{-}\overset{|}{\underset{|}{C}}{-}\overset{|}{\underset{|}{C}}{-}\underset{|}{C}{=}\underset{|}{C}{-}}} \qquad (92)$$

When a pure alkene is polymerized and only one form of the olefin is possible or when it is definitely known that only one molecular form is reacting under the polymerizing conditions, there is no need to decide which molecule is the acceptor and which the donor molecule. On the other hand, when two different olefins are copolymerized, it is necessary to decide (1) which one acts as the acceptor and which as the donor and (2) which hydrogen and carbon atoms in both are active in the reaction. A number of rules were set up in order to make possible the necessary decisions. Usually, the acceptor olefin will be the one with the more easily activated double bond; in other words, it will be the more readily polymerized olefin. In some cases the difference may not be marked and the two copolymerizing olefins may each act both as acceptor and donor molecules to approximately the same extent. In general, the ease of polymerization of alkenes increases from ethylene upward and is at the maximum with the pentenes. Tertiary alkenes are more readily polymerized than secondary alkenes. Dimers and trimers are far less easily polymerized than their monomers.

The active hydrogen atom in the donor molecule is the one which is attached to a terminal double-bonded carbon atom or to that double-bonded carbon to which the smaller alkyl group is attached. Thus, the active hydrogens in the following alkenes are indicated by asterisks:

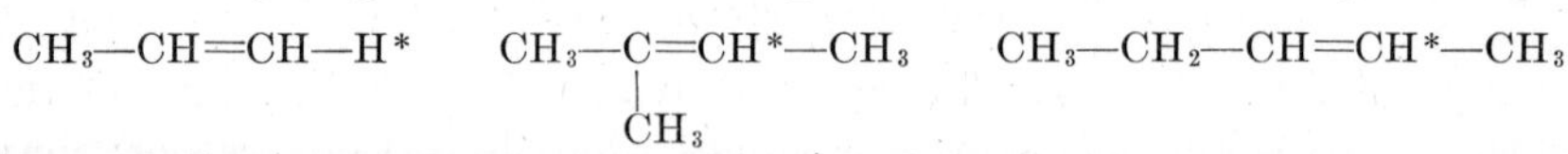

The carbon atom by means of which the alkene radical becomes attached to the acceptor molecule is, of course, the one to which the active hydrogen is attached.

The hydrogen atom adds to the terminal double-bonded carbon atom of the acceptor molecule or to that double-bonded carbon atom which has the smallest alkyl group; the olefin radical adds to the adjoining double-bonded carbon atom. When the acceptor olefin contains a tertiary double-bonded carbon atom, the olefin radical from the donor always becomes attached to it.

In accordance with these principles, formation of primary polymer and copolymers may be illustrated by the following examples:

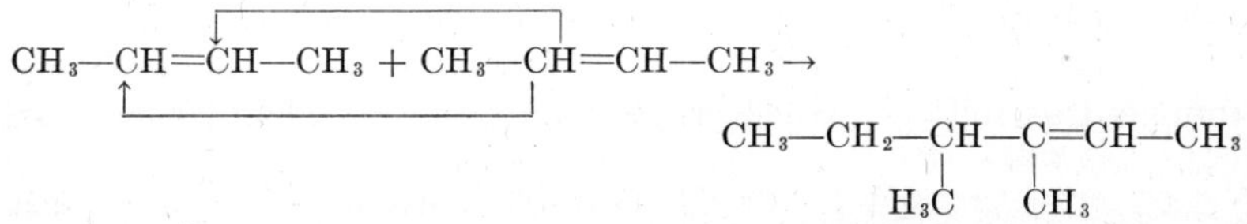
(93)

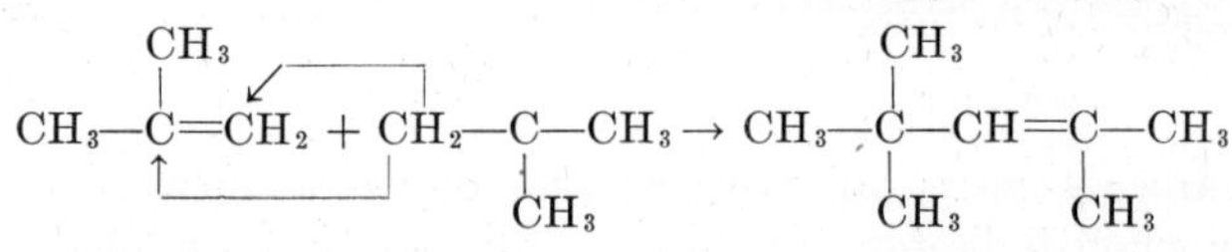
(94)

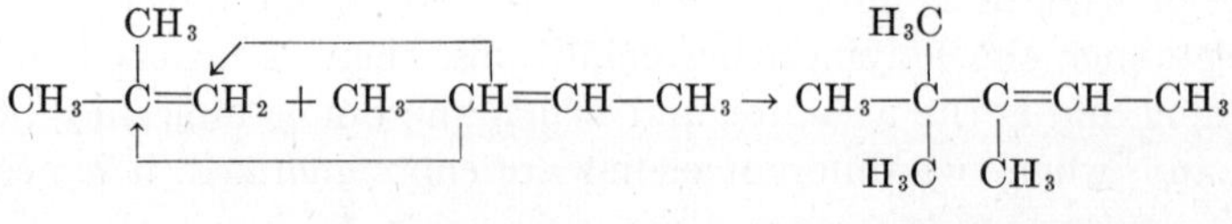
(95)

Since the catalysts which are active in polymerization are also catalysts for rearrangements of the reacting alkenes and their polymer products, these primary polymers may undergo rearrangement. The types of rearrangement were classified into four classes (Wachter, 62):

Rule A. Olefins having a terminal double bond (e.g., 1-butene) are less stable than straight chain olefins having an internal double bond (e.g., 2-butene) and tend to yield a latter type under the influence of heat, catalyst, and other methods of activation (Whitmore and Herndon, 65; Whitmore and Homeyer, 44).

Rule B. A secondary olefin will tend to rearrange to a tertiary olefin when a tertiary carbon atom is in alpha position to a double-bonded carbon atom. The rearrangement of 3-methyl-1-butene to 2-methyl-2-butene is well known (Norris and Reuter, 66).

Rule C. Tertiary olefins may isomerize to an equilibrium mixture of olefins differing in the position of the double bond at the tertiary carbon atom. This is related to the order of decreasing ease with which the different alkyl groups supply a hydrogen to form water in the dehydration of tertiary alcohols to olefins: isopropyl, ethyl, *n*-propyl, *n*-butyl, *n*-pentyl, methyl (Church *et al.*, 67). Thus, 3-methyl-3-heptanol dehydrates to give a mixture consisting of 55% 3-methyl-2-heptene (hydrogen from the ethyl group), 30% 3-methyl-3-heptene (hydrogen from the *n*-butyl group), and 15% 2-ethyl-1-hexene (hydrogen from the methyl group).

Rule D. When a quarternary carbon atom is the alpha position to a secondary or tertiary double-bonded carbon atom, a pinacol type of rearrangement involving migration of a methyl group may occur. Examples of this type of rearrangement include isomerization of 3,3-dimethyl-1-butene (*tert*-butyl-ethylene) to 2,3-dimethyl-2-butene (tetramethylethylene) and of 3,4,4-trimethyl-2-pentene to 2,3,3-trimethyl-1-pentene.

These rules, while they describe the various types of rearrangement that may occur to yield the products which are actually isolated from the polymerization of olefins, do not actually give an insight into the mechanism whereby the isomerization occurs. In this respect, the carbonium ion mechanism has an important advantage: as was shown in Section III, 1, the rearrangements mentioned in Rules A, B, and C may be explained as caused by the addition of a proton (from the catalyst) to one of the double-bonded carbon atoms followed by loss of a proton from a different alkyl group to yield the isomerized olefin. The rearrangements of the type of Rule D involve the addition of the proton to the olefin followed by migration of a methyl group with its electrons pair to the positive carbon atom and subsequent loss of a proton. In short, the carbonium-type mechanism offers a common basis for both the rearrangements and the polymerization steps.

The fact that tetramethylethylene which contains no hydrogen on either of the double-bonded carbon atoms undergoes polymerization to yield dimer might be considered as a means of choosing between the carbonium ion mechanism and the hydrogen separation mechanism. However, regardless of which mechanism is used, it is necessary to assume that the olefin first undergoes isomerization to *tert*-butylethylene.

A serious objection to the hydrogen separation mechanism is that it postulates that the hydrogen in vinylic position to the double bond is involved; it seems more reasonable to expect that the hydrogen in allylic position would be more readily separated. Furthermore, the mechanism gives no consideration to the role of the catalyst other than to state that it is the means whereby the bonds of the reacting molecules are activated. The postulation of the intermediate formation of carbonium ion or esters seems to be more plausible.

It should be noted that Wachter (62) has stated that the generalizations and working rules which he presents are supported by the evidence on the structure of the polymers and that they are put forward as working hypotheses of polymerization and not as a description of the actual mechanism. Their chief use is to aid in predicting the structure of the polymerization products.

5. *Hydrogen Transfer Mechanism*

A study of the isomerization of 1-butene to 2-butene in the presence of various catalysts, including radioactive phosphoric acid, has led to a hydrogen transfer mechanism to explain the catalytic isomerization of olefins, the alkylation of isoparaffins by olefins, cracking, and polymerization (Turkevich and Smith, 68). It was found that no exchange of tritium for hydrogen occurred when ethylene was treated with tritium phosphoric acid for 48 hours at 50°. Propylene and 1-butene on the other hand, underwent the exchange. It was shown further that the rate of isomerization of 1-butene in the presence of sulfuric and phosphoric acids is proportional to the amount of the catalyst when the latter is small and also that ester formation retards the rate of isomerization to 2-butene. This was taken as an indication that the rate determining step is the interaction of the acid with the butene and not the decomposition of an ester or other complex formed from the acid and the olefin. The fact that ethylene did not exchange with radioactive phosphoric acid was used as an objection to expressing the reaction in terms of carbonium ion or ester formation. Similarly, the failure of radioactive hydrogen chloride (a solution of hydrogen chloride in tritium water) to undergo exchange with 1-butene was taken further evidence that the tritium ion and, therefore, the hydrogen ion does not add readily to carbon-carbon double bonds. [It is the opinion of the present authors that these experiments merely indicate that ethyl phosphate is very stable under the reaction conditions, that its formation is not a reversible reaction, and that hydrogen chloride does not add readily to 1-butene under the experimental conditions, a fact which may readily be verified. Furthermore, it has been shown that when 2-butene is bubbled through radioactive sulfuric acid at such a rate that there is little absorption, hydrogen-tritium exchange occurs, demonstrating not only rapid exchange but also rapid reversible absorption of the olefin in the acid (Stewart and Denham, 69).]

The activation energy of the isomerization reaction was about 3.6 kcal. less than that for the exchange reaction. Comparison of the absolute rate of the two reactions at 27° showed that the rate of exchange was slower than the rate of isomerization and that the difference corresponded to a difference in activation rates for the two processes of 2.9 kcal., a result which is in rather good agreement with the value of 3.6 obtained from the temperature coefficients. This difference may be correlated with the difference in zero point energy of hydrogen and tritium. Hence, the two processes appear to follow the same mechanism, and the difference in activation energy for the two processes may be taken as an indication that the hydrogen and tritium are covalently bonded and do not react as hydrogen or tritium ions.

The mechanism which was proposed to explain isomerization of 1-butene and at the same time be consistent with the above-mentioned experimental results is based on the geometric relationships between the catalyst and the olefin. It was suggested that when a phosphoric acid or sulfuric acid molecule approaches a 1-butene molecule, the approach takes place in such a way that one of hydrogen atoms of the acid approaches the end carbon atom of the double-bonded pair while one of the oxygens atoms which has no hydrogen on it approaches the hydrogen atom on the third carbon atom in the 1-butene molecule. When the complex which is so formed decomposes, the acid may take with it the hydrogen atom from the third carbon atom and leave the hydrogen atom on the first carbon atom. The resulting product is 2-butene.

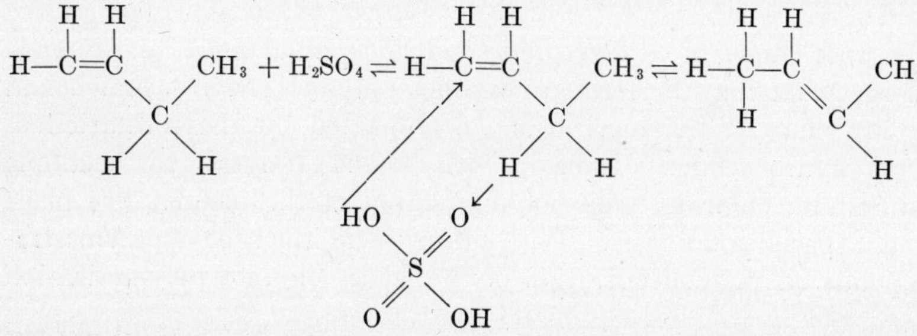

(96)

The activated complexes have no bond distortion in either the olefin or the catalyst molecule provided that the catalyst is able to furnish the hydrogen and accept another at a distance of approximately 3.5 A. This distance is the distance between the hydrogen on the first carbon and the third carbon atom in an aliphatic chain. This critical condition of the catalyst is satisfied by sulfuric acid, phosphoric acid, silicic acid, aluminosilicic acid, perchloric acid, moist aluminum chloride ("$HAlCl_4$"), and partially hydrogenated nickel.

Polymerization of isobutylene, for example, may be explained in terms of this hydrogen switch mechanism by assuming that the hydrogen switch occurs between two molecules of the olefin:

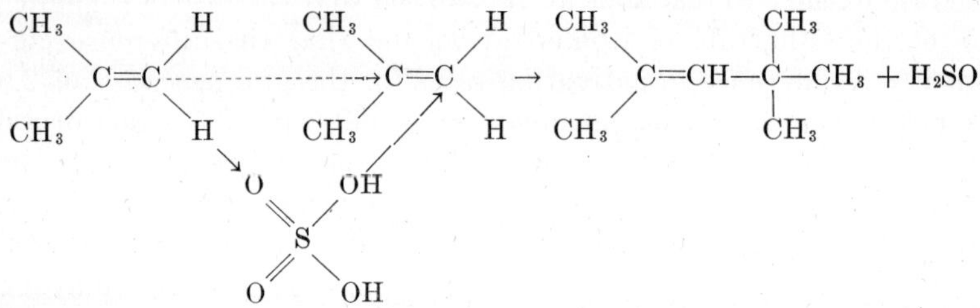

(97)

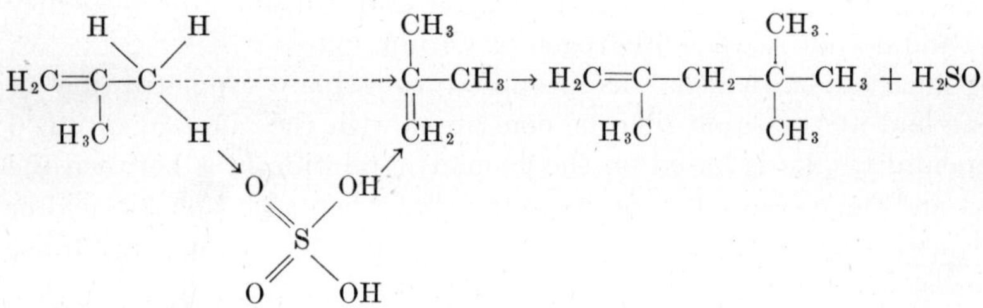

(98)

This mechanism does not seem to take into account the formation of products having a different carbon skeleton than that of the primary product; i.e., it does not explain the formation of those products which in accordance with the carbonium ion mechanism are formed by the migration of a methyl group.

IV. Conjunct Polymerization

By proper choice of catalyst and conditions, all alkenes can be made to undergo what may be termed conjunct polymerization (Ipatieff and Pines, 70), that is, polymerization accompanied by the formation of saturated hydrocarbons. Indeed, with some catalysts, for example with aluminum chloride, true polymerization of the olefin takes place only under special conditions. With sulfuric acid, the type of polymerization depends on the concentration of the acid; conjunct polymerization occurs with acid of greater than about 90% concentration while true polymerization occurs with more dilute acid. In the case of phosphoric acid catalysis, the type of polymerization is governed by the reaction temperature, conjunct polymerization occurring at temperatures above about 250–300°. A few typical examples of conjunct polymerization will be described here before the mechanism of the reaction is discussed.

1. Typical Examples

a. Aluminum Chloride as Catalyst. Ethylene did not react when treated with *pure* aluminum chloride even at superatmospheric pressure and 10–50° (Ipatieff and Grosse, 71). In the presence of hydrogen chloride or traces of impurities, on the other hand, an approximately exponential fall in pressure took place with the formation of two layers. The upper was water white and consisted of paraffin hydrocarbons while the lower was a viscous, dark red-brown oil and consisted of addition compounds of aluminum chloride and unsaturated cyclic hydrocarbons, resembling terpenes. The composition of the lower layer in an experiment in which 193 g. of ethylene was polymerized by 100 g. of aluminum chloride corresponded to the formula $C_nH_{2n-x}{\cdot}2AlCl_3$ ($x=2$ to 6). A maximum of about 10 moles of ethylene could be condensed per mole of aluminum chloride.

The polymerization of ethylene in the presence of aluminum chloride is fundamentally changed by the presence of metallic aluminum (Hall and Nash, 72). The product which was obtained at a reaction temperature of 100–200° under superatmospheric pressure was a mobile fuming liquid which was shown to contain diethyl aluminum chloride, a liquid spontaneously inflammable in air. Less conjunct polymerization occurred, the lower-boiling product consisting of olefins mixed with only minor amount of paraffins.

The diethylaluminum chloride was also a catalyst for the polymerization of ethylene. Very similar products were obtained in parallel polymerizations carried out at 300° with diethylaluminum chloride and with a mixture of aluminum chloride and aluminum. The distillation curves showed marked plateaus for C_6, C_8, and C_{10} hydrocarbons with complete absence of C_5, C_7 and C_9. The bromine numbers indicated that these fractions were mixtures of paraffins and olefins.

b. Sulfuric Acid as Catalyst. Although ethylene is not polymerized by concentrated sulfuric acid as such, it polymerized when treated with a solution of 5% copper sulfate and 2% mercurous sulfate in 95% sulfuric acid (Damiens and Lebeau, 73). The presence of the metallic salts permitted the solution of about one hundred times as much ethylene as was absorbed in their absence. After the solution had stood for some time, an upper hydrocarbon layer separated and a paste settled out. If a small quantity of the paste was immediately mixed with pure sulfuric acid, the mixture acquired the maximum absorbing capacity. This activity of the catalyst gradually decreased and was entirely lost in 24 hours. The hydrocarbon layer consisted of a mixture of saturated hydrocarbons, both paraffinic and naphthenic. The hydrocarbons tied up in the sulfuric acid layer consisted chiefly of unsaturated compounds resembling open-chain and cyclic terpenes (Boeseken and Max, 74).

When an olefin other than ethylene is treated with an excess of 96–98% sulfuric acid, it dissolves at first. On standing, two layers are formed: an upper, hydrocarbon layer and a lower, sulfuric acid layer (Ormandy and Craven, 75; Nametkin and Abakumovskaya, 76; Ipatieff and Pines, 77). When the reaction was carried out at 0° in the presence of 96% acid, the yield of hydrocarbon layer amounted to about 70% by weight of the olefin treated; the sulfuric acid layer showed an increase of about 20% in volume (Ipatieff and Pines, 77). About 32–40% of the product in the hydrocarbon layer boiled below 220° and was completely saturated. The higher boiling material consisted of a mixture of saturated, olefinic, and cycloolefinic hydrocarbons. The amount of unsaturated hydrocarbon in the fraction boiling above about 250° amounted to 30% of the total upper layer. The lower catalyst layer contained cycloolefins which were liberated on treatment with water. On the other hand, polymerization of isobutylene, for example, in the presence of 77% sulfuric acid at 0° yielded a completely olefinic product at least 84% of which boiled below 220°, di- and triisobutylenes being the chief products. It is interesting to note that a similar mixture of true polymers is obtained in the presence of alkanesulfonic acids at 80–100% concentration at 30–70° (Proell *et al.*, 78).

The ratio of acid to olefins treated has an effect on the nature of the products obtained. With isobutylene, for example, treatment with about 50% by volume of 96% sulfuric acid at 0° yielded a product of which the material boiling below 225–250° was completely paraffinic (Ipatieff and Pines, 77). When the isobutylene was treated with about 10% by volume of the concentrated acid, the polymer fraction boiling below 225° contained 50–60% of olefins.

The reaction of isobutylene with 91% sulfuric acid at 0° yielded a product which contained 63% of paraffin in the fraction boiling below 200°; when 87% sulfuric acid was used only traces of paraffins were found in the corresponding fraction, while with 77% acid no paraffins at all were produced. No polymerization of isobutylene occurred with 67% acid at 0°; at 35°, on the other hand, polymer composed of di- and triisobutylene was obtained.

Among the products which have been isolated from the polymer obtained by the reaction of isobutylene with concentrated sulfuric acid were octanes, dodecanes, hexadecanes containing about 15% of hexadecenes, eicosanes containing about 35% of eicosenes, and tetracosanes containing 42% of tetracosenes (Nametkin and Abakumovskaya, 76). Similarly, the polymer obtained from technical amylene boiling at 35–37° was found to contain decanes, pentadecanes, eicosanes, pentacosanes and triacontanes.

c. Phosphoric Acid as Catalyst. The polymerization of ethylene in the presence of 90% phosphoric acid at temperatures in the range 250 to 330° and under initial ethylene pressures of 50 to 65 atmospheres yielded a mixture of paraffinic, olefinic, naphthenic, and aromatic hydrocarbons (Ipatieff and Pines, 79). Depending on the reaction conditions, 31–46% by weight of the product boiled below 110°, 14–33% at 110–225°, 23–29% at 225–300° and 10–13% above 300°. The concentration of paraffins was highest in the lowest boiling fractions; the aromatics, on the other hand, appeared in the fractions distilling at 225° and higher. Unsaturated hydrocarbons were present in all fractions but one; no olefinic hydrocarbons were found in the product bioling below 60° obtained by the polymerization at 330°. Naphthenic hydrocarbons were present in the fractions boiling above 110°. Probably the most interesting fact about this polymerization was the formation of isobutane, the percentage of which increased with the temperature of the reaction. From 250 to 330° it varied from 2.5 to 18.8% by weight of the ethylene that reacted.

It is significant to note that the thermal polymerization of ethylene at 330° under the same conditions as those used for the phosphoric acid catalyzed polymerization yielded a product consisting of 8% paraffins, 68% olefins, and 24% naphthenes; aromatic hydrocarbons were not formed. Only 24% of the product boiled below 225° (Ipatieff and Pines, 79).

Polymerization of propene at 330° in the presence of 90% orthophosphoric acid under about 100 atmospheres initial pressure yielded a product consisting of paraffinic, olefinic, cycloparaffinic and cycloolefinic, and aromatic hydrocarbons (Ipatieff and Pines, 70). About 8% of the product boiled in the dimer (C_6) range and about 25% in the trimer (C_9) range. Isobutane was formed to the extent of more than 2% by weight of the total polymer.

The liquid product was shown to consist of 15% paraffins, 63% olefins, 10% cycloparaffins, 6% cycloolefins, and 6% aromatics. The olefins were present in the fractions boiling below, and the cycloolefins in those boiling above, 200°. The paraffins were concentrated in the lowest boiling fractions. The products boiling at 25–63°, representing 6.6% of the total polymer, consisted of 80% paraffin, the remainder being olefin. The fraction boiling at 75–85° contained 33% of paraffin and 67% of olefin, while the fraction boiling at 122–132° contained only 7% of paraffin, the remainder being olefin. Cycloalkanes appeared first in the product boiling at 155–165°, being present to the extent of 5%. Aromatic hydrocarbons were found only in the highest boiling fractions (132–145° at 8 mm.) where they formed 67% of the total fractions.

Propene also undergoes conjunct polymerization in the presence of dilute phosphoric acid at high temperatures and pressures (Monroe and Gilliland, 58). When propene was treated with 10–30% phosphoric acid at 260–305° and at 170–410 atmospheres pressure, the only operating variable which appreciably affected the composition of the polymer was the extent to which the feed was polymerized. At constant percentage reaction of the feed under these conditions, the temperature, pressure, and acid catalyst concentration had no effect on the product composition. At low conversions, the polymer consisted of nearly pure dimer; at 50% polymerization, two-thirds of the total was dimer and even when the feed was almost completely polymerized, the dimer fraction amounted to 35–40% of the total polymer. The dimer and trimer fractions obtained at temperatures of 305° or lower using a acid concentrations below 30% contained about 25% paraffins and little or no naphthenes or aromatic hydrocarbons.

While the composition of the products depend solely on the extent of polymerization of the feed when phosphoric acid concentrations below 30% are used, with higher concentrations of acid, the character of the polymer begins to change, an excess of heavier compounds forming at the expense of the yield of dimer. Polymers obtained with 40 and 50% acid showed dimer content 10–20% lower, and trimer content correspondingly higher, than were obtained at the same percentage of polymerization of the feed with lower concentrations of acid.

2. Mechanisms

a. Carbonium Ion Mechanism. Evidence in support of a carbonium ion mechanism to explain the formation of conjunct polymer was obtained in an investigation of the reaction of alkyl halides with isoparaffins (Bartlett *et al.*, 80). It was shown that hydrogen-halogen exchange occurs when for example, a mixture of isopentane and *tert*-butyl chloride is treated with aluminum chloride for a short time. The mechanism of this exchange appears to involve the removal from the isopentane of a hydrogen with its pair of bonding electrons by the *tert*-

butyl carbonium ion formed from the alkyl chloride, with the resulting formation of isobutane:

$$(CH_3)_3CCl + AlCl_3 \rightleftharpoons (CH_3)_3C^+ \, AlCl_4^- \qquad (99)$$
$$(CH_3)_3C^+ + C_2H_5(CH_3)_2CH \rightleftharpoons (CH_3)_3CH + C_2H_5(CH_3)_2C^+ \qquad (100)$$
$$C_2H_5(CH_3)_2C^+ + AlCl_4^- \rightleftharpoons C_2H_5(CH_3)_2CCl + AlCl_3 \qquad (101)$$

In a polymerization system, not only tertiary alkyl ions but also ions of the allyl type, because of their stabilizing resonance, would be formed readily. Hence, some hydrogenation and dehydrogenation of the primary polymer (e.g., $RCH_2CH_2CH{=}CHR'$) would occur in the following manner:

$$RCH_2CH_2CH{=}CHR' + HCl + AlCl_3 \rightleftharpoons RCH_2CH_2\overset{+}{C}HCH_2R' + AlCl_4^- \qquad (102)$$

$$RCH_2CH_2\overset{+}{C}HCH_2R' + RCH_2CH_2CH{=}CHR' \rightleftharpoons RCH_2CH_2CH_2CH_2R' + RCH_2\overset{+}{C}HCH{=}CHR' \qquad (103)$$

$$RCH_2\overset{+}{C}HCH{=}CHR' + AlCl_4^- \rightleftharpoons RCH{=}CH{-}CH{=}CHR' + HCl + AlCl_3 \qquad (104)$$

The final products are thus paraffins and polyolefins. The latter are present in the catalyst layer.

Olefin polymerization in the presence of acid catalysts may be indicated in similar manner. Thus,

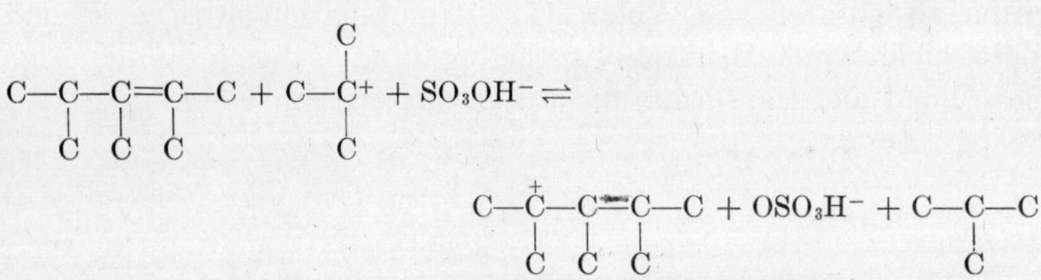

(105)

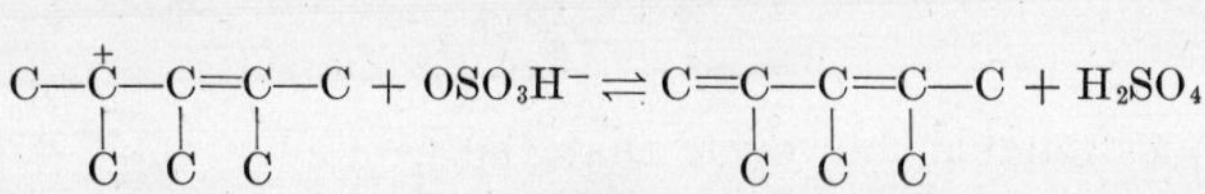

(106)

The polymerization of olefins in the presence of halides such as aluminum chloride and boron fluoride but in the absence of hydrogen halide promoter may also be described in terms of the complex "carbonium ion" formed by addition of the metal halide (without hydrogen chloride or hydrogen fluoride) to the olefin (cf. p. 28). These carbonium ions are apparently more stable than those of the purely hydrocarbon type; the reaction resulting in their formation is less readily reversed than is that of the addition of a proton to an olefin (Whitmore, 18). Polymerization in the presence of such a complex catalyst, may be indicated as follows (cf. Hunter and Yohe, 17):

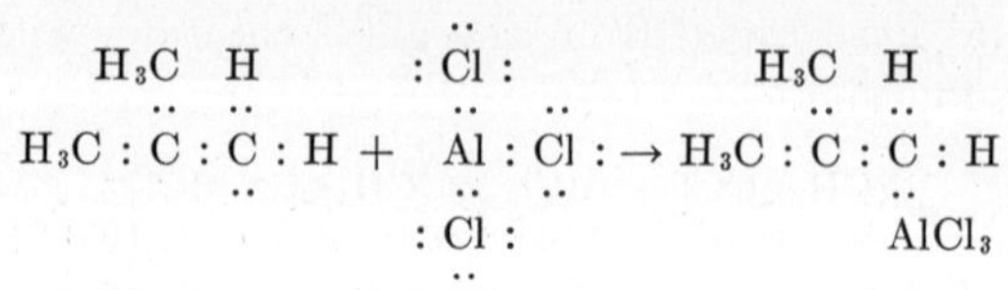

(107)

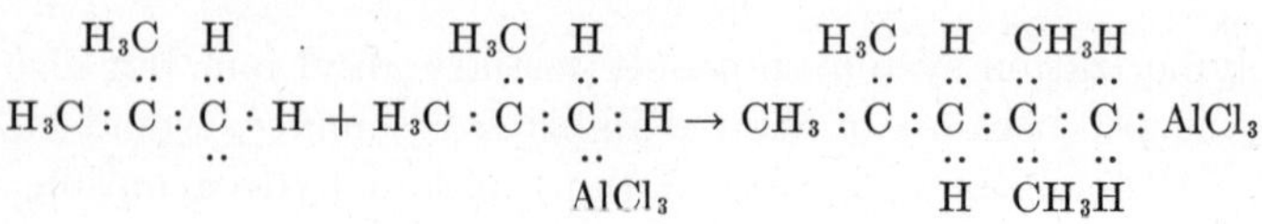

(108)

XLVI

The complex, XLVI, may add to another molecule of isobutylene to yield a higher polymer complex or eliminate aluminum chloride to yield the dimer; in the latter case intramolecular migration (a 1,5-shift!) of hydrogen must be postulated in order to form an olefin. On the other hand, cyclization may readily occur (particularly after a 1,2-shift of a proton from a methyl group) with the resultant formation of a naphthene.

It has been suggested that the carbon atom in the carbonium ion which is usually represented as containing only six electrons in its outer shell may actually contain six of its own electrons and two electrons which are donated by chlorine atom from aluminum chloride or, presumably, a fluorine from boron fluoride (N. V. Sidgwick, private communication to Hunter and Yohe, 17). Combination with a second molecule of olefin would then involve the breaking of the carbon-chlorine coordinate bond and the formation of a carbon-carbon bond.

It seems questionable, however, that the geometry of all metal halide catalysts would be such as to permit the formation of the carbon-halogen coordinate bond.

Formation of isobutane by the polymerization of ethylene in the presence of phosphoric acid may be represented as follows:

$$H_2C{::}CH_2 + PO(OH)_3 \rightleftharpoons \overset{+}{C}H_2CH_3 + OPO(OH)_2^- \quad (109)$$

$$H_2C{::}CH_2 + \overset{+}{C}H_2CH_3 \rightleftharpoons \overset{+}{C}H_2CH_2CH_2CH_3 \rightleftharpoons CH_3\overset{+}{C}HCH_2CH_3$$

$$\rightleftharpoons CH_3CH(CH_3)CH_2^+ \rightleftharpoons CH_3\overset{+}{C}(CH_3)CH_3 \quad (110)$$

Reaction of the *tert*-butyl carbonium ion with octene (ethylene tetramer), for example, will produce isobutane and an unsaturated carbonium ion which may form a diene by loss of a proton or which may cyclize to yield an ethylcyclohexyl carbonium ion:

$$\begin{array}{l} CH_3\overset{+}{C}CH_3 + CH_3CH_2CH_2CH_2CH_2CH_2CH{=}CH_2 \rightleftharpoons \\ \quad | \\ \quad CH_3 \end{array}$$

$$\begin{array}{l} CH_3CHCH_3 + CH_3CH_2\overset{+}{C}HCH_2CH_2CH_2CH{=}CH_2 \\ \quad\;\; | \\ \quad\;\; CH_3 \end{array} \tag{111}$$

$$\underset{\text{XLVII}}{CH_3CH_2\overset{+}{C}HCH_2CH_2CH_2CH{=}CH_2} \rightleftharpoons \underset{\text{XLVIII}}{\begin{array}{ccc} & H\;\; CH_2{-}CH_2 & \\ & |/ \qquad\quad \backslash & \\ CH_3CH_2C & & CH_2 \\ & \backslash \qquad\quad / & \\ & CH_2{-}\underset{+}{CH} & \end{array}} \tag{112}$$

$$\text{XLVII} \rightleftharpoons CH_3CH{=}CHCH_2CH_2CH_2CH{=}CH_2 + H^+ \tag{113}$$

$$\text{XLVIII} \rightleftharpoons \begin{array}{ccc} & H\;\; CH_2{-}CH_2 & \\ & |/ \qquad\quad \backslash & \\ CH_3CH_2C & & CH_2 \\ & \backslash \qquad\quad / & \\ & CH{=}CH & \end{array} + H_2 \tag{114}$$

$$\text{XLVIII} + RCH_2CH{=}CH_2 \rightleftharpoons \begin{array}{ccc} & H\;\; CH_2{-}CH_2 & \\ & |/ \qquad\quad \backslash & \\ CH_3CH_2C & & CH_2 \\ & \backslash \qquad\quad / & \\ & CH_2{-}CH_2 & \end{array} + R\overset{+}{C}HCH{=}CH_2 \tag{115}$$

Formation of aromatic compounds involves either the dehydrogenation (by way of reaction with carbonium ions) of the cyclohexane compounds or, less likely, the cyclization of á triolefinic carbonium ion.

b. Ester Mechanisms. Conjunct polymerization may be explained by the various ester type of mechanisms by assuming that hydrogen transfer between an ester and an olefin (or ester) occurs with the conversion of the ester to a saturated hydrocarbon and the formation of a new (unsaturated) ester or of an acid and a polyolefinic hydrocarbon. Such explanations, however, do not give an insight into the actual hydrogen transfer or show why the ester should act as the hydrogen acceptor and the olefin or other compound as the hydrogen donor.

In accordance with the "two ester" mechanism, the first step in the polymerization of ethylene in the presence of phosphoric acid is the formation of an ethyl phosphate (Ipatieff and Pines, 70). It was found that rapid absorption of ethylene took place when orthophosphoric acid was treated with etyhlene at 180° under an initial pressure of 50 atmospheres. The product was shown to be a monoethyl ester of phosphoric acid by converting it to a barium salt, $C_2H_5OPO_3Ba$. The ester decomposes when heated to 250° and 1-butene is formed as the primary product. Isomerization of the 1-butene to isobutylene then occurs; that 1-butene can isomerize to isobutylene under the experimental conditions was shown by heating pure 1-butene to 330° in the presence of phosphoric acid under 100 atmospheres initial hydrogen pressure. The product consisted of a liquid polymer and gas, the latter consisting of 50% isobutane, a small quantity of *n*-butane and normal butylenes. The amount of isobutane formed represented 6% of the reacting butene. Finally, six-membered ring cycloparaffins (formed by the cyclization of olefinic polymer) act as hydrogen donors and are converted to aromatic hydrocarbons while the isobutylene (as well as other olefins) is hydrogenated to isobutane (and other paraffins).

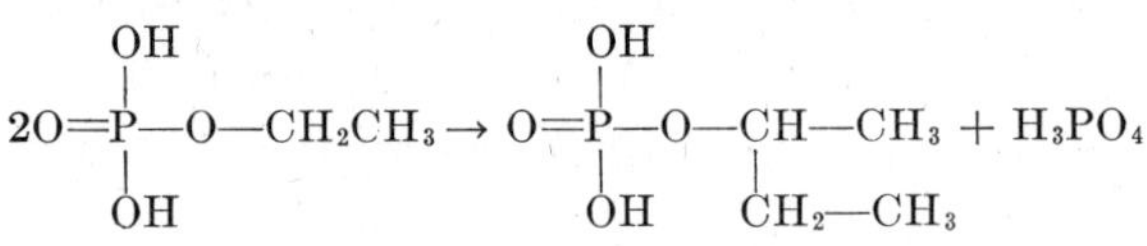
(116)

XLIX

$$CH_3CH{=}CHCH_3 + H_3PO_4 \leftrightharpoons \text{XLIX} \rightleftharpoons CH_2{=}CHCH_2CH_3 + H_3PO_4$$

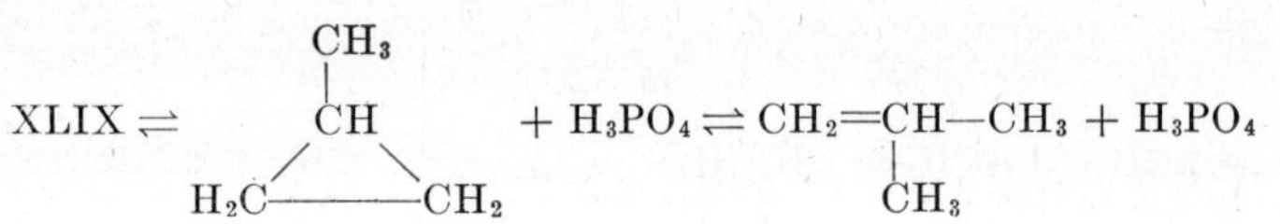
(117)

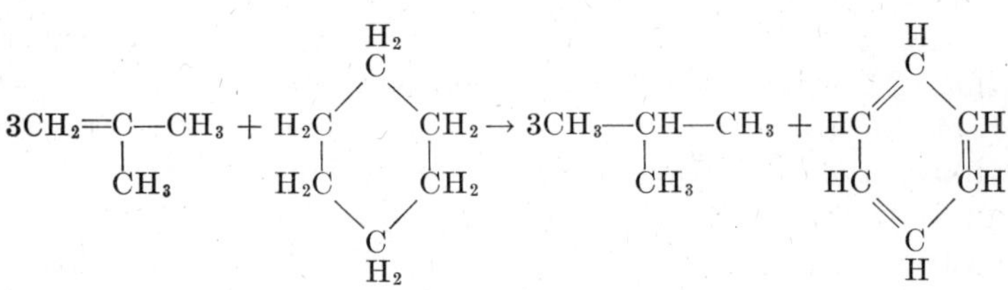
(118)

c. Other Mechanisms. Proponents of the other mechanisms (for example, the methyl and the hydrogen separation mechanism) do not seem to have attempted to account for conjunct polymerization.

V. Macropolymerization at Low Temperature

The low temperature polymerization of isobutylene (that is, polymerization at temperatures below about −70°) in the presence of Friedel-Crafts catalysts (particularly boron fluoride, aluminum chloride, and titanium tetrachloride, has been studied quite intensely. The reaction is commercially important because it yields a high molecular weight polymer having elastic properties (Thomas *et al.*, 81). The structure of the polymer offered an intriguing theoretical problem and the polymerization gained further interest and became the subject of much study following the discovery that a third substance (usually water) was necessary to cause the reaction to take place (Evans *et al.*, 82).

1. *Boron Fluoride*

The type of products depends largely on the reaction temperature (Thomas *et al.*, 81). At the boiling point of the isobutylene (−6°) polymerization in the presence of boron fluoride yielded an oil after a considerable induction period. On the other hand, if the isobutylene was precooled to −80° an immediate reaction occurred with almost explosive violence, producing a polymer of a very much higher molecular weight. The molecular weight of the polyisobutylene molecule increased from 10,000 to 200,000 when the temperature was decreased from −10 to −90°. Investigation of the structure of the polymers indicated that they were linear and that they were formed chiefly by head to tail polymerization. Each polymer molecule presumably contained one terminal double bond. Certain compounds are poisons for the reaction. For example, lower molecular weight products were obtained when the isobutylene contained *n*-butylene or di- or triisobutylene. The latter two compounds therefore cannot be intermediates in the polymerization to the high molecular weight polymers.

Hydrogen sulfide, mercaptans, and hydrogen halides are also poisons for the reaction. This effect of the hydrogen halides probably results from the formation of di- and triisobutylene by a mechanism similar to that which occurs at ordinary temperatures.

The catalytic activity of certain of the Friedel-Crafts catalysts was shown to decrease over a very wide range in the series boron fluoride, aluminum bromide, titanium tetrachloride, titanium tetrabromide, boron chloride, boron bromide and stannic chloride (Fairbrother and Seymour, mentioned in Plesch *et al.*, 83). When boron fluoride is added to isobutylene at dry ice temperatures, the olefin is converted to a solid polymer within a very few seconds. The time required for complete polymerization with aluminum bromide hardly extends to a few minutes while reaction times of hours are required with titanium chloride and periods of days with stannic chloride.

Pure isobutylene in the vapor phase at 50–100 mm. pressure did not polymerize when contacted with pure boron fluoride at about 50 mm. pressure (Evans and Polanyi, 84). Similar mixtures reacted instantaneously when water or *tert*-butyl alcohol was added. The boron fluoride was consumed in quantities approximately equivalent to the moles of third substance added. It was concluded that the catalyst does not form a stable complex with isobutylene and that the mechanism of the reaction involved the carbonium ion mechanism.

The gas phase polymerization of diisobutylene with boron fluoride also did not occur unless a third component was present. The addition of water or acetone caused the mixture to react rapidly, the boron fluoride combining instantaneously with the vapor of the third component in approximately equimolecular quantities. Certain substances, oxygen, hydrogen sulfide, and hydrogen chloride, did not produce a rapid polymerization of diisobutylene; these substances did not combine with the boron fluoride. In each of these cases, the final addition of water to the nonreacting mixture resulted in rapid polymerization. Ammonia formed an addition compound with the boron fluoride in approximately equimolecular quantities, but did not bring about the polymerization of the diisobutylene until water vapor was added, after which rapid reaction occurred (Evans and Weinberger, 85).

The experimental observations were explained by means of a mechanism which postulated that the catalyst acts as a proton transfer agent thereby converting an isobutylene molecule to the *tert*-butyl carbonium ion which is the chain initiator (Evans and Polanyi, 84):

$$BF_3 + HY \rightarrow F_3BYH \tag{119}$$

$$F_3BYH + CH_2{=}\underset{CH_3}{\underset{|}{C}}{-}CH_3 \rightarrow F_3BY^- + CH_3{-}\underset{CH_3}{\underset{|}{\overset{+}{C}}}{-}CH_3 \tag{120}$$

$$CH_3{-}\underset{CH_3}{\underset{|}{\overset{+}{C}}}{-}CH_3 + CH_2{=}\underset{CH_3}{\underset{|}{C}}{-}CH_3 \rightarrow CH_3{-}\overset{CH_3}{\overset{|}{\underset{CH_3}{\underset{|}{C}}}}{-}CH_2{-}\overset{CH_3}{\overset{|}{\underset{CH_3}{\underset{|}{C^+}}}} \tag{121}$$

$$CH_3{-}\overset{CH_3}{\overset{|}{\underset{CH_3}{\underset{|}{C}}}}{-}CH_2{-}\overset{CH_3}{\overset{|}{\underset{CH_3}{\underset{|}{C^+}}}} + CH_2{=}\underset{CH_3}{\underset{|}{C}}{-}CH_3 \rightarrow CH_3{-}\overset{CH_3}{\overset{|}{\underset{CH_3}{\underset{|}{C}}}}{-}CH_2{-}\overset{CH_3}{\overset{|}{\underset{CH_3}{\underset{|}{C}}}}{-}CH_2{-}\overset{CH_3}{\overset{|}{\underset{CH_3}{\underset{|}{C^+}}}}\text{ etc.} \tag{122}$$

The chain continues to grow by addition of isobutylene molecules until chain termination by proton transfer occurs (cf. discussion of titanium tetrachloride mechanism, p. 74). At higher temperatures, the loss or transfer of a proton occurs faster than does the addition reaction and lower molecular weight products are obtained.

Considerations involving the heats of reaction of head-to-tail, tail-to-tail, and head-to-head addition of the tertiary carbonium ion to the isobutylene indicated why the polymer chain is propagated chiefly by head-to-tail addition (Evans and Polanyi, 84).

When the polymerization of the isobutylene was carried out in liquid phase using pure liquid or a solution, reaction took place instantaneously on addition of the boron trifluoride, no third substance being necessary (Evans *et al.*, 86). Indeed, gas phase mixtures of pure isobutylene with boron trifluoride which showed no fast reaction in the gas phase always reacted when condensed by cooling in liquid air.

2. *Titanium Tetrachloride*

Unlike boron fluoride, titanium tetrachloride does not catalyze the liquid phase polymerization of isobutylene under anhydrous conditions (Plesch *et al.*, 83). The addition of titanium tetrachloride to a solution of the olefin in hexane at $-80°$ failed to cause any reaction. Instantaneous polymerization occurred when moist air was added. Oxygen, nitrogen, carbon dioxide, and hydrogen chloride had no promoting effect. Ammonia and sulfur dioxide combined with the catalyst; if these were added in small quantity only, subsequent addition of moist air permitted the polymerization to occur. Ethyl alcohol and ethyl ether, on the other hand, prevented the polymerization even on subsequent addition of moist air. They may be regarded as true poisons.

The addition of boron fluoride to a nonreacting mixture of isobutylene and titanium tetrachloride at -80° resulted in a rapid polymerization of isobutylene. In other words, as has already been mentioned, the liquid phase polymerization of isobutylene with boron fluoride catalyst apparently does not require the presence of water.

The action of water in the titanium tetrachloride catalyzed polymerization is paradoxical, since water at -60 to $-80°$ was present only in the solid phase; its solubility in hexane at these temperatures is in the order of 10^{-10} moles per liter (Plesch *et al.*, 83). It was found to be essential that the water be present as an extremely fine dispersion such as might result from the rapid bubbling of moist air through the liquid at the low temperature. Addition of liquid water which formed lumps of ice in the reaction mixture did not initiate polymerization. It may be concluded that a fine dispersion is necessary in order that reaction with titanium tetrachloride can occur and a chain reaction is initiated:

$$TiCl_4 + H_2O \rightarrow TiCl_4 \cdot H_2O^* \tag{123}$$

$$TiCl_4 \cdot H_2O^* + CH_2{=}\underset{\displaystyle CH_3}{\underset{|}{C}}{-}CH_3 \rightarrow CH_3{-}\underset{\displaystyle CH_3}{\underset{|}{\overset{+}{C}}}{-}CH_3 + TiCl_4OH^- \tag{124}$$

The asterisk is used to indicate that the initiating complex may be in an active state by carrying the heat liberated in its formation. The *tert*-butyl carbonium ion begins a polymerization chain which grows in the usual manner by addition of monomer molecule. Thus, if the isobutylene molecule is represented by M (for monomer) and the polymer molecule by M_n or M_{n+2}, the propagation of the polymerization chain proceeds as follows:

$$MH^+ + M \rightarrow M_2H^+ \quad (125)$$

$$M_2H^+ + nM \rightarrow M_{n+2}H^+ \quad (126)$$

Termination of the reaction chain may occur by either

$$M_{n+2}H^+ + TiCl_4OH^- \rightarrow M_{n+2} + TiCl_4 \cdot H_2O \quad (127)$$

or by

$$M_{n+2}H^+ + TiCl_4OH^- \rightarrow M_{n+2}H \cdot OH + TiCl_4 \quad (128)$$

Alternatively, transfer of the proton from the polymer carbonium ion to the monomer might occur:

$$M_{n+2}H^+ + M \rightarrow M_{n+2} + MH^+ \quad (129)$$

The carbonium ion formed in equation 129 would be able to start a new chain and each water molecule would thus cause the formation of more than one polymer molecule. If, on the other hand, the possibility of this chain transfer is excluded, and if it is assumed that only activated $TiCl_4 \cdot H_2O^*$ molecules can act as initiators, each water molecule will give rise to only one polymer molecule. The experimental data showed that the molar quantity of water added to initiate the reaction was of the same order as the number of molecules formed, but it does not seem probable, considering the experimental procedure, that each water molecule could enter into reaction.

The efficiency of titanium tetrachloride as a catalyst for initiating the polymerization of isobutylene may be dependent on the acid strength of the $TiCl_4 \cdot H_2O$ complex (Plesch *et al.*, 83). Furthermore, if a mechanism such as that outlined for titanium tetrachloride is applicable to the polymerization of isobutylene by Friedel-Crafts catalysts in general, it would appear that the activities of these catalysts will be directly proportional to the acid strengths of their hydrates. While reliable measurements of the acid strengths of the various Friedel-Crafts catalysts are not now available, investigations are in progress in a number of laboratories to obtain this information. It has been reported that the acid strengths of the strong catalysts such as aluminum bromide and boron fluoride are markedly greater than those of the weaker catalysts such as stannic chloride (Fairbrother, cited by Plesch *et al.*, 83).

Similarly, the efficiency of titanium tetrachloride as a terminator for the chain reaction should be related to the base strength of the complex anion $TiCl_4OH^-$; in the case of Friedel-Crafts catalysts in general, the terminator efficiency should be related to the base strength of the anion MX_nOH^-. Since this quantity is inversely proportional to the acid strength of the complex $MX_n{\cdot}OH_2$, the more active the catalyst as an initiator, the less active it will be as a terminator. Hence, the molecular weight of the polymer will be greater the more effective the catalyst, an observation which finds experimental confirmation.

3. Stannic Chloride

Water has also been shown to be essential for the liquid phase polymerization of isobutylene with stannic chloride as catalyst (Norrish and Russell, 87). The rates of reaction were measured by a dilatometric method using ethyl chloride as common solvent at −78.5°. With a mixture consisting of 1.15% stannic chloride, 20% isobutylene, and 78.8% ethyl chloride, the rate of polymerization was directly proportional to the amount of added water (up to 0.43% of which was added). A rapid increase in the rate of polymerization occurred as the stannic chloride concentration was increased from 0.1 to 1.25%; with higher concentrations the rate increased only gradually. It was concluded that a soluble hydrate is formed and functions as the active catalyst. The minimum concentration of stannic chloride below which no polymerization occurred was somewhat less than half the percentage of added water. When the concentration of the metal chloride was less than about one-fifth that of the added water, a light solid precipitated; formation of this insoluble hydrate which had no catalytic activity probably explains the minimum catalyst concentration. The addition of 0.3% each of ethyl alcohol, *tert*-butyl alcohol, diethyl ether, or acetone in the presence of 0.18% water reduced the rate to less than one-fifth of its normal value. On the other hand, no polymerization occurred on the addition of 0.3% of these substances in the absence of added water. The water-promoted reaction was halved when 1- and 2-butene were present in concentrations of 2 and 6%, respectively.

The rates of polymerization in systems containing very small quantities of added water (less than 0.1%) were subject to an acceleration in the later stages of the reaction, the acceleration occurring sooner and being more marked the lower the concentration of water. This was believed to be due to a decrease in the coefficient of chain termination as a result of the considerable increase in viscosity of the medium in the later stages of reaction: the chain-length and the consequent increase in viscosity were found to be greater the smaller the content of water.

It was suggested that the experimental observations could be explained by assuming that chain propagation is initiated both by positive and negative ions derived from the hydrated stannic chloride:

$$SnCl_4{\cdot}H_2O + 2(CH_3)_2C{=}CH_2 \rightarrow (CH_3)_3C^+ + SnCl_4OH{\cdot}C(CH_3)_2CH_2^- \quad (130)$$

Chain propagation occurs by successive additions of isobutylene to the ionized complexes so that there are both positive and negative growing chains. Termination results from the mutual neutralization of oppositely charged chains. This explains why the coefficient of termination decreases with increasing viscosity.

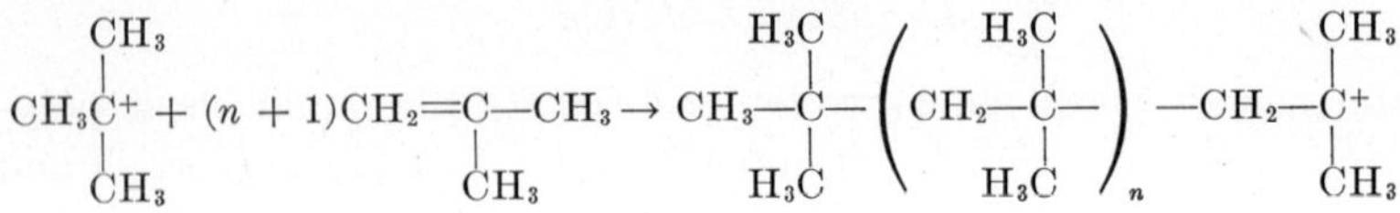

(131)

L

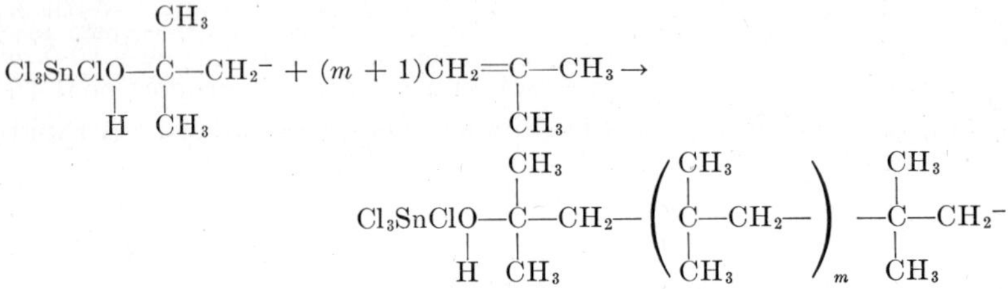

(132)

LI

L + LI → $Cl_3SnCl \cdot OH_2$ +

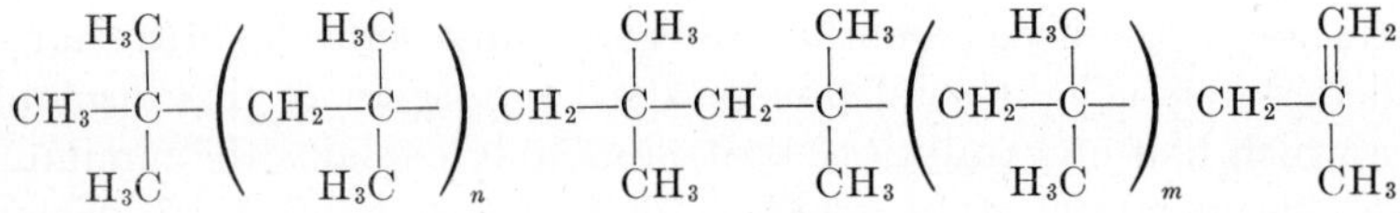

(133)

4. Aluminum Chloride

That the liquid phase polymerization of isobutylene in the presence of aluminum chloride or boron fluoride occurs even in the absence of a third component may be seen in the results of another worker (Houtman, 88). With solid aluminum chloride polymerization occurred at room temperature when isobutylene vapors were passed over the catalyst and at −80° when the isobutylene was in the liquid state. In both cases the reaction was incomplete probably because the catalyst was covered with a layer of polymer. The reaction at −80° was not however as explosive in character as when moisture was present. When boron fluoride was added to isobutylene at about −80°, "the polymerization started immediately in the gas phase above the liquid (owing to a somewhat higher temperature) and sometimes continued for a considerable time before reaction was noticeable in the liquid phase. There was no difference in behavior between the experiments in which the catalyst was led through or over the liquid." It was pointed out, however, that these results are not necessarily in contradiction with those of Polanyi and Evans and co-workers and that it is possible that the presence of moisture is necessary in certain cases, depending on the catalyst and olefin under investigation. It was suggested that the effect of moisture is not that it liberates hydrogen halide which then adds to the olefin as a proton. Instead, it is proposed that the metal halide adds to the pi electrons of the olefin to form a dipole which then attracts other monomer molecules resulting in the formation of the high molecular weight polymer. It was found that isobutylene polymers, prepared with boron fluoride, contained a noncombustible residue even after solution, filtration and precipitation. Addition of aluminum chloride and boron fluoride to the pi electrons of isobutylene occurs readily. With titanium tetrachloride there may be a question of steric hindrance due to the surrounding chlorine atoms; the addition of water may activate the catalyst by partial replacement of chlorine atoms by hydroxyl groups; thus diminishing the steric factor.

In ethylene the pi electrons are bound more strongly and presumably are not polarized. Hence, formation of the ethylene aluminum chloride complex requires a stronger activation. This may be obtained by the addition of water to the crystalline aluminum chloride which has the formula Al_2Cl_6 and in this state possesses a rather small ability to take up electrons. A molecule of water, it was postulated, splits the Al_2Cl_6 molecule with the formation of $AlCl_3$ (besides $AlCl_3 \cdot H_2O$) which will have a much greater attraction for electrons.

These considerations imply a mechanism which is similar to that proposed by Hunter and Yohe for the polymerization of olefins in the presence of aluminum chloride (see p. 68).

A more complete and detailed discussion of the low temperature polymerization of isobutylene cannot be given here. The interesting comments on the work of Evans, Polanyi, and co-workers (83 and 84) which have been published with the papers are well worth studying.

5. *Aluminum Bromide*

Evidence in support of a carbonium ion type of mechanism for low temperature polymerization was also obtained in an investigation of the kinetics of the homogeneous liquid phase polymerization of propene in the presence of aluminum bromide and hydrogen bromide at about $-78°$ (Fontana and Kidder, 89). The rate of reaction is approximately proportional to the concentration of the promoter, no polymerization occurring in its absence. During the main portion of the reaction, the rate is independent of the monomer concentration; toward the end, it decreases, due apparently to the low-concentration of the monomer, addition of more olefin resulting in an increase in the rate. It was concluded that the reaction involves an active complex, which may be regarded as a carbonium ion coupled with an anion:

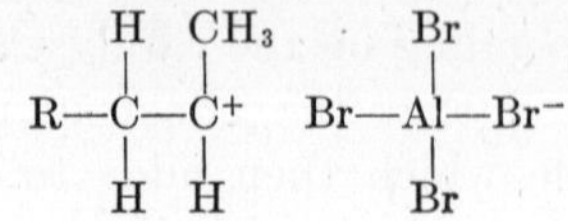

References

1. Baroni, E., in Schwab, G. M., Handbuch der Katalyse. Springer, Vienna, 1943, Vol. 7, Part I, pp. 344–479.
2. Burk, R. E., Thompson, H. E., Weith, A. J., and Williams, I., Polymerization. Reinhold, New York, 1937.
3. Egloff, G., The Reactions of Pure Hydrocarbons. Reinhold, New York, 1937, Chapter III.
4. Ellis, C., The Chemistry of Petroleum Derivatives. Reinhold, New York, 1934, Vol. I, Chapter 26; 1937, Vol. II, Chapter 26.
5. Ipatieff, V. N., Catalytic Reactions at High Pressures and Temperatures. Macmillan, New York, 1936, Chapter XI.
6. Thomas, C. A., Anhydrous Aluminum Chloride in Organic Chemistry. Reinhold, New York, 1941, Chapter 18.
7. Staff Report, *Chem. Inds.* **60,** 219 (1947).
8. Staudinger, H., *Trans. Faraday Soc.* **32,** 97 (1936).
9. Flory, P. J., *J. Am. Chem. Soc.* **59,** 241 (1937).
10. Price, C. C., *Ann. N. Y. Acad. Sci.* **44,** 351 (1943).
11. McKinley, J. B., Stevens, D. R., Baldwin, W. E., *J. Am. Chem. Soc.* **67,** 1455 (1945).
12. Frey, F. E., and Hepp, H. J., *Ind. Eng. Chem.* **28,** 1439 (1936).
13. Whitmore, F. C., *Ind. Eng. Chem.* **26,** 94 (1934).
14. Evans, A. G., and Polanyi, M., *J. Chem. Soc.* **1947,** 252.
15. Brown, H. C., and Pearsall, H., Paper presented at the New York meeting of the American Chemical Society, September, 1947.
16. Fairbrother, F., *J. Chem. Soc.* **1937,** 503; **1941,** 293.
17. Hunter, W. H., and Yohe, R. V., *J. Am. Chem. Soc.* **55,** 1248 (1933).
18. Whitmore, F. C., *Chem. Eng. News* **26,** 668 (1948).
19. Luder, W. F., and Zuffatni, S., *Chem. Revs.* **34,** 345 (1943).
20. Whitmore, F. C., and Meunier, P. L., *J. Am. Chem. Soc.* **63,** 2197 (1941).
21. Kazanskii, B. A., and Rozengart, M. I., *J. Gen. Chem. U.S.S.R.* **13,** 304 (1943); cf. translation by J. G. Tolpin in *Natl. Petroleum News* **36,** R-643 (1944).
22. McCubbin, R. J., and Adkins, H., *J. Am. Chem. Soc.* **52,** 2547 (1930).
23. Whitmore, F. C., and Church, J. M., *J. Am. Chem. Soc.* **54,** 3710 (1932).
24. Whitmore, F. C., Rowland, C. S., Wrenn, S. N., and Kilmer, G. W., *J. Am. Chem. Soc.* **64,** 2970 (1942).
25. Whitmore, F. C., and Laughlin, K. C., *J. Am. Chem. Soc.* **55,** 3732 (1933).
26. Whitmore, F. C., and Rohrmann, E., *J. Am. Chem. Soc.* **63,** 2033 (1941).
27. Whitmore, F. C., Wilson, C. D., Capinjola, J. V., Tongberg, C. O., Fleming, G. H., McGrew, R. V., and Cosby, J. N., *J. Am. Chem. Soc.* **63,** 2035 (1941).
28. McCubbin, R. J., *J. Am. Chem. Soc.* **53,** 356 (1931).
29. Lebedev, S. V., and Kobliansky, G. G., *Ber.* **63,** 103, 1432 (1930).
30. Drake, N. L., and Veitch, Jr., F. P., *J. Am. Chem. Soc.* **57,** 2623 (1934).
31. Lebedev, S. V., and Orlov, S. M., *J. Gen. Chem. U.S.S.R.* **5,** 1581 (1935).
32. Orlov, S. M., *J. Gen. Chem. U.S.S.R.* **7,** 923 (1937).
33. Brooks, B. T., *J. Am. Chem. Soc.* **56,** 1998 (1934).

34. Hoog, H., Smittenberg, J., and Visser, G. H., 2me Congres Mondial du Petrole, Paris, 1937. Vol. 2, p. 489.
35. Whitmore, F. C., and Mosher, W. A., *J. Am. Chem. Soc.* **63,** 1120 (1941).
36. Drake, N. L., Kline, G. M., and Rose, W. G., *J. Am. Chem. Soc.* **56,** 2076 (1934).
37. Whitmore, F. C., and Mosher, W. A., *J. Am. Chem. Soc.* **68,** 281 (1946).
38. Whitmore, F. C., and Stahly, E. E., *J. Am. Chem. Soc.* **55,** 4153 (1933); **67,** 2158 (1945).
39. Brunner, H., and Farmer, E. H., *J. Chem. Soc.* **1937,** 1939.
40. Laughlin, K. C., Nash, C. W., and Whitmore, F. C., *J. Am. Chem. Soc.* **56,** 1395 (1934).
41. Cook, N. C., Ropp, W. S., and Whitmore, F. C., Paper presented at the New York meeting of the American Chemical Society, September, 1944.
42. Ipatieff, V. N., and Schaad, R. E., *Ind. Eng. Chem.* **37,** 362 (1945).
43. Brooks, J. W., Paper presented before the Petroleum Division of the American Chemical Society, at the Chicago meeting, April, 1948, *Ind. Eng. Chem.* **41,** 1694 (1949).
44. Whitmore, F. C., and Homeyer, A. H., *J. Am. Chem. Soc.* **55,** 4194 (1933).
45. Whitmore, F. C., Laughlin, K. C., Matuszeski, J. F., and Surmatis, J. D., *J. Am. Chem. Soc.* **63,** 756 (1941).
46. Schmerling, L., *J. Am. Chem. Soc.* **67,** 1152 (1945).
47. Whitmore, F. C., and Mixon, L. W., *J. Am. Chem. Soc.* **63,** 1460 (1941).
48. Schmerling, L., *J. Am. Chem. Soc.* **67,** 1778 (1945); **68,** 275 (1946).
49. Price, C. C., *Chem. Revs.* **29,** 37 (1941).
50. Bloch, H. S., Pines, H., and Schmerling, L., *J. Am. Chem. Soc.* **66,** 1422 (1944).
51. Bergmann, E., *Trans. Faraday Soc.* **35,** 1025 (1939).
52. Lwow, J., *J. Russ. Phys. Chem. Soc.* **1884,** 462.
53. Kondakow, I., *J. prakt. Chem.* (2) **54,** 442 (1896).
54. Katsuno, M., *J. Soc. Chem. Ind. Japan* **44,** 831 (1941); *Chem. Abstracts* **42,** 2226 (1948).
55. Berthelot, M. P. E., Die Chemische Synthese. Brockhaus, Leipzig, 1877, p. 84.
56. Ipatieff, V. N., *Ind. Eng. Chem.* **27,** 1067 (1935).
57. Norris, J. F., and Joubert, J. M., *J. Am. Chem. Soc.* **49,** 873 (1937).
58. Monroe, L. A., and Gilliland, E. R., *Ind. Eng. Chem.* **30,** 58 (1938).
59. Farkas, A., and Farkas, L., *Ind. Eng. Chem.* **34,** 716 (1942).
60. Kline, G. M., and Drake, N. L., *J. Research Natl. Bur. Standards* **13,** 705 (1934).
61. Kharasch, M. S., and Marker, R., *J. Am. Chem. Soc.* **48,** 3130 (1926).
62. Wachter, A., *Ind. Eng. Chem.* **30,** 822 (1938).
63. Sparks, W. J., Rosen, R., and Frolich, P. K., *Trans. Faraday. Soc.* **35,** 1040 (1939).
64. Morikawa, K., Trenner, H. R., and Taylor, H. S., *J. Am. Chem. Soc.* **59,** 1103 (1937).
65. Whitmore, F. C., and Herndon, J. M., *J. Am. Chem. Soc.* **55,** 3428 (1933).
66. Norris, J. F., and Reuter, R., *J. Am. Chem. Soc.* **49,** 2624 (1927).
67. Church, J. M., Whitmore, F. C., and McGrew, R. V., *J. Am. Chem. Soc.* **56,** 176 (1934).
68. Turkevich, J., and Smith, R. K., *J. Chem. Phys.* **16,** 466 (1948).
69. Stewart, T. D., and Denham, H., *J. Am. Chem. Soc.* **68,** 1135 (1946).
70. Ipatieff, V. N., and Pines, H., *Ind. Eng. Chem.* **28,** 684 (1936).
71. Ipatieff, V. N., and Grosse, A. V., *J. Am. Chem. Soc.* **58,** 915 (1936).
72. Hall, F. C., and Nash, A. W., *J. Inst. Petroleum Technol.* **23,** 679 (1938).
73. Damiens, A., and Lebeau, P., *Bull Soc. chim.* (4) **33,** 76 (1923).
74. Boeseken, J., and Max, N., *Rec. trav. chim.* **48,** 486 (1929).

75. Ormandy, W. R., and Craven, E. C., *J. Soc. Chem. Ind.* **47,** 317T (1928).
76. Nametkin, S. S., and Abakumovskaya, L. N., *J. Gen. Chem. U.S.S.R.* **6,** 1166 (1936); *Chem. Abstracts* **31,** 1002 (1937).
77. Ipatieff, V. N., and Pines, H., *J. Org. Chem.* **1,** 464 (1936).
78. Proell, W. A., Adams, C. E., and Shoemaker, B. H., *Ind. Eng. Chem.* **40,** 1129 (1948).
79. Ipatieff, V. N., and Pines, H., *Ind. Eng. Chem.* **27,** 1364 (1935).
80. Bartlett, P. D., Condon, F. E., and Schneider, A., *J. Am. Chem. Soc.* **66,** 1531 (1944).
81. Thomas, R. M., Sparks, W. J., Frolich, P. K., Otto, M., Muller-Cunradi, M., *J. Am. Chem. Soc.* **62,** 275 (1942).
82. Evans, A. G., Holden, D., Plesch, P. H., Polanyi, M., Skinner, H. A., and Weinberger, M. A., *Nature* **157,** 102 (1946).
83. Plesch, P. H., Polanyi, M., Skinner, H. A., *J. Chem. Soc.* **1947,** 257.
84. Evans, A. G., and Polanyi, M., *J. Chem. Soc.* **1947,** 252.
85. Evans, A. G., and Weinberger, M. A., *Nature* **159,** 437 (1947).
86. Evans, A. G., and Meadows, G. W., and Polanyi, M., *Nature* **158,** 94 (1946).
87. Norrish, R. G. W., and Russell, K. E., *Nature* **160,** 543 (1947).
88. Houtman, J. P. W., *J. Soc. Chem. Ind.* **66,** 102 (1947).
89. Fontana, C. M., and Kidder, G. A., *J. Am. Chem. Soc.* **70,** 3745 (1948).

Early Studies of Multicomponent Catalysts

ALWIN MITTASCH

Heidelberg, Baden, Germany

Translated into English by W. Frankenburg*

CONTENTS

The study of catalysis has expanded with increasing speed since about 1900, both in its technical and in its scientific aspects. In the course of this development, use has been made to a large extent, and with considerable success, of multicomponent catalysts (Mehrstoff-catalysts), that is, of catalysts which contain mixtures of various chemical constituents rather than one single chemical element or one single chemical compound.

It may be of interest to describe the first stages of the development which led to the thorough investigation and to the technical use of these catalysts.

* Written by the author mainly from his private notes and with but little literature available, this paper represents the early development of multicomponent catalysts, a field in which he played a decisive role. A few insertions on theoretical points and several references have been added by the translator where it seemed indicated. The theoretical investigations and practical applications of multicomponent catalysts after about 1920 are not included in this paper.

I. The Use of Multicomponent Catalysts before the Development of the Ammonia Catalyst

Although a truly systematic study of multicomponent catalysts did not start prior to 1900, the older literature shows that catalysts of this type were used occasionally in earlier times.

More than a hundred years ago Johann Wolfgang Döbereiner, the first explorer of catalytic phenomena, described the catalytic activity of certain combined substances, and he emphasized the increased efficiency of some of such combinations compared with that of their components (1). In 1824, he reported that the catalytic action of metallic iridium on a hydrogen-oxygen mixture is enhanced by the combination of this metal with osmium. He further observed in 1844 that platinum sponge to which alkali has been added is more active in catalyzing the reaction between hydrogen and oxygen than untreated platinum sponge. In 1852, Fr. Wöhler and Mahla (2) described the considerable catalytic activity of a copper-oxide-chromium oxide catalyst for the oxidation of sulfur dioxide. In the 1870's, H. Deacon, investigating the oxidation of hydrogen chloride by oxygen, studied the effect of added substances to his copper catalyst such as of sodium chloride, sodium sulfate and magnesium sulfate, and expressed the opinion that these additions are "accelerating substances" (3). In 1899, G. Bredig observed that the catalytic action of platinum on the decomposition of hydrogen peroxide is enhanced in the presence of sodium hydroxide (4).

As to catalytic reactions in homogeneous media, Moritz Traube found in his studies of the oxidation of hydrogen iodide by hydrogen peroxide in aqueous solution, that the catalyst ferrous sulfate is activated by copper sulfate (5). As to the magnitude of such effects, Price stated in 1898 (6) that the simultaneous action of iron and of copper compounds on the reaction between persulfate and hydrogen iodide causes an unexpected acceleration of the reaction, which is more than twice as great as the acceleration calculated as an additive effect of the two single catalysts. However, effects were also observed of the opposite kind, i.e., deactivations of catalysts by added substances. An example of this type is the effect described by Döbereiner in 1832 as a "depotentiating action of ammonia" on a platinum catalyst for the hydrogen-oxygen reaction (7).

At the turn of the century, studies were made by Johannes Brode (in Ostwald's Institute, Leipzig) on "combined catalysts" (8). Continuing the work of Price with catalysis in solutions, this author demonstrated the existence of re-enforcing and of weakening influences exerted on a given catalyst by added compounds. He showed that a catalyst can be considerably improved even by such substances which, by themselves, do not exert any catalytic action, or only a negligible one.

Some more observations were made on solid catalysts for gas reactions between 1900 and 1910. Rudolf Knietsch of the B. A. S. F.* found that effective catalysts for the oxidation of gaseous sulfur dioxide to sulfur trioxide can be prepared from iron oxide and from similar oxides by coating them with platinum (9). Hölbling and Dietz reported for the same reaction that the efficiency of cerium oxide is increased by the oxides of didymium, yttrium, zirconium, etc. (10). Bastian and Calvert reported that a combination of platinum and palladium represents a catalyst which is superior to each of the pure components, for the ignition of inflammable gas mixtures (11). In 1906 Fokin reported that in catalytic hydrogenations, the action of nickel is increased by a second "galvanically acting" metal such as zinc or magnesium (12).

We shall omit the description of the numerous cases in which a decrease or a complete annihilation of catalytic action by a second substance was observed, an effect which has been called "catalyst poisoning" since the investigations of Knietsch and Bredig (13). Effects of this type have been encountered again and again since the first report of Döbereiner; they were often due to the unintentional presence of harmful substances in the catalysts or to contaminations of the reacting components. There are, of course, also numerous cases in which added substances exert no detectable influence on a catalyst.

II. Initial Work of the Author at the Badische Anilin und Soda Fabrik Started in 1904

From 1900 to 1902 the author studied the chemical dynamics of the formation and decomposition of nickel carbonyl (14) under the supervision of M. Bodenstein in Wilhelm Ostwald's Physical Chemical Institute in Leipzig. In establishing the equilibrium conditions for the carbonyl, it was found that oxygen retarded and that sulfur increased the reaction velocities. The action of sulfur was sufficient to counterbalance the detrimental presence of oxygen. Furthermore, an astonishingly large additional acceleration of the equilibration was achieved by mercury vapor. Therewith, a typical case of multicomponent catalysis was established, the same system being accessible to the influence of the negative catalyst, oxygen, and to that of the positive multicomponent catalyst, sulfur plus mercury.

Shortly after the author had joined the B. A. S. F. in 1904 the chemist C. Bosch gave him, as his assistant, the assignment of working out a technical method for the chemical fixation of elementary nitrogen in the

* Badische Anilin and Soda Fabrik, Ludwigshafen am Rhein, Germany.

form of cyanides and nitrides or related compounds, at temperatures below that of the electric arc, i.e., without the use of electrical energy. Having become familiar with the concept and importance of catalysis, the author attempted to use catalytic methods for this problem. It seemed possible that catalysts would prove to be of value in all the various reactions, the aim of which was to fix elementary nitrogen of the air chemically in the form of various metal compounds. The more promising reactions which were studied successively, were the following ones:

1. Barium cyanide and barium cyanamide from barium carbonate, carbon and nitrogen at about 1400°C. (15).

2. Titanium nitride (or titanium carbonitride) previously called cyano-nitrogen titanium from titanium oxide, carbon and nitrogen, again at about 1400°C. (16).

3. Silicon nitride from silica, carbon and nitrogen, at about 1500°C.

4. Aluminum nitride from bauxite, carbon and nitrogen (17), and the formation of silica-aluminum-nitride mixtures from various clays at about 1500°C. and higher temperatures.

In all these studies which were carried out over the five-year period, 1904–1909, the effects of catalysts on these fixations of nitrogen were investigated. For the barium compounds, iron oxide acted as a catalyst. For the titanium nitrides, various other metal oxides as well as a number of inorganic salts proved to be effective. For the silicium nitrides and aluminum nitrides, again metal oxides and salts had beneficial effects but in other proportions and to another extent than found for the titanium nitrides. Often, two or more catalysts were added to the reaction mixtures with beneficial effects, but no systematic quantitative comparisons were carried out at this time. Table I lists some of these experiments.

Simultaneously, a general study was made on the chemical fixation of elementary nitrogen to many other chemical elements including the reaction between nitrogen and hydrogen, and the optimal conditions were investigated for the formation of many nitrides, including ammonia.

In an attempt made by the author in 1905 to synthesize ammonia from nitrogen and hydrogen at 1000°C. with very pure iron as a catalyst, the reacted gases contained 0.004 volume per cent of ammonia. It is clear today that the high temperature and the ordinary pressure employed in this experiment had to result in a very low yield. During the period between 1905 and 1908 many substances were tested as catalysts for such an ammonia synthesis under, as we know today, unfavorable temperatures and pressures. We worked with pure iron (mostly prepared from iron oxalate), with iron carbide, tantalum, uranium, thorium, nickel, palladium, titanium, molybdenum and tungsten. As a rule, we passed a mixture of pure nitrogen and hydrogen, at normal pressures and at temperatures between 600 and 1000°C. through quartz tubes containing the catalyst. Also, mixtures were tested as catalysts such as for example, iron powder plus 3% calcium fluoride plus 3% mixed alkali chlorides (1909). Occasionally, we obtained measurable quantities of ammonia, particularly with iron, uranium and tantalium, but the amounts produced remained insignificant for the same reason as mentioned above.

TABLE I

Amounts of Nitrogen Fixed by Silicium and by Aluminum, as a Function of Catalysts added. All Experiments Carried out by Treating the Solid Phase with a Current of Nitrogen at 1000°C.

Substance used for the fixation of nitrogen	Silicium							Aluminum	
Catalyst	None	5% KCl	5% Na_2CO_3	5% Borax	5% CaF_2	5% NaF	5% NaCl	None	5% Na_2CO_3 + K_2CO_3
Amount of N_2 fixed, as per cent of solid phase obtained.	1.1	3.6	4.2	7.0	19.6	20.5	29.7	6.7	12.3

Repeatedly, an intermittent method was employed by treating the catalyst alternately with nitrogen and then with hydrogen in an attempt to enforce a formation and subsequent reduction of nitrides.

A technical exploitation of these experiments appeared unattractive, until 1909. Other methods of the chemical fixation of nitrogen, particularly the electric arc process for the formation of nitrous oxides (18) and the calcium cyanamide process (19) appeared far superior at that time.

This picture changed completely as, in February 1908, F. Haber approached the B. A. S. F. with the proposition to obtain their financial assistance for an intensified continuation of his research on the catalytic synthesis of ammonia, following a suggestion of professor K. Engler of Karlsruhe, a member of B. A. S. F.'s advisory board. The catalytic synthesis of ammonia had been investigated previously, but without having met with any practical success.

The following data give a short historic survey on the first steps toward a synthesis of ammonia from hydrogen and nitrogen. In 1823, Döbereiner claimed to have achieved an ammonia synthesis from the elements (20). His experiments proved to be erroneous. In 1884, Ramsay and Young (21) showed that the catalytic decomposition of ammonia does not proceed quantitatively, a first indication for the existence of measurable equilibria for this system. In 1901 Le Chatelier (22) had filed for a French patent in which, hypothetically, a synthesis was claimed of ammonia at high gas pressures. In 1904, Perman and Atkinson (23) claimed again to have obtained traces of ammonia from hydrogen and nitrogen, in the presence of catalysts at ordinary pressure. Haber and van Oordt reported on similar observations in the same year (24). In 1907, Nernst and Jost published a paper (25) in which they reported that considerable quantities of ammonia were formed from hydrogen and nitrogen in the presence of catalysts (platinum foil, iron and manganese) and at pressures of up to about 75 atmospheres. Since 1908, following the advice of Nernst, Haber had continued his studies with Le Rossignol using pressures between 100 and 300 atmospheres (26) and osmium and uranium as catalysts.

As to efforts to carry out ammonia synthesis in a technical direction, studies along that line had been started in the B. A. S. F. after Wilhelm Ostwald had suggested such work in 1900. In laboratory experiments considerable yields of synthetic ammonia had been obtained by W. Ostwald (27). However, all attempts to reproduce these yields on a larger scale were futile, and finally Ostwald had to admit that in his original experiments, ammonia had probably been erroneously introduced into the synthesis reactor with a foreign source, presumably in form of an iron nitride, which had been formed by a previous treatment of the iron catalyst with ammonia.

Numerous studies relating to Ostwald's proposition were carried out in the B. A. S. F. in 1900 by the chemist C. Bosch who had entered the services of the company shortly before 1900. He used pressures of one up to a few (about 5) atmospheres and catalysts such as iron wire, as recommended by Ostwald. Calculations carried out by W. Gaus indicated that elevated pressures should lead to better yields. However, no definite success was achieved at this stage.

Haber's interest in the ammonia synthesis originated mainly from a discussion between him and Nernst concerning the correct values for the chemical equilibrium $N_2 + 3H_2 \rightleftarrows 2NH_3$. In order to prove their points, Haber as well as Nernst undertook experiments independently with the purpose of establishing the equilibria, using high pressures and catalysts. After his attempts of a synthesis with pressures of 100 atmospheres and above had proved to be successful, Haber realized the great technical potentialities of such a process, and offered his rights, in February 1908 to the B. A. S. F., in a synthesis process to be carried out under high pressure, and as a new feature, in form of a recycle operation.

From this time on, and after further progress made by Haber in the year 1909, the aim of C. Bosch and his associates in the B. A. S. F. was to transform Haber's findings into an economic technical large-scale process. One of the main points in this program was the replacement of the costly osmium and uranium catalysts used by Haber by equally effective, but less sensitive and commercially acceptable materials. The catalytic laboratory experiments which were now started on a new basis in the B. A. S. F. were, following Haber, carried out at a pressure of about 100 atmospheres.

III. Transition from the Nitride Studies to the Use of Multicomponent Catalysts for the Ammonia Synthesis

His observations during the earlier work with nitrides, induced the author to start a systematic search for multicomponent catalysts for the ammonia synthesis. As a result of these studies, a highly effective ammonia catalyst was finally developed early in 1910. This catalyst consisted of pure iron containing relatively little (a few per cent) aluminum oxide and a still smaller amount of potassium oxide. This result, however, was not achieved by following the lines of research started by Döbereiner, Price, Brode and others which were mentioned above. Rather, the course of our work moved on a detour.

In February 1909, the results of the experiments on nitride formation had led to the outline of a patent application which covered the preparation of metal nitrides in the presence of auxiliary substances. Following a hypothetical concept of the action of these additions, they were defined as "flux promoters." This draft of an application ended with the following sentence: "Finally, it is also advantageous to add a flux promoter to metals or alloys which serve as catalysts for the ammonia synthesis." This statement was made in view of the early catalytic experiments in which we had observed the synthesis of traces of ammonia in the presence of catalysts similar to those which acted favorably for the nitride formations.

The reasoning which led the author to make this first "shot in the dark" regarding the usefulness of combinations of solid compounds as ammonia catalysts was as follows: If we assume that a labile iron nitride is an interminate in the catalytic ammonia synthesis, every addition to the iron which favors the formation of the iron nitride ought to be of advantage. In other words, the hypothesis was used that surface catalysis acts *via* the formation of intermediate compounds between the catalyst and one or more of the reactants. An experimental support for this theory was the fact that a stepwise synthesis *via* the formation and successive hydrogen reduction of nitrides had been carried out with calcium nitrides (Haber), and cerium nitrides (Lipski). Later, the author found molybdenum nitride as being the best intermediate for such a stepwise synthesis.

Thus, the idea of preparing multicomponent catalysts originated actually from the experience gathered in the nitride experiments. Similarly, our future progress did not move along a straightforward course, but rather in winding ways due to initial setbacks.

We can readily understand these setbacks today if we consider the high sensitivity of iron as an ammonia catalyst toward numerous catalyst poisons. In those early years, this fact was unknown to us. Specifically, no one suspected the harm which is done to the catalyst by substances such as sulfur and sulfur compounds. Even Haber had never discussed a catalyst poisoning by impurities, because he had been able, apparently to avoid the presence of catalyst poisons in his small scale experiments.

One disappointment followed another as the author, with the

assistance of G. Stern and H. Wolf, in the summer of 1909, started to carry out the ammonia synthesis at gas pressures of about 100 atmospheres and with catalysts which were prepared by following the multicomponent principle. The reasons are clear today: As a rule, the metallic or oxidic iron preparations used as raw materials for the catalysts contained traces of sulfur and other impurities and were, accordingly, inactive for the ammonia synthesis from the very start.

The main emphasis was laid, in this initial work, on Haber's catalysts, e.g., osmium and uranium compounds, as well as on a series of iron catalysts. Some other metals and their compounds which we tested are, as we know today, less accessibble to an activation by added substances than iron. Therefore, they showed no improvement or only small positive effects if used in the form of multicomponent catalysts. Finally, the substances which we added to the metal catalysts in this early stage of our work were mostly of the same type as those which had proved to favor the nitride formation, e.g., the "flux promoting" chlorides, sulfates, and fluorides of the alkali and alkaline earth metals. Again, we know today that just these compounds do not promote, but rather impair the activity of ammonia catalysts.

Thus, the first months of our efforts to obtain satisfactory promoted catalysts led to discouraging results. As an example, an iron preparation called "Crocus Martis" ("Colcothar," an iron sulfate or a natural or synthetic iron sulfide, decomposed at high temperatures) showed by itself considerable catalytic activity, but was not improved by the addition of potassium-, calcium-, or magnesium-compounds. This effect was obviously due to the fact that Crocus Martis contained in itself oxides of the alkali- and earth-metals as impurities, and moreover, usually traces of sulfur which acted as catalyst poisons*. Similarly, the poor showing of various pyrites ashes which we tested as catalysts was in all probability caused by traces of sulfur which had remained in these samples. A series carried out with a Crocus Martis preparation passing a mixture of one part of nitrogen and three parts of hydrogen over the catalyst at 100 atmospheres pressure, 500°C. and with a velocity of 15 to 18 liters per hour, yielded the precentages of ammonia in the exit gas as shown in Table II, depending on the pretreatment of the original Crocus Martis sample.

Ferrum purissimum (Merck) yielded, under the same working conditions, initially, 0.4% ammonia, but after a relatively short time this yield decreased to zero. An addition of 1.7% of sodium hydroxide to this pure iron resulted in yields up to 1.8% ammonia, additions of 0.6%

* A typical analysis of a reduced Crocus Martis powder was: about 95% of Fe; 1.42% CaO; 0.73% Na_2O; 0.45% MgO; 0.25% SiO_2; 0.1% S; 0.06% Cl.

potassium hydroxide gave, at best, 2.7% ammonia and additions of 3.5% potassium hydroxide led to 4% of ammonia as the highest yields.

The results remained irregular during this entire period and lacked reproducibility. An important progress however was that during this time my associate Stern developed a laboratory scale reactor which made it possible to work easily at pressures of a hundred odd atmospheres, a reactor of which we soon had two dozens and more in continuous

TABLE II

Pretreatment of Catalyst

	None	Extracted with water	Moistened & pressed into rods	5% of $NaNO_3$ added	Fused in the hydrogen-oxygen flame
Percentage of NH_3 in exit gas	1.0%	1.5%	1.1%	1.5%	1.8%

operation. This type of equipment facilitated greatly all our further studies.

An encouraging surprise came finally as Wolf introduced into the catalytic test oven, on November 6, 1909, a sample of magnetite from Gällivare (Sweden) which had stood for years on our shelf. With this substance as a catalyst, the exit gas contained for a relatively long period of operation, about 3% of ammonia compared with an average of 1% which was obtained heretofore over extended periods. What was the cause for this unexpected favorable effect?

Still following the "macro-structural" hypothesis which we favored at that time, we abondoned the idea of a specific favorable influence of flux promoters and assumed instead that the cause for the success of the magnetite experiment was the "compact porous" structure of the iron sponge which was formed in the test oven by the reduction of the Swedish ore. An apparent support of this idea was that contrary to the favorable action of the dense iron sponge obtained from magnetite, catalysts of a looser structure such as, e.g., iron asbestos preparations had always been particularly ineffective.

At first this new assumption seemed confirmed by the enhanced activities of catalysts which were obtained by pressing loose powders of iron oxide into dense tablets. Soon, however, it was found that magnetites of another origin than the Swedish specimen yielded no ammonia, or only negligible amounts, in spite of the dense structure of their reduction products.

In addition to the commercially available types of iron and iron oxides, we then tested a great number of native iron ores as to their catalytic activities parallel with analytical studies of their composition. I mention **Iron Glance** (Merck), **Hematites** (Hungary), **Limonites** (Saxony), **Siderites** (Rhineland), **Iron Reds** and **Chalybdites** (Spain), **Iron Reds** (Ruhr), **Fibrous Hematites,** and **Magnetites** of various origins. An example of pure synthetic iron compounds which we tested is an iron-acetyl acetonate which was prepared by Dr. Villiger in our Ludwigshafen laboratory.

The result of all these experiments was that the idea of a particularly beneficial "macro-compact porous" state of the catalyst had to be abandoned, although patents had already been filed in which this had been claimed. Instead, we returned still more convinced of its correctness, to the previous more general hypothesis of the desirability of multi-component catalysts.

With the aim of becoming independent of the individual origins and compositions of the various materials, we attempted to prepare, synthetically, an iron-containing catalyst equivalent in structure and in composition to the Magnetite of Gällivare (Sweden) which had proved, in repeated tests, to be an efficient and dependable catalyst.

IV. Systematic Experiments with Activated (Promoted) Catalysts, 1910–1912

After we had become convinced once again that added substances were of essential importance for the ammonia catalyst,* the work in this field developed more and more into an increasingly systematized search.

The following rules were laid down for this search:

(1) The use of iron preparations of the highest obtainable purity (both commercial and laboratory preparations) as starting materials. Principally employed were "Ferrum reductum, Kahlbaum," and later, as an ultra pure material "Iron Kahlbaum" (cost of 1 kg. about $80.00).

(2) The development of special methods of preparation for obtaining intimate mixtures of catalyst and added substances.

(3) The exclusive employment of a synthesis gas, which consisted of highly purified and dried hydrogen and nitrogen. For a long period, electrolytically produced hydrogen and oxygen-free nitrogen were prepared for this purpose.

As to the method of preparing our catalyst samples, considerable work was devoted to the various methods of preparation by which the ferroferrioxide (Fe_3O_4) can be obtained from metallic iron, in view of the

* First patent application in which this point was clearly stated was filed January 8, 1910 (28).

outstanding properties of native magnetite, the composition of which is identical or very close to Fe_3O_4. As a particularly convenient method, we frequently used the oxyfusion of metallic iron, initially with perchlorate as the oxidant, and later, more advantageously, in form of a fusion under pure oxygen, the process being started with a glowing iron wire. A batch of iron, as a rule about 20 g. was fused in the hemispherical cavity of a small steel block. Soon, further methods of preparation were taken up in addition to the oxyfusion, such as the thermodecomposition of nitrate mixtures, the coprecipitation of iron together with other components, and also a mere mixing of iron powders with added substances as well as the impregnation of iron powders with solutions containing such added substances.

On the one hand, we attempted to imitate the Gällivare magnetite as closely as possible, based on thorough analytical investigations of this ore. For instance, a synthetic catalyst was prepared by the oxyfusion of 20 parts of iron, 0.2 parts of calcium oxide, 0.4 parts of aluminum oxide, 0.1 part of magnesium oxide and 0.1 part of chromium oxide, corresponding to the magnetite analysis. This synthetic catalyst acted, after reduction, catalytically just as well as one of the best specimens of native magnetite.

On the other hand, it became increasingly more important to know which substances were actually responsible for good or poor catalytic activities of iron catalysts. To obtain information on this point, only one single component was added at a time to pure iron powders previous to the oxyfusion. For instance, the following single compounds were tested as additions to a given amount of iron: potassium oxide, sodium oxide, magnesium oxide, calcium oxide, aluminum oxide, sulfur, phosphorus, lead, tin, copper, etc. The amounts of these additions were varied. After oxyfusion, the product was broken into small pieces and tested in laboratory scale reactors. Later on, not only single elements and compounds, but various mixtures were added before the oxyfusion. After a relatively short while, it became clear that a "macro-compact porous" state of the catalysts, although favoring good and peristent yields of ammonia (per unit of catalyst volume) is not of decisive importance but rather the kind and amount of the added activators (promoters).* Instead of attempting to describe here the vast number of promoter tests carried out in 1910 (29), we shall report briefly the main

* These promoter effects were strikingly new and unexpected at that time. In a letter (January 17, 1910) in which F. Haber congratulated C. Bosch and his coworkers on their achievement, he wrote: "It is strange how matter shows again and again new peculiarities. Thus, it is the iron, a substance with which Ostwald was the first to work, and which we have tested several hundred times in a pure state but not a single time in an impure condition!"

conclusions reached after this period of intensified systematic work and of considerable progress in the approach toward a technically satisfactory catalyst.

These conclusions were:

(1) Certain substances, if intimately mixed in very fine dispersion with iron, increase the catalytic efficiency of the latter for the ammonia synthesis to a large degree, with a simultaneous extension of its useful life. Specifically, oxides of very high melting points and of great resistance against reduction are, even in small amounts, efficient promoters. Of such oxides, alumina and magnesia are particularly favorable, and to a lesser degree calcium oxide and barium oxide.

(2) Other substances decrease or annihilate, even in traces, the catalytic properties of iron. Such "catalyst poisons" had already been known as a nuisance in the catalytic oxidation of sulfur dioxide. With the ammonia catalysis several elements, particularly sulfur proved to be harmful, even in amounts of 1/100 of one per cent. Chlorine, phosphorus and arsenic showed a similar behavior (30), just as certain metals, such as lead, tin and zinc.

Both these final conclusions were equally important because they indicated clearly the way in which to proceed further if dependable results were to be achieved.

(3) The addition of some further metals and oxides to iron catalysts proved to be of no effect, such as of copper, silver, titanium dioxide, etc.

(4) Promoters can, to a certain extent, counterbalance the presence of small amounts of poisons. Thus the oxides of the alkali metals, which in themselves do not improve the activity of pure iron, have an indirect favorable action by binding traces of sulfur.*

(5) As a rule, further improvements can be achieved by producing catalysts from mixtures of more than two components. For instance, the combination of iron plus alumina can be favorably influenced by small amounts of potash.

(6) Special properties are observed with mixtures of iron with other catalytically active metals such as cobalt, molybdenum, tungsten and uranium. These systems represent cases of mutual activations.

(7) Metals other than iron which catalyze the ammonia synthesis such as osmium, ruthenium, uranium, and molybdenum can also be promoted by added substances. However, several of these metals do not show improvements of the same magnitude as does iron. Further-

* Remark of the translator: More recent experience does not agree with this explanation of the promoting action of alkali metal oxides. On the one hand the sulfur retention by alkalinized iron is only weak; on the other hand, the promoter action by alkali oxides is also observed in the sulfur free system.

more, type and amounts of promoters have to be varied from one metal to the other. Favorable multicomponent catalysts are, e.g., osmium-potassium oxide* and nickel-molybdenum.

(8) A "support-effect" such as demonstrated in the classical case of platinum asbestos is of negligible importance with ammonia catalysts except with osmium and ruthenium.

In October, 1910, the author had established the fact that certain magnetites as well as synthetic iron catalysts can be as effective as the uranium carbide catalyst which, in the hands of F. Haber, had proved to be of outstanding activity. At 500°C. and at a pressure of 100 atmospheres and a gas velocity of 50 liters per hour, 5 volume % of ammonia were obtained in the exit gases with 2 g. of a magnetite and 4.5% with a synthetic iron catalyst, the latter being operated at 550°C.

An important property of the multicomponent catalysts was that with highly purified gases, they remained active over long periods, a fact which greatly helped their practical usage for the technical ammonia synthesis.

Soon it was recognized that the iron-alumina-potash combination was the best suitable catalyst for the large scale industrial synthesis. Again and again, it was proved that this particular catalyst gave the most dependable yields both in the pilot plants and in large scale operation.

The various possibilities for modifications of catalyst activities in two component systems are shown schematically in Figure 1. On the abscissa, the relative proportion of two catalyst components A and B are plotted, and on the ordinate the catalytic efficiencies, measured under constant conditions, in arbitary units. As an example for Case I, we mention the system Fe-Al_2O_3, and for case IV, the system Fe-Mo.

A brief survey of the positive or negative effects of various elements (viz., of their oxides) on the activity of iron as an ammonia catalyst follows:

(a) Positive: Al, Ba, Be, Ca, Ce, Cr, Er, K, La, Li, Mg, Mn, Mo, Na, Nb, Nd, Rb, Sm, Sr, Ta, Th, Ti, U, V, W, Y, Zr.
(b) Negative: As, B, Bi, Br, C, Cd, Cl, F, I, P, Pb, S, Sb, Sn, Te, Tl, Zn.
(c) Indifferent or dubious: Au, Co, Cu, Hg, Ir, Ni, Os, Pd, Pt, Si.

A decision in which of these three classes to list a given element depends to a certain extent on the proportion in which it is added to iron and also on the question whether a given element such as, e.g., Si, Ti, and B is employed in its elementary form or as its oxide. Actually, various tables would be required to give a complete picture.

* Potassium osmiate can be used, advantageously, as the starting material.

In February 1912, the author prepared for C. Bosch a survey on the characteristics of the more promising ammonia catalysts which had been tested so far. The following remarks may be of interest:

"No sharp limit exists either between the promoting and the indifferent components nor between the indifferent and the harmful ones. Of decisive importance are the proportions employed, the structures of the solid products obtained, and some further factors. As a rule, the

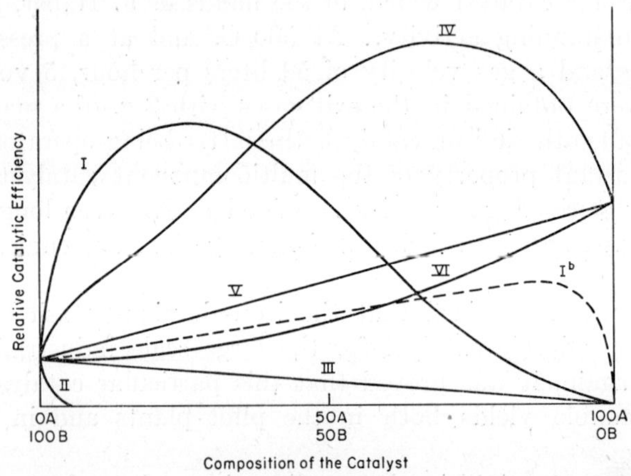

FIG. 1.

I. Simple activation
I^b. Support effect
II. Poisoning
III and V. Additive effect
IV. Mutual activation
VI. Deactivation.

hydrides, nitrides, amides, carbides, and if easily reducible, even the oxides of metallic elements are equivalent to the metals proper in their effects on an iron catalyst."

"It is of importance to draw a distinction between additions which are sensitive toward oxygen and those which are not sensitive toward oxygen. For instance, metallic potassium, under the influence of oxygen, forms potassium oxide, and the promoting effect of the latter differs from that of the alkali-metal. The same element can act on the catalyst differently, depending on whether it is present as the oxide, chloride, or other compound."

"A comparison of the various catalysts proves the high specifity of promoter actions. The same promoter does not necessarily act in the same sense and to the same extent on different metals. Whereas, e.g., potassium oxide is an excellent promoter for osmium and ruthemium, it has no, or little, effect on uranium and molybdenum."

The following description refers to comparative experiments carried out with various metallic catalysts using 2 g. of catalyst, average gas velocity 60 liters per hour, 550°C. 100 atmospheres.

Iron: High efficiency of the purest iron sponge as long as its high porosity is maintained. This can be achieved by producing a highly dispersed skeleton of earth metal oxide throughout the metal. Without promoter, about 2% ammonia and decrease with time; with promoter, 4 to 5% ammonia for a long period.

Cobalt: Similar but weaker action than iron; less susceptible to promoter effects. Pure metal, about 0.2% ammonia; promoted up to 1.5%.

Nickel: Still less efficient than cobalt, almost ineffective.

Molybdenum: In its pure form, without additions, it is the most efficient catalyst of all the easily obtainable and reducible substances, and it is less easily poisoned than iron. It catalyzes in another way than iron, insofar as it forms analytically easily detectable amounts of metal nitrides (about 9% nitrogen content) during its catalytic action, whereas iron does not form, under synthesis conditions, analytically detectable quantities of a nitride. In this respect, molybdenum resembles tungsten, manganese and uranium which all form nitrides during their operation as ammonia catalysts. Molybdenum is clearly promoted by nickel, cobalt and iron, but not by oxides such as alumina. Alkali metals can act favorably on molybdenum, but oxides of the alkali metals are harmful. Efficiency, as pure molybdenum, 1.5%, promoted up to 4% ammonia.

Tungsten: Similar to molybdenum but of weaker activity. 0.4% ammonia as the pure metal, and 1.2%, if promoted.

Osmium: This metal had already been found by Haber to be an excellent ammonia catalyst. Its activity is further increased by alkali metal oxides, especially by potassium hydroxide. As the pure metal 2% ammonia, promoted, 4%.

Manganese: In its pure form a fairly good catalyst; it can be promoted by certain oxides and metals. However, it is very easily oxidized and rendered inactive by water vapor. As the pure metal, 0.8%, promoted, up to 2.5%.

Similarly sensitive toward oxygen or water vapor are the following metals:

Uranium: (studied at an early stage by Haber). It can be promoted by certain metals. 1.0% to 2.5% ammonia.

Cerium has but little activity in the form of its hydride, nitride or amide. It can also be used in the form of its carbide. 0.3% to 1% ammonia.

As for the remaining metals and their hydrides and nitrides, numerous combinations were found to possess catalytic activities. We may mention here a few pairs of metals (31):

Al-Mn	Cr-Mg	Li-Nb
Ba-Cr	Cr-Zr	Li-V
Ba-Os	Cr-Na	Mg-V
Ca-Mn	K-V	Mn-V
Ce-Mg	Li-Mn	V-Zr.

Again and again it was confirmed that, as R. Willstätter expressed it, "multicomponent catalysts can be regarded, in their activities as equivalent to new specific substances."

V. Multicomponent Catalysts for Reactions Other than the Ammonia Synthesis

Still during our experiments with ammonia catalysts, other technically important reactions were studied which promised to be accessible to the catalytic method, and particularly to the influence of multicomponent catalysts. As this further development has been repeatedly described elsewhere, only a short survey will be presented here.

1. Catalytic Production of Hydrogen

W. Wild studied, in the B. A. S. F., the production of hydrogen from water gas at relatively low temperatures according to the shift reaction

$$H_2O + CO \rightarrow H_2 + CO_2$$

Strangely enough, a combination similar to the ammonia catalyst, iron oxide plus alumina, yielded particularly good results (32). Together with Ch. Beck, the author found that other combinations such as iron oxide with chromium oxide, zinc oxide with chromium oxide, lead oxide with uranium oxide, copper oxide with zirconium oxide, manganese oxide with chromium oxide, and similar multicomponent systems were quite effective catalysts for the same reaction (33).

2. Catalytic Hydrogenation of Carbon Monoxide to Methane, and of Unsaturated Organic Compounds

The reactions studied were: the catalytic formation of methane from carbon monoxide and hydrogen (according to Sabatier (34), normal pressure), the catalytic hydrogenation of unsaturated hydrocarbons and also of unsaturated fatty acids ("fat hardening" according to Normann (35)). Here again, a certain analogy was established between the catalysts suitable for these types of reactions and the ammonia catalyst: the typical hydrogenation catalysts such as nickel, cobalt, etc., were made more effective by promoters such as magnesia, alumina, thoria, zirconia, uranium oxide, chromium oxide and others, generally by oxides of the earth metals and of related elements.

In this special field, earlier work had been done in other laboratories, such as by the Schering Company, Berlin (36), and by Ipatieff (37) in connection with his work on the hydrogenation of camphor and of other organic compounds. At both places, the favorable effect of alkali oxides and earth alkali oxides on nickel, cobalt and copper has been investigated. Similarly, Paal and his coworkers (38) have used a palladium-aluminum hydroxide catalyst in 1913 for the hydrogenation of double bonds. Bedford and Erdman (39) had reported that the catalytic action of nickel oxide is enhanced by the oxides of aluminum, zirconium, titanium, calcium, lanthanum, and magnesium.

However, the hydrogenation catalysts can be promoted also by compounds which are ineffective for the iron ammonia catalysts, e.g., by silica and silicates, silicofluorides, borates, and phosphates. An interesting type of hydrogenation catalysts was found in the form of zeolites which by ion exchange were impregnated with heavy metal salts.

3. Production of Various Organic Substances by the Catalytic Reduction of Carbon Monoxide with Hydrogen under Pressure

In the aforementioned studies, our aim was to find particularly effective catalysts for such reactions which were already known but which had not yet been subjected to a systematic search for the most effective catalysts. A new field of catalytic reactions, however, was entered as we followed a suggestion of C. Bosch to attempt the reduction of carbon monoxide with hydrogen at high pressures with the aim of obtaining, as reduction products, not methane, but organic compounds such as higher hydrocarbons, alcohols, etc. Thermodynamic calculations proved that the formation of a large number of organic substances from carbon monoxide and hydrogen lay in the range of possibilities. The question was whether catalysts and operating conditions could be found which would permit one to direct the reduction of carbon monoxide along highly selective reaction paths. It is understandable that we based much of our work in this new field on our previous favorable experience with multicomponent catalysts in the hope that we would be able to achieve, with the help of such catalysts, the needed specificity of catalytic action. In investigating this field, high pressures were used for reactions which hitherto had been studied with normal pressures only, and the proportions of the reacting gases were varied, within wide limits.

Starting in 1913, these studies went through various stages.

(a) *Work carried out together with Ch. Schneider.* The catalysts used for the pressure reduction of carbon monoxide with hydrogen consisted mainly of mixtures of cobalt oxide or of osmium with sodium hydroxide, and later on, of iron oxide with zinc oxide and alkali oxide. As a rule, the products obtained were of a twofold nature as they separated into an oily layer of various hydrocarbons and into an aqueous layer containing methyl alcohol and higher alcohols, aldehydes, esters and other oxygen-containing organic compounds (40).

(b) *Work aiming at the preferential formation of alcohols.* For this specific aim, the leading viewpoint concerning the catalyst was to avoid iron and iron compounds, which, according to our previous experience, favored the formation of hydrocarbons. A start was made in this direction by C. Müller (1919) who used material consisting mainly of alkali metal compounds as catalysts. Later on (1923), M. Pier and K. Winkler obtained highly satisfactory results with catalysts of the type zinc oxide plus chromium oxide which yielded, as the product of the carbon monoxide reduction, practically pure methanol. Addition of alkali compounds to catalysts of the zinc oxide-chromium oxide type changed the selectivity of the reaction with the result that higher alcohols, particularly isobutyl alcohol, were the main products (41). Similar results were obtained by O. Schmidt and K. Ufer with copper combined with oxides of other metals as catalysts (42).

As is well known, starting in 1923, Franz Fischer, Tropsch *et al.* developed the synthesis of hydrocarbons by carbon monoxide reduction to a technical process. They developed highly efficient multicomponent catalysts of the cobalt, nickel and iron type for this synthesis (43).

4. Catalytic Oxidations

A technically important reaction is the oxidation of ammonia with air to nitrous oxides which are then converted into nitric acid. In this catalysis, losses are mainly caused by the formation of nitrogen as an oxidation product.

For this reaction, our aim was to improve the known catalytic action of iron oxides to an extent that the high activity of the generally used platinum catalyst would be equalled. Together with Ch. Beck, the author finally found iron oxide-bismuth oxide combinations, and iron oxide-manganese oxide-bismuth oxide combinations which were outstanding catalysts. An iron oxide-lead oxide catalyst proved to be less satisfactory (44).

It may be mentioned briefly that also for other catalytic oxidations, e.g., for those of ethylene and of anthracene, multicomponent catalysts proved to be very effective. For the anthracene oxidation, the following mixtures were used with good success: Uranium oxide-alumina, copper oxide-bismuth oxide, nickel oxide-bismuth oxide, manganese oxide-chromium oxide, manganese oxide-uranium oxide, and iron oxide-chromium oxide.

Rather unique multicomponent catalysts are those in which one of the components, being volatile, has to be replenished continuously. The oxidation of dilute hydrogen sulfide to elementary sulfur by air with active carbon as a catalyst is such an example. Small amounts of ammonia, if added to the gases entering the reactor, favor and complete this catalytic reaction: the ammonia leaves the reactor unchanged (45).

We shall omit the numerous other examples of successful uses of multicomponent catalysts. Since about 1920, they have been employed to an ever increasing extent in chemical industries all over the world, and it may be considered rather an exception than the rule if a single component compound serves as a catalyst. Nevertheless, it appears that we have merely entered this vast field, particularly in view of further catalytic possibilities in organic chemistry and in biochemistry.

VI. Definition of Multicomponent Catalysts

To a certain extent the expression "multicomponent catalysts" is an arbitrary one. There is no doubt that the pure chemical elements and pure chemical compounds have to be called "single component catalysts." It is, however, questionable whether a material such as steel should be classified as a single component system or as a multicomponent system. Some of the multicomponent catalysts, for instance, the iron-alumina catalyst consist of two separate solid phases but it would be misleading to accept the presence of more than one phase as the decisive criterion for multicomponent catalysts. The more than additive catalytic action of Cu-ions and Fe-ions in an homogeneous aqueous medium represents obviously a case of multicomponent catalysis, although it occurs in a single-phase system. As to solid multicomponent catalysts, they usually consist of more than one single phase, but there are exceptions to this rule, such as in cases in which mixed crystals or solid solutions are formed from the components.

A purely practical decision whether two substances form, if combined, a multicomponent catalyst, can be reached by investigating if mixtures of both components, made in various proportions, show a change of catalytic activities as a function of their relative amounts (as shown in Fig. 1).

VII. Theoretical Remarks

In some older hypotheses, catalytic processes have been set apart as essentially different from "normal" chemical reactions. However, these ideas have been abandoned in favor of the concept that catalytic processes are caused and controlled by the same forces of chemical affinity as they are effective in bringing about "regular" chemical reactions.

There can be, however, no doubt that in catalytic processes, purely physical factors play an important role, in addition to the chemical valence forces. This is particularly true for the solid catalysts of heterogeneous reactions for which the properties of surfaces, as the seats of catalytic action are of prime importance. The total surface areas, the fine structure of the surfaces, the transport of reactants to and from surfaces, and the adsorption of the reactants on the surfaces, can all be considered as processes of a predominantly physical nature which contribute to the catalytic overall effect. Any attempt, however, to draw too sharp a line between chemical and physical processes would be futile. This is illustrated clearly by the fact that the adsorption of gases on surfaces can be described either as a mere "physical" condensation of the gas molecules "on top" of the solid surface, as well as the result of chemical affinities between adsorbate and adsorbent. Every single case of adsorption may lie closer to either one of the hypothetical extremes of a "purely physcial" or of a "purely chemical" adsorption, and it would be misleading to maintain an artificial differentiation between physical and chemical factors.

In order to act as a catalyst, a given substance has to bear its specific affinities on the reacting molecules. Obviously, a modification of "the field of force" of these specific affinities can be achieved by the presence of additional components in a catalyst, either via a combination of the affinities of both components at phase boundaries, or via a gradual change of the affinities of the main catalyst due to the formation of solid solutions, alloys, or mixed crystals with the accessory component.

Further, a stabilization of the total surface of the main catalyst by added substances may explain some promoter effects, but this explanation holds only for a few multicomponent catalysts. For the iron-alumina catalyst, a beneficial stabilizing effect of the promoter alumina on the fine structure of the iron has to be accepted as a partial explanation. The fact that highly dispersed pure iron sinters at temperatures above 300°C. to a considerable extent, and that sintering practically does not occur with iron of the same high dispersion which contains 1 to 2% of alumina, is a strong qualitative support for this concept.* In a quantitative way, the work of P. H. Emmett (47) and his associates has proved this point beyond any doubt; it gives similarly valuable

* The oldest observations on this effect date back to the studies carried out by G. Magnus in 1825 on pyrophoric iron (46).

information about the true surface values of unpromoted and promoted iron preparations. As is to be expected, the alumina containing samples show, according to Emmett, for equal weights, much higher surface values than unpromoted iron samples. Further, these large specific surfaces are maintained at higher temperatures in the promoted catalysts whereas the pure iron samples show a sharp surface shrinkage if heated to equally high temperatures. Even before these valuable quantitative results had been obtained, the action of alumina on iron as a re-enforcing agent for its fine structure was anticipated. Thus, in one of our earlier patents (48), the outstanding qualities of alumina-containing iron catalysts were interpreted as follows: "The afore-mentioned substances (alumina and related oxides) are dissolved in the very hot metal oxide, in a highly dispersed form, and constitute, after the reduction of the catalyst, an almost invisible skeleton within the bulk of the metal, thus preventing a decrease of its surface and of its catalytic properties."

It is strange and typical for the erratic path of some laboratory work, that this final concept of the promoter action on ammonia catalysts was just opposite to our initial working hypothesis according to which flux promoters were considered to be essential for good catalytic activity.

Although the "fine structure hypothesis" of the promoter action of alumina on iron appears well founded, it is conceivable that, in addition to the conservation of the iron surface, the beneficial effect of alumina involves also a modification of the valence forces at the borderlines between alumina (or the iron-alumina spinel $Al_2O_3 \cdot FeO$), and the crystallites of metallic iron (49).

A favorable combination of valence forces of both components seems to be the basic principle of the nickel-molybdenum ammonia catalyst. It has been found (50) that an effective catalyst of this type requires the presence of two solid phases consisting of molybdenum and nickel on the one hand and an excess of metallic molybdenum on the other. Similar conditions prevail for molybdenum-cobalt and for molybdenum-iron catalysts: their effectiveness depends on an excess of free metal, molybdenum for the molybdenum-cobalt combination and iron for the molybdenum-iron combination, beyond the amounts of the two components which combine with each other. A simple explanation for the working mechanism of such catalysts is that at the boundary lines between the two phases, an activation takes place. In the case of the nickel-molybdenum catalyst, the nickel-molybdenum phase will probably act preferentially on the hydrogen and the molybdenum phase on the nitrogen.

Generally speaking, the many complicated and specific factors underlying catalytic action were, and still are today, inaccessible to a satisfactory theoretical and quantitative interpretation. Thus, a purely empirical search for suitable catalysts still has to be employed for any new problem instead of a method which would permit one to predict such catalysts directly and in a dependable way. It is due to this situation that the author and his associates had to carry out about 20,000 small scale tests and to investigate some 3000 different preparations as potential catalysts for the ammonia synthesis. The more promising substances were tested under varied conditions and in combination with different promoters. Accordingly the number of single test runs exceeds the number of samples.

Whereas some knowledge has been obtained about the working mechanism of ammonia catalysts (51), this does not apply to the same extent to catalysts used for many other processes. However, a few typical cases of multicomponent catalysts have been investigated both in the author's laboratory and by others. The main conclusion to be drawn from these studies is that it would be wrong to seek one universal explanation for the promoter effects in solid catalysts. As outlined above, structural as well as chemical effects may cause the improvements which are observed after certain substances have been added to a given catalyst.

In many cases, those promoters which seem to act by virtue of a combination of their chemical affinities with those of the main catalysts do not merely increase the activity of the unpromoted catalyst, but they also cause the catalytic reaction to proceed in a more specified direction. The application of promoters to guide reactions selectively toward the formation of desired product is, from a practical viewpoint, often more important than the achievement of an overall acceleration of the catalytic process.

One of the simplest examples for such effects is the oxidation of ammonia with iron oxide-bismuth oxide as a catalyst. Here, the addition of bismuth oxide results in the formation of nitrous oxides as the main product whereas an iron oxide catalyst without bismuth oxide yields nitrogen almost exlcusively. Selectively guiding catalysts become increasingly important in the synthesis of organic compounds, e.g., in the hydrogenation of carbon monoxide where the type of obtainable product can be varied, within wide limits, by the kinds of catalysts and promoters which are employed.

Since the time when the author retired some twenty years ago from practical work in catalysis, the use and the study of catalysts and particularly of multicomponent catalysts has extended beyond expectations. A great many theoretical (52) and practically important catalytic reactions have been discovered and investigated in all their aspects. This domain of catalytic chemistry will extend over an even wider range when we shall have learned more about the deeper reasons of catalytic action and if we shall be able to build catalysts which resemble, in efficiency and in complexity, the catalytic agents of biochemical systems.

References

1. Döbereiner, J. W., *Ann.* **1,** 29 (1832); *Gilb. Ann.* **74,** 269 (1823); **76,** 102 (1824).
2. Möhler, F., and Mahla, A., *Ann.* **81,** 255 (1852).
3. Deacon, H., *J. Chem. Soc. London* (2) **10,** 725 (1872); **20,** 1062 (1887); British Pat. 505 (1873); 906, 1632 (1875).
4. Bredig, G., *Z. physik. Chem.* **31,** 301 (1899).
5. Traube, M., *Ber.* **17,** 1062 (1884); **18,** 1885, 1890 (1885).
6. Price, Th. S., *Z. physik. Chem.* **27,** 474 (1898).
7. Döbereiner, J. W., *Schweiggers J.* **34,** 91 (1822); **38,** 321 (1823).
8. Brode, J., *Chem. Z.* **24,** 1116 (1901); *Z. physik. Chem.* **37,** 257, 290 (1901); **49,** 209 (1904).
9. German Patent 140,353 (1901).
10. German Patents 142,144, 149,677 (1902).
11. British Patent 6448 (1905).
12. Fokin, G., *Z. Elektrochem.* **12,** 747 (1906).
13. Bredig, G., and Ikeda, K., *Z. physik. Chem.* **37,** 7 (1899).
14. Mittasch, A., *Z. physik. Chem.* **40,** 1 (1902).
15. Margueritte, F., and Sourdeval, H., *Compt. rend.* **50,** 1100 (1860).
16. Wöhler, F., *Ann.* **74,** 217 (1850); see also Bichowsky, V., U.S. Patent 1,570,802 (1924).
17. Serpek, O. German Patents 181,991, 181,992, 216,746 (1905).
18. Birkeland and Eyde, German Patents 170,585, 179,882 (1905); Schönherr, O., and Hessberger, German Patent 201,279 (1905); *Z. angew. Chem.* **21,** 1633 (1908); Pauling, German Patent 198,241 (1905); U.S. Patent 887,230 (1907).
19. Frank, A., and Caro, N., German Patent 88,363 (1895); 108,971 (1898).
20. Döbereiner, J. W., *Ann. Chim. Phys.* (2) **24,** 91,880 (1823).
21. Ramsay, W., and Young, C., *J. Chem. Soc. London* **45,** 88 (1884).
22. LeChatelier, H., French Pat. 313,950 (1901); Compt. rend Acad. Sci. **164,** 588 (1917).
23. Perman, W., and Atkinson, G., *Proc. Roy. Soc. London* **74,** 110 (1904); **76,** 167 (1905).
24. Haber, F., and van Oordt, G., *Z. anorg. Chem.* **43,** 111 (1904); **44,** 341 (1905).
25. Nernst, W., Jost, W., and Jellinek, G., *Z. Elektrochem.* **13,** 521 (1907); **14,** 373 (1908); *Z. anorg. Chem.* **57,** 414 (1908).
26. Haber, F., and Le Rossignol, R., *Ber.* **40,** 2144 (1907); *Z. Elektrochem.* **14,** 181 (1908); **16,** 244 (1910); **19,** 53 (1913). Haber, F., *Naturwissenschaften* **10,** 1041 (1922). Haber, F., German Patents 223,408, 229,126 (1909), assigned to B. A. S. F.
27. Ostwald, W., Lebenslinien. Vol. 2, Leipzig, 1926–27, 279 ff.
28. German Patent 249,447 (1910); 254,437, 258,146, 262,823 (1910).
29. A list of these later patents for catalysts of the ammonia synthesis is given in Ullmann, Enzyklopädie der Technischen Chemie, 2d edition, Urban and Schwarzenberg, Berlin, Wien, 1928, p. 420–426.
30. German Patents 254,344, 263,612 (1910).
31. U.S. Patents 1,094,194, 1,052,951, 1,068,966 to 1,068,969, 1,118,628, 1,152,930, 1,225,725 (1910–1915).
32. German Patent 292,615 (1913).
33. German Patents 279,582 (1913), 293,585 (1914); U.S. Patent 1,330,772 (1918).

34. Sabatier, P., and Senderens, R., *Compt. rend.* **132,** 1257 (1901); **133,** 321 (1901); **134,** 514 (1902); **135,** 226 (1902); **137,** 301 (1903). French Patents 35,462, 35,590 (1905).
35. Normann, W., German Patents 141,029 (1902); 211,669, 221,890 (1907).
36. German Patents 219,043, 219,044 (1908); 271,157 (1909).
37. Ipatieff, V. N., *Ber.* **34,** 596 (1901); **40,** 1281 (1907); **42,** 2089 (1909); **43,** 3387 (1910); **45,** 3205 (1912); *J. prakt. Chem.* **87,** 479, 482 (1913); *J. Russ. Phys. Chem. Soc.* **46,** 470 (1914).
38. Paal, H., German Patent 236,488 (1910); 298,193 (1913); *Ber.* **38,** 1406 (1905).
39. Erdmann, E., German Patent 211,669 (1907). Bedford, R., German Patent 221,890 (1907).
40. German Patent 293,787 (1913); 295,202 (1914); U.S. Patent 1,201,850 (1914).
41. German Patents 415,686, 441,433, 462,837 (1923); U.S. Patents 1,558,559, 1,569,775 (1923).
42. Schmidt , O., and Ufer, H., German Patents 571,355, 571,356, 580,905 (1928).
43. Fischer, F., and Tropsch, H., *Brenn.* **4,** 276 (1923); **7,** 97, 299 (1926); **8,** 1, 165, 226 (1927); *Ber.* **59,** 830, 832, 923 (1926); **60,** 1330 (1927). German Patents 411,216 (1922); 484,337 (1925); 524,468 (1926).
44. German Patent 283,824 (1914); U.S. Patents 1,207,706 to 1,207,708, 1,211,394 (1914/15).
45. German Patents 303,862 (1914); 338,829 (1918).
46. Magnus, G., *Pogg. Ann.* **3,** 81 (1825).
47. Brunauer, S., and Emmett, P. H., *J. Am. Chem. Soc.* **57,** 1754 (1935); Emmett, P. H., and Brunauer, S., *J. Am. Chem. Soc.* **59,** 310, 1553, 2682 (1937); *Trans. Electrochem. Soc.* **71,** 383 (1937); Brunauer, S., Emmett, P. H., and Teller, E., *J. Am. Chem. Soc.* **60,** 309 (1938).
48. German Patent 254,437 (1910); see also Mittasch, A., and Keunecke, E., *Z. Elektrochem.* **38,** 666 (1932); Brill, R., *Z. Electrochem.* **38,** 669 (1932).
49. Frankenburger, W., *Z. Elektrochem.* **39,** 45, 97, 269, 819 (1933).
50. Mittasch, A., *Z. Elektrochem.* **36,** 567 (1930); Keunecke, E., *Z. Electrochem.* **36,** 690 (1930); Mittasch, A., and Keunecke, E., *Z. physik. Chem.*, Bodenstein Festband 574 (1931).
51. Frankenburger, W., Synthetic Ammonia, in F. Ullmann, Enzyklopädie der Technischen Chemie. 2d ed., Urban and Schwarzenberg, Berlin, Wien, 1928, vol. 1, p. 383–400, vol. 6, p. 436; Messner, G., and Frankenburger, W., *Z. physik. Chem.* Bodenstein Festband, 593 (1931); Mittasch, A., and Frankenburger, W., *Z. Elektrochem.* **35,** 920 (1929). Mittasch, A., *Z. Elektrochem.* **36,** 569 (1930). Almquist, J. A., Black C. A., *J. Am. Chem. Soc.* **48,** 2314 (1926). Almquist, J. A., *J. Am. Chem. Soc.* **48,** 2820 (1926). Emmett, P. H., Hendricks, S. B., and Brunauer, S., *J. Am. Chem. Soc.* **52,** 1456 (1930). Emmett, P. H., *J. Chem. Education* **7,** 2571 (1930). Brunauer, S., Jefferson, M. E., Emmett, P. H., and Hendricks, S. B., *J. Am. Chem. Soc.* **53,** 1778 (1931). Wyckoff, W. G., and Crittenden, E. D., *J. Am. Chem. Soc.* **47,** 2866 (1925). Mayer, C., *Z. Krist.* **70,** 383 (1929); Brill, R., *Z. Krist.* **68,** 379 (1928); *Z. Elektrochem.* **38,** 669 (1932); Emmett, P. H., and Brunauer, S., *J. Am. Chem. Soc.* **55,** 738 (1933); **56,** 35 (1934); **62,** 1732 (1940). Jorris, G. G., and Taylor, H. S., *J. Chem. Phys.* **7,** 893 (1939). Guyer, R. F., Jorris, G. G., and Taylor, H. S., *J. Chem. Phys.* **9,** 287 (1941). Emmett, P. H., "Synthetic Ammonia" in Fixed Nitrogen. Ed. by H. Curtis, The Chemical Catalog Co., New York, 1932, p. 150. Mittasch, A., *Elektrochem. Z.* **35,** 16, 96 (1929).
52. Schwab, G. M., Catalysis from the Viewpoint of Chemical Kinetics Berlin, 1931, p. 203. Hinshelwood, C. H., *J. Chem. Soc. London* **1939,** 1203.

Catalytic Phenomena Related to Photographic Development

T. H. JAMES

Research Laboratories, Eastman Kodak Co., Rochester, New York

CONTENTS

I. INTRODUCTION

The photographic process had its inception in the discovery by Daguerre (1835, although not announced until 1839) of the development of the latent image. He found by accident that a superficially iodized silver plate which had been exposed to sunlight for a time insufficient to produce a visible image yielded a good image upon subsequent fuming with mercury vapor. A short time later, and without knowledge of Daguerre's discovery, the Rev. J. B. Reade and, independently, Fox

Talbot found that an invisible latent image produced by the action of light on paper impregnated with silver chloride could be developed by brushing the paper with a mixture of gallic acid and silver nitrate. The process worked out by Fox Talbot is the forerunner of the present-day photographic process.

1. Nature of the Photographic Sensitive Layer

The light-sensitive layer of the present-day photographic material consists essentially of a large number (e.g., 10^8 per square centimeter) of tiny crystals of silver halide embedded in a layer of gelatin. The tiny crystals, or grains as they are commonly called, of the most sensitive photographic materials are composed of silver bromide, a small percentage of iodide, and a very small but very important amount of silver sulfide (Sheppard, 1) or possibly silver (Carroll and Hubbard, 1a) or both. The halide in the less sensitive materials may be simply bromide, chloride, or mixtures of the two.

2. Latent Image Formation

The sensitive layer shows no visible change in appearance when it is exposed to light in the normal process of "taking a picture." However, the reactivity of some of the silver halide grains towards certain reducing agents has been altered, and if the material is immersed in a developing solution for a suitable time, the visible developed image emerges. Some of the silver halide grains have been reduced to metallic silver. The number of such reduced grains per unit area is larger, the greater the amount of actinic light absorbed in that area (providing the material has not been over-exposed). The optical density of the silver deposit is roughly proportional to the number of developed grains under most working conditions, and hence the developed image gives a representation of the tone values of the object photographed.

The identity of the latent image material produced by the action of the exposing light has not been established analytically because of the extremely small amounts of photolytic product involved, but the indirect evidence that this material is metallic silver is strong. The silver is believed to be formed by the reduction of silver ions at pre-existing "sensitivity specks" on the grain surface or in the interior. (In the usual negative material the specks are mostly on the surface.) The surface specks are believed to be silver sulfide, silver, or both. Imperfections in the lattice structure also may act as sensitivity specks, particularly in the interior of the grain.

If the photolytic silver is added on to the material of the speck, the nuclei which are responsible for making a grain developable consist of the pre-existing sensitivity speck plus the photolytic silver. The latter may amount to no more than a few atoms of silver per speck on the average. The amount of photolytic silver may be as little as 10^{-8} of the silver formed when the grain is developed.

J. W. Mitchell (1b) has suggested that the latent image nucleus is composed not of metallic silver but of F-centers in a two-dimensional aggregate of suitable size and configuration. (An F-center consists of an electron trapped at a vacant anion lattice site.) He believes that development takes place in two stages. In the first stage the F-center aggregate is enlarged to such a size that it breaks away from the silver halide matrix and forms a colloidal silver particle. In this stage, according to Mitchell, electrons pass from the developing agent to the F-center aggregate and a corresponding number of vacant anion sites migrate to the aggregate to form F-centers with these electrons. Once colloidal silver is formed, development proceeds by another mechanism. If Mitchell's views are correct, the discussion of the mechanism of development in Section V of this chapter applies only to this second stage of development.

3. Development

If the exposed photographic material is developed for only a very short time, the image will consist largely of partially reduced grains, particularly in the higher exposure areas. If development is prolonged, very few partially reduced grains can be found. The great majority of the developed grains have been largely or completely reduced to silver. For a given exposure, the number of developed grains increases with time of development. However, some reduced grains also appear in the areas of the material which have not been exposed to light, and some silver may deposit as a nearly uniform haze of very fine particles. The silver which has been formed without reference to the light exposure is termed "fog." In practical work, development is usually terminated at some point before the fog formation becomes objectionable for the particular use which is to be made of the photograph.

For theoretical purposes, the problem of determining whether a given grain is developable is not always easy. In general, the problem must be treated statistically. Exposure to light has increased markedly the rate at which some of the silver halide grains are reduced by the developer. If the exposure has been large, but not in the solarization* region, the

* The developable density of many photographic materials passes through a maximum at a sufficiently high exposure and subsequently decreases with further increase in exposure. This loss of developability with increasing exposure is termed "solarization," and probably is caused by a superficial rehalogenation of the surface latent image centers (cf. J. H. Webb, *J. Optical Soc. Am.* **30,** 445, 1940).

difference between exposed and unexposed grains is sharply defined. The difficulty arises when the exposure is small, particularly when it is near the threshold of sensitivity of the photographic material in question. Reduction of a grain must progress to a point where a microscopically visible amount of silver is formed before development of the grain can be detected. The time elapsing between the immersion of the exposed sensitive layer in the developer and the point at which development of a grain can be detected is not the same for all grains, and may vary widely according to conditions. Hence, a grain which does not appear developable within a certain time may develop at some longer time. But unselective reduction of the silver halide grains becomes more and more prominent as the time of immersion in the developer increases, and no sharply defined border exists between fogging reduction and the development of a grain which has received an exposure lying in the sensitivity threshold region. Viewed on a macro scale as, for example, in terms of mass of reduced silver per unit area or in terms of optical density, image development gradually merges into fog formation as development is prolonged.

In the study of the mechanism of development, it is generally desirable to compare the rate of reduction of grains which have been given a relatively large exposure to light (avoiding, however, over-exposure leading to solarization) with the rate of reduction of completely unexposed grains. The differentiation in rate is then large. A definition of threshold of developability cannot be avoided, however, when the minimum number of quanta which must be absorbed before a grain becomes "developable" or the minimum number of silver atoms which will serve as a nucleus to promote development is under consideration. Then, some particular time of development under the particular set of conditions must be chosen. The time at which the rate of increase of optical density in the image region becomes equal to the rate of increase of density in the unexposed region is a convenient criterion for the "maximum degree of development." The number of developable grains is the number of grains which then contain a visible amount of reduced silver, corrected by the number of similarly reduced grains in the unexposed area. (If, however, much silver has been deposited at some distance from the original grains, this enumeration of developable grains becomes impractical.)

Silver catalysis of the reduction of silver ions appears to be a necessary condition for normal development. The reaction of developing agents including several types of chemical compounds, e.g., hydroquinone, *p*-aminophenol, hydroxylamine, catechol, and *p*-phenylenediamine, are known to exhibit this catalysis to a high degree.

4. Types of Development

There are two basic methods by which a developed silver image can be obtained. The one commonly employed in practice is termed "chemical" or, better, "direct" development. The exposed sensitive layer is placed directly in a suitable reducing solution, and the silver image is derived from the reduction of the silver halide grains. This is the type already considered to some extent in the preceding section.

In the second type, the silver halide is dissolved out after the exposure but before development. An alkaline solution of sodium thiosulfate is suitable for this "fixing" process. The latent image silver remains behind in the gelatin layer. Development is carried out in a solution which contains both a reducing agent and a soluble silver salt. The developed image is built up by silver derived from the developing solution, not from the original silver halide. This type is termed post-fixation "physical" development, a misnomer which, however, is firmly established in the photographic literature.

In a modified and more complicated form (pre-fixation physical development) the exposed sensitive layer is placed directly in the developing solution without first dissolving out the silver halide. The developing solution contains a silver halide solvent in addition to the soluble silver salt, and part of the developed silver in this process comes from the original silver halide. Some direct development also occurs.

Post-fixation physical development is simpler in mechanism than direct development. The fundamental reaction is the reduction of silver ions from a solution of silver salt. This reaction is accelerated by the presence of silver nuclei, and the mechanism of the development is the mechanism of this catalytic process.

II. Physical Development and the Reduction of Silver Ions from Solution

1. Sheppard's Mechanism

Piper (2) in 1908 suggested that development is an "action of a catalytic nature, the latent image being the catalyzer," but he suggested no specific mechanism. Sheppard (3) in 1919 discovered that the reduction of silver ions in a solution of silver nitrate and sodium sulfite is catalyzed by silver. The silver ions in this solution are largely bound in the form of the soluble complex, $Ag(SO_3)_2^{\equiv}$, and Sheppard thought that the reaction was an autoreduction occurring within this complex. He suggested that the reaction occurring in development followed a similar course, i.e., that a complex is first formed between the developing agent and a silver ion of the silver halide and that the autoreduction of the silver ion in this complex is accelerated by silver.

Sheppard's mechanism was formulated primarily in reference to direct development (Sheppard, 3; Sheppard and Meyer, 3a) and he assumed that the complex was formed in the act of or as a result of adsorption of the developing agent by the silver halide. Since he did not specifically suggest application of the basic mechanism to physical development, consideration of his mechanism will be deferred until direct development is treated in detail.

2. Oxygen Oxidation of Developing Agents

Volmer, on the basis of some experiments on the oxygen oxidation of developing agents, suggested (1921) that development is a catalytic process. In his experiments (Volmer, 4) he divided a solution of 0.02 g. of developing agent in 10 cc. of 1 *N* sodium carbonate into two equal parts, added finely divided silver to one, and passed air through each in a constant stream. He used the comparative rates of coloration of the solutions as a measure of the relative rates of oxidation, and concluded that the silver accelerated the oxidation by from two- to sixfold. He suggested that silver would similarly catalyze the reduction of silver salts by the developing agents. Thus, Volmer treats the catalysis as an activation of the developing agent and as being nonspecific with respect to the oxidizing agent, whereas Sheppard's mechanism is specific for the reduction of silver (or metal) salts.

Weissberger and Thomas (5) have questioned the validity of Volmer's use of color as a measure of the rate of primary oxidation of the developing agent, since the colored products are largely formed as a result of secondary reactions. They measured the actual rate of uptake of oxygen by solutions of the same developing agents as Volmer used. The experimental conditions were such that the solutions were kept saturated with oxygen during the measurements. The results show that colloidal silver accelerates to only a very small degree the oxygen oxidation of hydroquinone, catechol, *p*-aminophenol, and *p*-hydroxyphenylglycine. Weissberger and Thomas conclude that a catalysis of the type suggested by Volmer is of no significance for photographic development by these particular agents. On the other hand, they found a marked acceleration (as much as one hundred fold) of the oxygen oxidation of *p*-phenylenediamine, an agent not tested by Volmer.

3. Reduction of Silver Ions by Hydroquinone

James has investigated in some detail the kinetics of the reduction of silver ions by several developing agents. Without exception, the reduction was markedly catalyzed by metallic silver.

From the kinetic viewpoint, the simplest reaction between a typical developing agent and silver ions is that involving hydroquinone in slightly acid solution. The silver-catalyzed reaction can be studied conveniently in the presence of gelatin or gum arabic, using colloidal silver or gold as the initial catalyst. The nuclei of the added sol grow in size as the reaction proceeds and, under suitable conditions, the number remains substantially constant. With a nuclear gold sol, and over the concentration ranges used in the kinetic studies (James, 6) a linear relation exists between the mass of silver and the optical density.

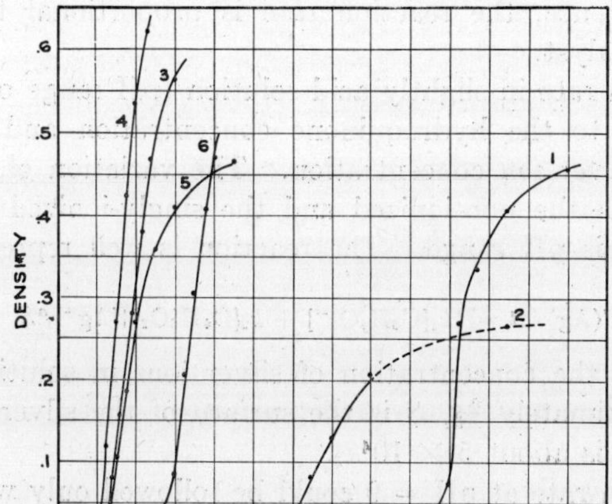

FIG. 1. Reduction of silver ions by hydroquinone: 1, no catalyst added; 2, quinone added; 3, silver sol added; 4, gold sol added; 5, palladium sol added; 6, silver sulfide sol added. Gelatin (1 %) was present as stabilizer.

Figure 1 shows experimental reaction curves obtained in the absence of added catalyst and in the presence of various colloidal catalysts. The reaction has a pronounced induction period in the absence of added catalyst, and the curve shows the typical shape expected for an autocatalytic reaction. Addition of colloidal silver, gold, palladium, and silver sulfide markedly decrease the magnitude of the induction period.

The most extensive work on the catalyzed reaction was carried out with a nuclear gold sol (phosphorus reduction) added as the initial catalyst. The gold particles soon become coated with silver, so the major portion of the measured reaction is actually silver-catalyzed. The reaction rate is proportional to the surface area of the catalyst. This is shown by the following experimental results.

For any particular silver ion and hydroquinone concentration, the rate is proportional to the amount of nuclear sol added. When reaction proceeds to a point where the original gold particles are covered with silver, the proportionality still holds. This result could mean either proportionality to the mass of the catalyst or to its surface. However, as reaction proceeds in the presence of a fixed amount of nuclear sol the rate is well represented by the equation

$$d(\text{Ag}) = k(\text{Ag})^{2/3} \quad (1)$$

as long as the change in silver ion and hydroquinone concentrations are relatively small. Since the nuclei grow in roughly spherical form, the surface is approximately proportional to the two-thirds power of the silver mass. Hence, the reaction rate is proportional to the surface area of the catalyst.

The reaction rate in slightly acid solution (pH range of 5.15 to 6.27) is proportional to the hydroquinone concentration and to the ⅔rds power of the silver *ion* concentration. The variation of rate with pH shows that both the non-ionized and the singly-ionized hydroquinone are active in this pH range. The reaction is well represented by the equation.

$$d(\text{Ag})/dt = \{k_1[C_6H_5O_2^-] + k_2[C_6H_6O_2]\}[\text{Ag}^+]_s^{\alpha}S \quad (2)$$

where $[\text{Ag}^+]_s$ is the concentration of silver ions in solution, α has the value of approximately ⅔, S is the surface of the silver catalyst, and the constant k_1 is about $5 \times 10^4 k_2$.

The reaction rate at pH = 9 could be followed only when the silver ion concentration was kept to a low value. This was accomplished by using as a source of silver ions the soluble silver sulfite complex ion which has a dissociation constant of about 3×10^{-9} at 25°. The undissociated complex itself is not involved in the reaction to any significant extent. The reaction rate varies as about the half power of the silver ion concentration under these conditions. The dependence upon the hydroquinone concentration, as indicated by the data in Table I, is somewhat greater than a direct proportionality (James, 7).

TABLE I

Variation of Rate with Hydroquinone Concentration; pH = 9.0

Hydroquinone	R_1	R_2
millimoles per liter		
5.0	0.026	
10.0	0.062	0.047
20.0	0.15	0.102
40.0	0.37	0.24

R_1: $AgNO_3$, 5.0 millimoles; Na_2SO_3, 50.0 millimoles; gold, 20 ml.; gelatin, 0.5%; experiments in the presence of air. Total volume, 1 liter.

R_2: $AgNO_3$, 2.0 millimoles; Na_2SO_3, 50.0 millimoles; gold, 40 ml.; gum arabic, 0.5%; experiments under nitrogen. Total volume, 1 liter.

The variation of rate with a fractional power of the silver ion concentration can be explained if adsorption of the silver ions by the silver occurs prior to reaction and if the adsorption follows a Freundlich isotherm. The reaction rate then would be directly proportional to the concentration of adsorbed silver ions, since the concentration of adsorbed ions would be related to that in solution by the equation

$$[Ag^+]_a = k'[Ag^+]_s^\alpha \qquad \text{where} \qquad \alpha \geqq 1 \tag{3}$$

Direct evidence for the adsorption of silver ions by silver and gold has been obtained by several investigators (Euler, 8; Proskurnin and Frumkin, 9; Veselovsky, 10; Euler and Zimmerlund, 11; cf. James, 12). Veselovsky, who was careful to work with oxide-free surfaces, found that adsorption begins at a silver ion concentration of only 1×10^{-13}. The concentration used in the kinetic experiments just discussed are greater than this by several orders of magnitude.

Some experiments by Rabinovich and his coworkers (Rabinovich *et al.*, 13; Rabinovich and Peisakhovich, 13a) seemed to show that hydroquinone is strongly adsorbed by silver. Rabinovich and his coworkers found, for example, a loss of 0.0178 g. hydroquinone from a 0.005 *M* solution when the solution was shaken at 25° with a Kohlschütter silver sol containing 1.12 g. of silver, and the silver was separated out by ultrafiltration. Calculations based on their reported values for the silver surface (calculated from particle count data, and hence smaller than the true specific surface) indicate that 1 molecule of hydroquinone is adsorbed for every 10 sq. A. of calculated surface. Since the van der Waals' projection area value for hydroquinone is 43 sq. A. for flat orientation and 35 sq. A. for edgewise orientation, complete monolayer adsorption is indicated if the specific surface of the silver is no more than four times the calculated surface. However, the accuracy of the adsorption data is questionable. Oxygen was not excluded in the experiments, and a large loss of hydroquinone through oxidation occurred. Adequate corrections for this oxidation were not made. The formation of silver oxide in sols exposed to oxygen is always a dangerous source of error in such experiments, and some oxygen oxidation of the hydroquinone probably took place during the filtration operation.

A more careful investigation was made by Perry *et al.* (14). In their experiments, a buffered silver sol prepared by dextrose reduction of ammoniacal silver nitrate was mixed with a solution of hydroquinone in a hydrogen atmosphere. Both solutions were deaerated before mixing. The mixed solution was stirred with hydrogen for one-half hour, then pressed through an ultrafilter under hydrogen pressure. The filtrate was analyzed for hydroquinone, and correction was made for retention of hydroquinone by the filter membrane. A small amount of hydroquinone (0.0225 milliequivalents or 0.00124 g. for a 0.68 g. silver sample) remained unaccounted for and possibly adsorbed by the silver. The missing hydroquinone, however, was within the experimental error in determining hydroquinone analytically, so it is doubtful if any adsorption occurred. Certainly, any real adsorption must be far smaller than that reported by Rabinovich and his coworkers. (Assuming that the average amount of unaccounted for hydroquinone in three experiments represents adsorbed compound, Sheppard (15) calculated that this still could represent 60 to 84% coverage of the silver surface by hydroquinone. However, errors in his calculations make these values too high by a factor of approximately 10, the correct values being 7 to 8%. Moreover, the area of the silver surface was calculated from size-frequency data obtained from electron microscope photographs of the colloidal particles, and the calculated area is certainly smaller than the true one.)

Thus, the available evidence indicates that little or no adsorption of hydroquinone by silver occurs. Rabinovich's data are unacceptable because of the large experimental errors involved. The possible amount of adsorption indicated by the data of Perry, Ballard, and Sheppard does not exceed the limits of error in their analytical determination of hydroquinone and could not under any circumstances cover more than a small fraction of the silver surface. The kinetics of the reaction between hydroquinone and silver ions do not indicate adsorption of the reducing agent, although the first-order dependence of rate on concentration is not incompatible with weak adsorption. It seems unlikely, accordingly, that adsorption of hydroquinone by silver plays a role of any consequence in the silver catalysis of the reaction between hydroquinone and silver ion.

There are indications that another type of catalysis is present in the reaction between hydroquinone and silver ions in alkaline solution. The increase of rate with increasing hydroquinone concentration is greater than direct proportionality. This situation is similar to that observed in the oxygen oxidation of durohydroquinone (tetramethylhydroquinone) (James and Weissberger, 16) where the quinone formed in the reaction catalyzes subsequent oxidation. A direct check on quinone catalysis of the hydroquinone-silver ion reaction was not made, since quinone is unstable in alkaline solution, particularly in the presence of sulfite which reacts with it. Experiments were made, however, on the reaction between durohydroquinone and silver ion. This reaction shows the same dependence of rate upon the square root of the silver ion concentration as the hydroquinone reaction does. Addition of duroquinone to the reaction mixture produces a definite acceleration, as shown in Table II.

The mechanism of the quinone catalysis probably is the same as that operative in the oxygen oxidation. The quinone reacts with the divalent quinonate ion to yield a highly reactive semiquinone. Quinone catalysis was not observed in the reduction of silver ions in acid solution when silver was present initially. The concentration of divalent quinonate ion probably is too small for quinone catalysis to compete with silver catalysis. The addition of quinone to the reaction mixture in the absence of nuclear sol, however, did shorten the induction period (see Fig. 1). This suggests a quinone catalysis of the homogeneous reaction.

Bagdasar'yan (17) has proposed an alternative explanation of the kinetics of the reduction of silver ions by hydroquinone. He treats the reaction as being fundamentally a discharge of silver ions at a static

TABLE II

Duroquinone Catalysis

Durohydroquinone, 2.5 millimoles; $AgNO_3$, 0.80 millimole; Na_2SO_3, 40 millimoles; temperature, 25°C. Total volume, 1 liter

Duroquinone	R_m	R_i
millimoles		
0.00	0.044	0.028
0.25	0.048	0.034
0.50		0.046
1.25	0.078	0.072

R_m: Slope at mid-point of reaction.
R_i: Slope at one-tenth of reaction course.

electrode, and he assumes that the slowest stage in the reaction is the transfer of the electron from the molecule or ion of developing agent to the silver nucleus. Following Frumkin's formulation of the electrode process (Frumkin, 18) he obtains the rate equation

$$d(\mathrm{Ag})/dt = k[\mathrm{C_6H_6O_2}]e^{\alpha FV/RT}S \tag{4}$$

where V is the potential drop at the border, silver/solution, S is the area of the interface, and α is a fraction. V can be expressed by the equation

$$V = (RT/F) \log [\mathrm{Ag^+}] \tag{5}$$

Substitution of (5) in (4) gives the rate equation

$$d(\mathrm{Ag})/dt = k'[\mathrm{C_6H_6O_2}][\mathrm{Ag^+}]^\alpha S \tag{6}$$

Thus, an equation in agreement with the experimental data for the hydroquinone-silver ion reaction can be derived either on the basis of the assumption that adsorption of silver ions by the silver is a prelude to the reaction, or on the basis of the assumption that the rate-controlling step in an electrode process is the rate of transfer of electrons to the silver electrode. The first mechanism carries with it the assumption that a silver ion adsorbed by silver is more easily reduced than an ion in solution. It is assumed that the free energy of formation of the activated complex involving the adsorption site (Hq..Ag^+..Ag)‡ is lower than that of the complex (Hq..Ag^+)‡ by a sufficiently large amount to allow the heterogeneous reaction to proceed at a higher net rate (cf. Laidler *et al.*, 19). Bagdasar'yan's mechanism, on the other hand, does not require direct interaction between the silver ion and the hydroquinone. According to him, an electron acquired at one spot on the silver surface can reduce a silver ion at any point where the latter may contact the surface.

4. Reduction of Silver Ions by Hydroxylamine

Hydroxylamine, although not commercially used as a developing agent, shows good selectivity under proper conditions. Its use in the study of reaction mechanism offers the distinct advantage that the rate of the reaction can be followed by measuring the rate of evolution of nitrogen or nitrous oxide.

The oxidation products of hydroxylamine depend upon the nature of the oxidizing agent and the conditions of oxidation. Acid solutions of hydroxylamine react with ferric oxide to give quantitative yields of nitrous oxide and water (Knorre and Arndt, 20; Bray *et al.*, 21); alkaline solutions of hydroxylamine react with cupric hydroxide to give 95–96% nitrous oxide and no nitrogen (Knorre and Arndt, 20). On the other hand, silver salts oxidize acid or neutral solutions of hydroxylamine with a yield of 99% or more nitrogen (James, 22). In alkaline solution the reaction yields pure nitrogen or a mixture of nitrogen and nitrous oxide according to experimental conditions (Nichols, 23; James, 22, 24, 25, 26).

In the oxidation of hydroxylamine by silver salts and mercurous salts, the nature of the reaction product apparently depends upon the extent to which catalysis participates in the total reaction. This is illustrated by some results obtained with mercurous nitrate as oxidizing agent. The reaction is strongly catalyzed by colloidal silver, and is likewise catalyzed by mercury. The reaction of 0.005 *M* mercurous nitrate with 0.04 *M* hydroxylamine at pH 4.85 proceeds rapidly without induction period. The mercury formed collects at the bottom of the vessel in the form of globules when no protective colloid is present, so the surface available for catalysis is small. Under these conditions the yield is largely nitrous oxide. Addition of colloidal silver accelerates the reaction and increases the yield of nitrogen. Some data are given in Table III.

A nitrogen yield of 98% was obtained by adding solid mercurous nitrate crystals to the hydroxylamine solution in the presence of 5 ml. silver sol. These and other experiments with mercury and silver salts suggest that the oxidation product of the uncatalyzed reaction is largely or wholly nitrous oxide, whereas that of the metal catalyzed reaction is entirely nitrogen (James, 22).

The reduction of silver ions by hydroxylamine from acid or slightly alkaline solution in the presence of colloidal silver proceeds with virtually a quantitative yield of nitrogen, and the reaction rates measured in terms of the amount of silver formed are identical within the limits of experimental error with those measured in terms of the amount of nitrogen evolved. The reaction rate varies as approximately the two-thirds power of the silver ion concentration at pH 4.16 and as approximately the half power at pH 8.54, in good agreement with the results

TABLE III

Oxidation of Hydroxylamine by Mercurous Nitrate at 20°; 0.07% Carey Lea Dextrine Ag Sol Used; Total Volume 50Ml.

Catalyst added (ml. soln.)	N_2%	N_2O%
None	13	87
0.1	30	70
2.0	51	49
10.0	63	34

obtained in the hydroquinone reaction. Unlike the latter, however, the rate variation is less than proportional to the hydroxylamine concentration; it changes from a dependence on slightly less than the first power at pH 3.8 to a dependence on about the 0.3 power at pH 8.54. The rate is proportional to the surface of the catalyst.

The same considerations can be applied to the mechanism of the hydroxylamine-silver ion reaction as have been given already for the hydroquinone-silver ion reaction. Bagdasar'yan's equation adequately expresses the dependence of rate on the silver ion concentration and on the surface of the catalyst, but not the dependence on the hydroxylamine concentration. The latter dependence, on the other hand, would be in agreement with the assumption that hydroxylamine is adsorbed by the silver prior to reaction.

5. *Reduction of Silver Ions by p-Phenylenediamine*

The kinetics of the reduction of silver ions by *p*-phenylenediamine differ in important respects from those of the reduction by hydroquinone and hydroxylamine. Once more, the silver catalysis is marked and the reaction rate varies directly as the area of the catalyst surface, but the rate is directly proportional to the silver ion concentration (James, 7).

The dependence of rate upon the concentration of reducing agent is determined by the experimental conditions. When gum arabic was used as protective colloid, the rate varied as the square root of the *p*-phenylenediamine concentration over a range of 120 fold; when gelatin was used as protective colloid, however, the rate varied as the quarter power of the concentration over a twenty-fold range, but at higher concentrations the power steadily increased. Data for the reactions, plotted on a logarithmic scale, are given in Fig. 2.

Unlike the hydroquinone and hydroxylamine reactions, the *p*-phenylenediamine reaction shows a strong positive salt effect. The pH depend-

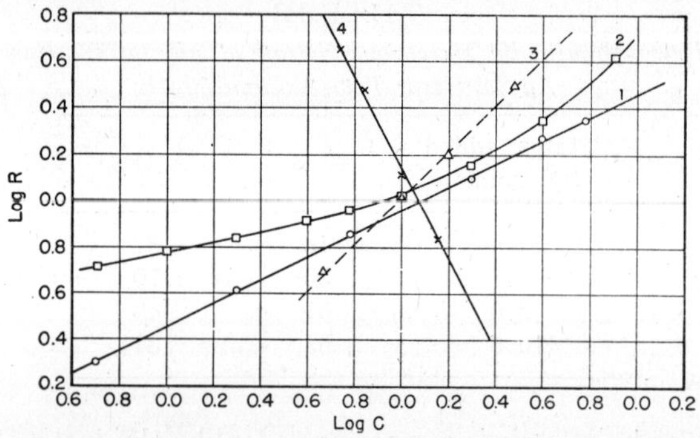

FIG. 2. Kinetics of the silver-catalyzed reduction of silver ions by *p*-phenylenediamine: 1, variation of rate with *p*-phenylenediamine concentration in presence of gum arabic; 2, same in presence of gelatin; 3, variation with silver ion concentration; 4, variation with sulfite ion concentration.

ence is curious, changing in magnitude and even in direction with changing salt content. For example, the rate of the reaction in the presence of gelatin increased with pH increasing from 9 to 13 when no neutral salt was added. In the presence of 0.2 *M* potassium nitrate, however, the rate *decreased* steadily with increasing pH.

The data for the *p*-phenylenediamine-silver ion reaction are not accounted for by Bagdasar'yan's treatment. On the basis of an adsorption mechanism, the data would suggest that the important phase for the catalyzed reaction is adsorption of the *p*-phenylenediamine by the silver catalyst. The extent of the adsorption would depend upon the surface conditions of the catalyst, which apparently depend on changes in the protective colloid or in the salt concentration. A catalytic mechanism involving activation of the *p*-phenylenediamine by the catalyst would be consistent with the observation of Weissberger and Thomas that colloidal silver markedly catalyzes the oxygen oxidation of *p*-phenylenediamine.

6. *Physical Development by p-Phenylenediamine*

A transition step between the reaction system just considered (silver ions, *p*-phenylenediamine, silver sol) and actual physical development of a fixed photographic plate or film is supplied by the investigations of Arens (27, 28). He used a series of silver, gold, and silver sulfide sols which had been coated on glass plates, using gelatin as the binding material. In this way he obtained plates in which colloidal particles

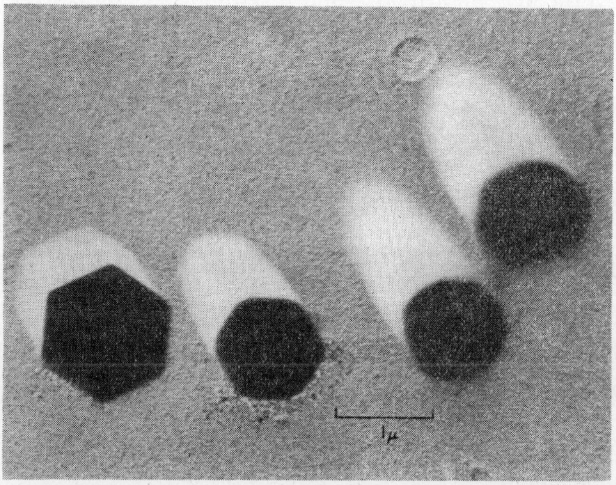

FIG. 3. Electron micrographs of colloidal silver nuclei after 21 hours' physical development. Gold shadowing at a 20° angle was used to indicate the thickness of the crystals.

were suspended in a thin layer of set gelatin, thus similating with artificial silver nuclei the conditions existing in the exposed, fixed-out photographic material. The developer solution was closely analogous in composition to Lumiere's standard *p*-phenylenediamine formula. Under such conditions, the nuclei grow as compact and often well-formed silver crystals in the gelatin layer as shown by the electron micrograph reproduced in Fig. 3 (Schoen, 29).

Arens found that the relative rate of growth of the artificial nuclei during physical development depended in the early stages upon their size and concentration in the gelatin layer. For a given number of particles per unit area, the relative rate of growth decreased with increasing size, while for a given particle size it increased with decreasing numbers. For longer development times, the number of nuclei was of primary importance in determining the amount of silver deposited. When the original particle size did not exceed about 10^{-15} g., the amount of silver developed in 2 hours per unit area was substantially independent of the size, the method of preparation, or the chemical nature (i.e., Au, Ag, or Ag_2S) of the nuclei. The dependence on number was linear, however, only at low particle concentrations (10^7 to 10^8 nuclei/sq. cm.). Beyond this, the ratio: (developed Ag)/(number of nuclei) decreased steadily with increasing number of nuclei. Arens showed that the same general relation between developed silver and number of nuclei held for physical development of the latent image.

The results just described, while useful in the interpretation of the photographic results obtained by physical development of a latent image, do not yield much information on the ultimate mechanism of physical development. Arens' conditions correspond closely to those obtaining during physical development of a photographic material, but the rate of this process is dependent on the rate of agitation of the developing solution (Vanselow and Quirk, 30) and hence is at least partially diffusion controlled.

Replacement of the latent image silver by gold in the exposed photographic sensitive layer increases the developability (James *et al.*, 31). Since the mass of developed silver depends only upon the number of developed nuclei, the gold treatment has obviously increased the number of active nuclei. Hence, smaller gold nuclei (in terms of numbers of atoms) than silver can initiate development, unless the treatment used to effect a replacement of silver by gold has resulted in something more than a simple replacement.

Growth of the nuclei during the gold treatment is a possibility. The exposed, fixed photographic film in these experiments was immersed for from 1 to 5 minutes in a solution prepared by heating to boiling 40 cc. of 0.1% potassium chloraurate with 0.5 g. potassium thiocyanate, cooling, and diluting to 1 liter. The gold in this solution is in the form of the aurous thiocyanate complex. The maximum effect of the treatment upon development is attained within 5 minutes, and further treatment up to an hour or more results in no further increase in developability. If the treatment is prolonged for several days, however, a visible image appears. The gold solution itself acts as a physical developer, depositing gold at the latent image nuclei and building up gold nuclei which eventually become large enough to be detected with the electron microscope (James, 32) (see Fig. 6). This development is autocatalytic, however, so it seems unlikely that more than simple replacement of silver by gold occurs within the first few minutes.

7. Reduction of Silver Ions by Other Agents

The reduction of silver ions by hydrazine (a photographic developer, but one of poor selectivity) is catalyzed by silver and, to a smaller degree, by gold (Jablczynski and Kobryner, 33; James, 34). The important factor in this catalysis is the adsorption of the hydrazine rather than of the silver ions. The reaction is strongly catalyzed by copper ion.

The reduction of silver ions by catechol (a good developer) is silver-catalyzed. This reaction has been studied to only a limited extent (James, 7, 35) but the mechanism appears to be quite similar to that of the hydroquinone reaction. The rate is directly proportional to the catechol concentration at a pH of 7.58.

The reduction of silver ion in a sulfite solution is catalyzed by silver, (Sheppard, 3; James, 12) but the catalysis is less pronounced than that operating in the reactions of the developing agents already considered. The sulfite reaction is very slow at 20°, even in the presence of large amounts of colloidal silver. Data obtained at 59.8° indicate a dependence of rate upon the 0.75 power of the silver ion concentration and upon the first power of the sulfite ion concentration. The reaction, even in the presence of silver catalyst, is further catalyzed by copper salts.

The autoacceleration of the silver-catalyzed reaction in the early stages is greater than would be expected from a simple increase in catalyst surface by growth of the silver nuclei orginally present. Apparently new nuclei are readily formed during the reaction. A possible mechanism which could lead to this result is:

$$SO_3^{=} + Ag^+ \rightarrow -SO_3^- + Ag \qquad \text{(catalyzed)} \qquad (7)$$

$$-SO_3^- + Ag^+ + H_2O \rightarrow Ag + SO_4^{=} + 2H^+ \qquad (8)$$

The reaction between the monothionate radical and silver ions should occur readily in solution without catalysis, and the silver atoms formed in this way could condense to form new catalyst nuclei. The cupric ion catalysis of the reduction of silver ions by sulfite can be explained in a similar way, the monothionate radical being formed by the reaction

$$SO_3^{=} + Cu^{++} \rightarrow -SO_3^- + Cu^+ \qquad (9)$$

Direct development by sulfite ion does not take place, probably because the rate of reduction of silver ions by the sulfite is much smaller than the rate of solution of the silver halide (see Section VI of this chapter). The writer has obtained physical development, however, with a solution of silver nitrate and sodium sulfite. A specially hardened gelatin film which was able to withstand the action of the solution at 70° for 40 minutes was used. The developing solution contained 1.7 g. silver nitrate and 13 g. sodium sulfite per liter. Fog formation was relatively high, as might be expected.

Sodium stannite (which is not a developer) reacts very rapidly with silver nitrate or with the silver sulfite complex, and the rates have not been measured. The reaction with the silver thiosulfate complex ion proceeds without measurable silver catalysis (James, 12) and the complex ion rather than free silver ion is involved. Reduction of silver salt in a solution prepared by dissolving silver iodide in a concentrated potassium iodide solution, however, proceeds with marked silver catalysis.

Desylamine and furoin* (not developers in the usual sense) reduce silver ions without catalysis (James, 12). The reaction rate varies with the first power of the reducing agent concentration, and is independent of the silver ion concentration. Moreover, the rate of reaction with oxygen is substantially the same as that of reaction with silver ion. These facts, together with the linear dependence of rate upon the hydroxyl ion concentration, show that the rate controlling process is the enolization of the reducing agent (cf. Weissberger *et al.*, 36).

The reduction of silver ions by resorcinol is catalyzed by silver (Krishna and Ghosh, 37). Resorcinol generally is not considered to be a developing agent. However, it will develop a simple silver bromide photographic film under extreme conditions, and the development process appears to be a normal one. For example, a solution of 11 g. resorcinol and 28 g. potassium hydroxide per liter at 25° will develop an image of good density on a strip of specially hardened motion picture positive film in 4 hours. Without the special hardening, the gelatin layer decomposes in the strongly alkaline solution before development becomes noticeable.

8. *The Critical Size of the Nucleus for Physical Development*

It is not known with any degree of certainty what is the lower limit to the size of a nucleus for development. Calculations based upon the amount of light absorbed by the silver halide in forming the latent image lead to only very rough figures. The amount of light absorbed determines the amount of photolytic silver formed when conditions are such as to prevent recombination of the silver and halogen, but the distribution of this silver is not known. Moreover, the latent image nuclei very probably are formed at places where "sensitivity" nuclei of silver, silver sulfide, or both existed before the exposure, and the photolytic silver has added on to such nuclei.

* The formula for desylamine is

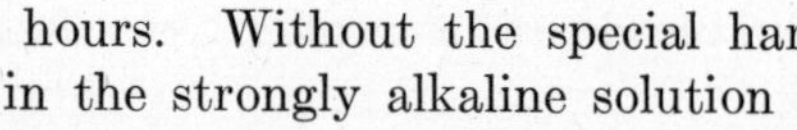

that for furoin is

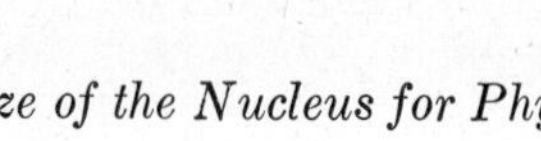

Vogler and Clark (38) suggested that the minimum size of a development nucleus is determined by the number of silver atoms in a group required "to form a strong space lattice." Svedberg (39) and Clark (39a) suggested that this number is of the order of 300 atoms, drawing upon analogy to Zsigmondy's evidence that gold particles of about this size are required to serve as nuclei for the preparation of gold and silver sols of larger particle size. However, Zsigmondy's value is too high. Baker and Usher (40) in a study of the formation of gold sols have demonstrated the existence of stable and active nuclei of less than one-tenth this size. Their smallest active nuclei were approximately the size of a unit cell of crystalline gold (14 atoms).

An attempt at a direct determination of the number of silver atoms in a nucleus just able to initiate physical development was made by Reinders and his co-workers (Reinders and Hamburger, 41; Reinders and deVries, 42). Silver was deposited on glass plates by sublimation in a high vacuum. The source of the silver was a molybdenum wire coated with silver and heated electrically to 500–600°. The glass plates were carefully cleaned and outgassed before use.

Plates containing varying numbers of silver atoms per unit area were immersed for 5 minutes in a hydroquinone-silver nitrate physical developer which gave no spontaneous deposition of silver within 30 minutes. The number of developed centers per unit area was counted under dark field illumination at a magnification of 500×.

Reinders and his coworkers assume that the silver atoms sublimated onto the glass plate adhere to it at the point of contact. They then calculate on a probability basis the number of atom pairs, triplets, etc. expected per unit area from the total number of atoms impinging upon that area. They assume that a pair is formed only if the second atom strikes the first or if it strikes the glass at a distance from the first not greater than the normal distance between adjacent atoms in the silver crystal. Experimentally, they find that the number of developed nuclei corresponds closely with the calculated number of initial centers containing four or more atoms. A recalculation of their data based on a more accurate probability treatment (Berg, 43) does not materially change the result.

The accuracy of Reinders' conclusions depends upon the validity of his assumption that migration of silver atoms along the condensing surface is of little significance in determining the number and size of nuclei formed. This assumption is not on firm ground. The only evidence given for a lack of mobility of silver atoms along the glass surface is that when an object was interposed between the glass plate and the source of silver atoms, and the plate subsequently developed, a sharp shadow of the object was obtained. This in itself is not conclusive. Cockcroft (44) obtained similar shadows in his experiments with cadmium. When a wire was interposed between the cadmium beam and a freshly cleaned plate, no film formation took place in the shielded area, but when the plate was first "sensitized" by preliminary deposition of cadmium on a cooled surface, film formation occurred in the shielded area.*

Cockcroft's results are in accord with Frenkel's theory of condensation (Frenkel, 45) i.e., that (a) atoms on striking the surface stay on that surface for a definite time during which they move about like a two-dimensional gas, and (b) when they collide with other atoms of the condensing vapor the pair forms a nucleus with a much longer mean life on the surface than the single atoms. These "doublet" atoms act as centers of condensation for other atoms, etc. Experiments by Hass (46) in which he showed that silver films of 5–10 mμ thickness formed at $-175°$ undergo rapid recrystallization on warming up to room temperature attest to the high mobility of freshly deposited silver atoms.

9. *The Mechanism of Physical Development*

Some physical development may result from a selective deposition upon the latent image nuclei of silver formed in the homogeneous reaction between silver ions and developing agent. Under the usual conditions of development, however, catalysis of the actual reduction of silver ions is the important factor.

The probable mechanism of physical development can be summarized briefly as follows: The latent image silver nuclei (and later silver formed in the reaction) catalyze the reaction between silver ions and the developing agent. Adsorption of silver ions by the catalyst is the essential step when the developing agent is not itself appreciably adsorbed (e.g. hydroquinone, catechol); adsorption of the developing agent plays a dominant role when that agent is strongly adsorbed (e.g., *p*-phenylenediamine); adsorption of both silver ions and developing agent are involved in intermediate cases (e.g., hydroxylamine). The homogeneous reaction involves an activated complex consisting of only silver ion and developing agent, and reaction is relatively slow. The heterogeneous reaction involves an activated complex consisting of silver ion, developing agent, and one or more adsorption sites located on the surface of the nuclei. Both complexes decompose into silver and oxidized developing agent. Because of distortion of the complex consisting of

* Data on the critical temperature of deposition of silver on glass are unsatisfactory. Knudsen (47) gives the critical temperature as above 575°. Cockcroft, however, reports critical temperatures ranging from -90 to $-15°$, depending on the filament temperature, for deposition of silver on mica.

silver ions and developing agents by the adsorption forces of the catalyst, the heterogeneous reaction occurs with considerably smaller activation energy than the homogeneous reaction and, in the presence of sufficient catalyst (nuclei) the rate of the heterogeneous reaction is much greater than that of the homogeneous reaction.

III. Reduction of Solid Silver Salts

An investigation of the kinetics of reduction of silver halides by developing agents evidently should give results which are more directly applicable to the interpretation of direct development than investigations on the reduction of silver ions from solution. Work with the solid salts is more difficult experimentally, however. Luther and Leubner (48) attempted such a program, using freshly precipitated silver bromide as the halide and hydroquinone as the reducing agent. Their results were inconclusive, however, because they were not able to devise an adequate experimental procedure.

1. Reduction of Silver Chloride by Hydroxylamine

Hydroxylamine is a much easier developing agent to work with than any of the organic agents, since the reaction rate can be followed continuously by following the evolution of the gaseous reaction product. A series of investigations of the reduction of solid silver salts by hydroxylamine have been carried out by James. The apparatus used consisted of a modified form of that devised by Weissberger *et al.*, (36) for autoxidation studies. The silver salt and the hydroxylamine solution were kept in a state of vigorous agitation by shaking the reaction vessel, and the evolution of gas was followed by water displacement in a jacketed burette.

The reduction of silver chloride, precipitated in the presence of excess chloride ion, yielded the S-shaped curve typical of an autocatalyzed reaction (James, 25). The initial reaction rate, measured in terms of the reciprocal of the time required to complete 5% of the total reaction, varied directly as the hydroxylamine concentration and inversely as the chloride ion concentration when the latter was relatively large. The specific surface of the freshly prepared precipitate, as measured by dye adsorption, decreased with aging, and the reaction rate decreased proportionately.

The reaction rate was diminished by shaking the precipitate in a gelatin solution for some time prior to addition of the hydroxylamine. The rate decreased sharply as the amount of gelatin increased until a minimum rate was attained. Further addition of gelatin had no measurable effect. Gelatin also decreased the rate of reduction of silver ions from the soluble silver sulfite complex, but the relative effect was much smaller, and no minimum rate was attained even at high gelatin concentration. Evidence from other sources shows that gelatin is rather strongly adsorbed by silver halides (Sheppard *et al.*, 49) and the effect of the gelatin on the rate of reduction of silver chloride must be associated with such adsorption.

The cyanine dye, 3,3′-diethyl-9-methylthiacarbocyanine chloride, had a much greater effect than gelatin in decreasing the reaction rate of the silver chloride. The rate of reduction of silver chloride varied linearly with the amount of silver chloride surface *not* covered by the dye, and the rate attained at complete coverage was of the order of one-thousandth that for the undyed precipitate. The dye exerted scarcely any effect upon the reduction of silver ions from silver sulfite complex solution.

The reduction of silver chloride probably is initiated at certain unprotected spots on the crystal surface. These spots may be imperfections in the crystal structure or may be simply silver ions unprotected by adsorbed chloride ions. With the fairly high chloride ion concentration used, the number of such silver ions might be expected to vary inversely as the chloride ion concentration. If active spots corresponding to impurities or imperfections in the crystal surface are involved, these spots do not selectively adsorb the cyanine dye or gelatin, since the reaction rate is directly proportional to the amount of uncovered surface. (In the reduction of mercurous chloride the greatest effect of dye was observed during coverage of the first 10% of the surface (James, 26).) Moreover, the number of active spots on the silver chloride precipitate decreased proportionately with the surface when the latter was changed by aging. This result would not be expected if the active spots consisted of imperfections in the crystal surface.

Exposure of the silver chloride precipitate to actinic light nearly eliminated the induction period of reduction (James, 50). This effect, together with the regression which occurs when sufficient time elapses between exposure and reduction, is illustrated in Fig. 4. The regression effect is not associated with a change of specific surface of the silver chloride, since aged precipitates which had already attained minimum surface were used. The maximum reaction rate of the freshly exposed precipitates was reached within the first 5% of the total reaction course, whereas the maximum rate for the unexposed precipitates was not reached until 25–30% of the reaction had taken place. This suggests that many more nuclei are formed by the action of the light than are formed during direct attack of the hydroxylamine on the silver chloride.

The kinetics of the silver-catalyzed reduction of silver chloride were studied on two types of preparations. In the first the nuclei were obtained by exposure of the precipitate to light. In the second, nuclei were formed by reaction with an amount of hydroxylamine sufficient to reduce only 25 or 50% of the silver chloride. The hydroxylamine nucleation was carried out under one set of carefully controlled conditions so that, in subsequent measurements of the rates of the catalyzed reaction, the initial state of the catalyst was always the same.

The reduction of nucleated precipitates varied directly as the 0.8 power of the hydroxylamine concentration and inversely as the 1.25

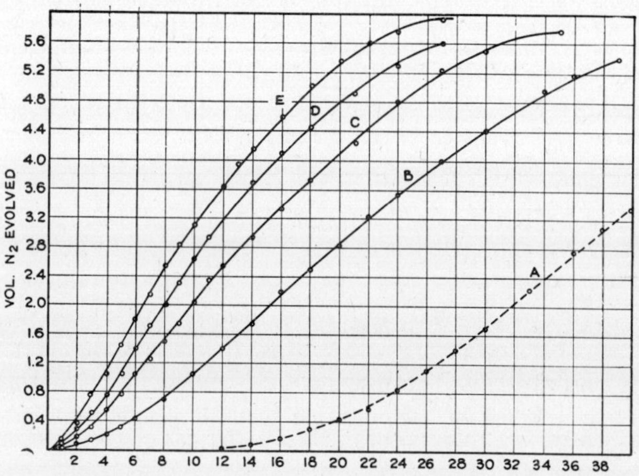

FIG. 4. Effect of age of exposure on reduction of silver chloride by hydroxylamine: A, no exposure; B, exposure 17 hours old; C, 6 hours old; D, 2 hours old; E, 20 minutes old.

power of the chloride ion concentration. The effect of gelatin and the cyanine dye on the reaction rate of the nucleated silver chloride was similar to that on the initial rate of reduction of the unnucleated precipitates. A minimum rate was attained at quite small additions of gelatin. Complete coverage of the surface with dye reduced the rate to a very low value.

The effect of the cyanine dye and of gelatin on the reaction rate shows that reduction of silver ions from solution is not the rate-controlling process. These influences of adsorbed components on the reaction rate speak against the concept that solution of the silver halide is the rate controlling process. Hence, the silver catalyzed reduction of silver chloride by hydroxylamine takes place substantially at the solid silver/silver halide interface.

The temperature coefficient of the catalyzed reaction is smaller than that of the initial reaction of the unnucleated precipitates. Some data are given in Table IV. R_i is the rate of the initial reaction of the unnucleated precipitate measured in terms of the reciprocal of the induction period; R_c is a measure of the rate of the silver-catalyzed reaction beyond the induction period; R is the rate of the reaction in the nucleated precipitates. The results are in the direction expected if the catalysis is characterized by a decrease in activation energy.

TABLE IV

Temperature Coefficients of the Reduction of Silver Chloride by Hydroxylamine

Temperature	Unnucleated precipitates				Nucleated precipitates			
					By NH_2OH		By light	
	R_i	Temperature coefficient	R_c	Temperature coefficient	R	Temperature coefficient	R	Temperature coefficient
°C.								
12.00	0.022		0.0090		0.058		0.145	
		4.46		3.35		3.25		2.80
20.03	0.073		0.0237		0.149		0.332	
		3.78		3.15		3.12		2.87
29.80	0.27		0.073		0.454		0.925	

At high pH the rate of reduction of silver chloride by hydroxylamine becomes too fast to be followed by the experimental procedure used. However, the composition of the gaseous reaction product gives a clue as to the course of the reaction. Other things remaining constant, the percentage of nitrogen in the gaseous product decreases with increasing pH. At pH 7.2 (that used in most of the work just described) the yield of nitrogen is nearly 100%; at pH 10.3 it is only 63%; at pH 12.7 it is only 5%. The reaction of hydroxylamine and silver bromide shows a similar pH effect. At pH 10.3 it yields 100% nitrogen; at pH 12.7 only 65% nitrogen (James, 22). Thus, at sufficiently high pH, the uncatalyzed reduction of silver halide can occur as fast or faster than the catalyzed reaction.

Exposure of the silver chloride or bromide to light results in an increase in the nitrogen yield. This is to be expected, because the action of the light supplies nuclei for the catalyzed reaction. On the other hand, exposure of silver thiocyanate, which is relatively insensitive to the action of light, has little or no effect on the amount of nitrogen obtained on subsequent reduction with hydroxylamine.

2. *Reduction of Silver Chloride by Hydrazine*

The reduction of silver chloride by hydrazine shows some points of similarity to the action of hydroxylamine, but also some important points of difference (James, 34). An induction period was obtained with the unnucleated precipitates which, under some conditions, was relatively large. However, exposure of the precipitate to actinic light had only a small effect upon the induction period and upon the subsequent course of the reaction. Previous nucleation of the precipitate by the action of hydroxylamine decreased the induction period without eliminating it, and produced little or no effect upon the subsequent course of the reaction. Addition of the dye, 3,3′-diethyl-9-methylthiacarbocyanine chloride, produced no effect until the surface of the precipitate was more than half covered. Further increase in the amount of dye added produced an irregular decrease in the reaction rate. Gelatin decreased the reaction rate, but to a smaller extent than in the hydroxylamine reaction, and a minimum rate was not attained. As the gelatin concentration increased, more and more reduced silver appeared in colloidal form in the solution.

A large part of the reduction of silver chloride by hydrazine evidently takes place by a different mechanism from that of the reduction by hydroxylamine. The effect of gelatin and dye on the process, together with the appearance of colloidal silver in the solution when gelatin is present to stabilize it, shows that the reaction involves dissolved silver chloride to a greater degree than the hydroxylamine reaction. Indeed, if the reaction rate is plotted against a silver ion concentration calculated on the assumption that a saturated solution of silver chloride is maintained, the same relation is obtained as is found for the reduction of silver ions from a solution of the sulfite ion complex.

The essential difference between the hydroxylamine reaction and the hydrazine reaction appears to be that silver nuclei are formed in the solution much more readily by hydrazine than by hydroxylamine. At sufficiently low pH and in the absence of copper, hydroxylamine does not readily form nuclei in the solution, and the catalytic reduction of the silver chloride occurs essentially at a solid interface with the silver nuclei. Hydrazine, on the other hand, readily forms nuclei in the solution and an important fraction of the total reaction involves the catalytic reduction of dissolved silver chloride. This would account for the well-known photographic properties of the two agents. Hydroxylamine is a clean-working developer which, under proper conditions, yields little fog. Hydrazine shows much less selectivity and, although it develops an image, it also yields a relatively high fog density.

The autocatalysis in the reduction of silver ions by hydrazine may be a kind of heterogeneous chain reaction. The net reaction is given by the equation

$$4Ag^+ + N_2H_4 \rightarrow 4Ag + N_2 + 4H^+ \tag{10}$$

but the reaction probably takes place in several steps, each involving the formation of an intermediate oxidation product. Such products may be much more reactive than the hydrazine molecule itself and, if they should become desorbed from the silver before they react with silver ions at the surface, could react to form metallic silver in solution. New catalytic nuclei could thus be formed in the solution.

The action of an active intermediate oxidation product would explain another feature of the reaction. The reduction of silver ions by hydrazine is extremely sensitive to the presence of small amounts of copper. For example, a solution containing a mixture of silver nitrate, sodium sulfite and hydrazine which normally showed no sign of reduced silver for several minutes underwent almost immediate reaction when merely stirred with a clean copper rod. In the presence of gum arabic as stabilizer, streamers of colloidal silver passed out from the copper surface. Similarly, the addition of small amounts of cupric sulfate to a hydrazine solution eliminated the induction period of the reaction with silver chloride.

Cupric sulfate exerts an effect on the silver chloride-hydroxylamine reaction similar in kind to that which it exerts on the hydrazine reaction, but in a smaller degree. If sufficient cupric sulfate is added to the hydroxylamine solution, the character of the reduction of silver chloride shifts towards that shown by the hydrazine reaction, e.g., the effect of gelatin becomes less pronounced, a minimum rate at a small gelatin addition is not obtained, and significant amounts of colloidal silver appear in the solution.

A reactive intermediate may be responsible for the copper catalysis of the hydroxylamine reaction. The intermediate formed in the silver-catalyzed reaction, if it has any real existence, is not further oxidized but breaks down into nitrogen and water. Oxidation of hydroxylamine by cupric ion, on the other hand, yields predominately nitrous oxide. The intermediate formed by the removal of a single electron from the hydroxylamine in this reaction must be further oxidized to yield the final product. Such an intermediate may react readily with silver ions in solution.

IV. Reduction of the Photographic Grain

Development of the photographic material represents a more complicated process than any of those considered up to this point, but many of the results already discussed are directly applicable to the formulation of its mechanism. The direct studies of development have followed, in general, three paths: (1) thermodynamic studies, notably by Reinders and his co-workers (cf. Mees, 51); (2) microscopic and kinetic studies of the development of individual grains; and (3) kinetic studies of the development of the photographic material as a whole. Of these, the first will not concern us here, since the results obtained have no demonstrated bearing on the reaction mechanism.

It is important to keep in mind that, in the development of the sensitive layer as a whole, we are dealing with an ensemble of reaction units where reaction may or may not proceed in a parallel fashion among the many units. Under certain conditions the kinetics of development of a typical single grain can be inferred directly from a measurement of the overall rate of formation of silver, but this is not true as a general proposition. Studies of development of the individual grains are of fundamental importance, since the grain is the real unit of development.

Microscopic examinations of developing grains show that the reaction starts at discrete spots on the grain surface, and proceeds from there throughout the entire grain. Unfortunately, little work which can be applied to mechanism considerations has been done with individual grains. Most of that work has been on relatively large grains of low sensitivity. Even with such grains the experimental difficulties are great.

Rabinovich and his coworkers (Rabinovich, 52; Rabinovich *et al.*, 53) have followed the development of silver bromide grains of 15–17 μ diameter by making photomicrographs at suitable time intervals. The reactants were contained on a thermostated microscope stage specially constructed for the purpose. Five types of development were observed depending upon the character of the preparation and the composition of the developing solution. "Regular development" began usually from one of the edges of the grain and rapidly spread over the surface in a more or less circular zone. "Irregular development" began at multiple points. In places where development had started the grain became ulcerated, protuberances were formed, and the darkening spread in various directions. "Mixed development" presented various transitions from the first type to the second. Usually it began as irregular, then at some stage changed to a rapid regular development. "Explosion development" was an extreme case of irregular development. From one point on the crystal a strong protuberance was thrown out and the whole grain usually moved in the opposite direction. The area of the developed grain was much larger than that of the original silver bromide. "Diffuse development" began at once in many points on the crystal surface, which was soon covered by multiple black spots. These spots coalesced as development proceeded.

Evaluation of the data obtained for "regular" development showed that the results of experiments involving several different emulsions and developer compositions could be expressed by the relation

$$R = R_0 e^{kt} \tag{11}$$

where R is the radius of the blackened zone and R_0 is a constant which Rabinovich and coworkers interpret as the critical size of a "viable" nucleus. R_0 has a value of approximately 4 mμ, which corresponds to about 20,000 silver atoms. Rabinovich suggests, however, that the latent image nuclei may be considerably smaller.

Equation (11) does not give an accurate relation between the rate of formation of silver and the extent of the silver/silver halide interface.

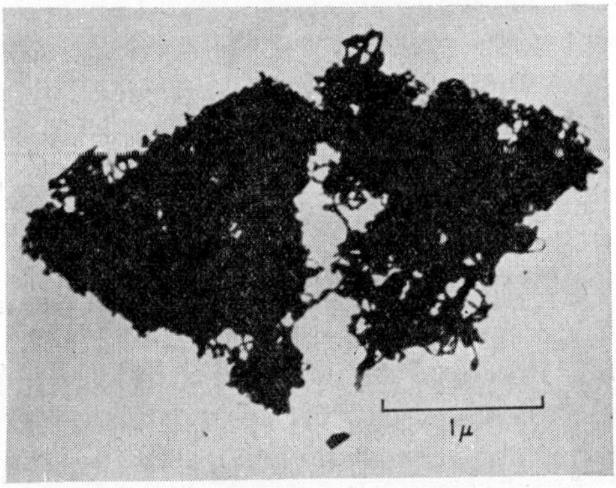

FIG. 5. Electron micrograph of two silver grains obtained by developing silver bromide grains in a "distortionless" hydroquinone developer.

Electron microscope photographs show that the propagation of the interface is irregular even under conditions such that it appears regular when observed under the optical microscope. The developed grain consists of a tangled mass of silver filaments of irregular form. This structure is readily seen when the grain is developed on the microscope slide (Ardenne, 54; Hall and Schoen, 55; Mees, 51). When the grain is developed in its normal surroundings in the sensitive layer, the electron microscope shows a filamentary structure around the edges even when the outlines of the developed grain appear under the optical microscope to be the same as those of the original silver halide grain. Figure 5 shows an electron micrograph of the silver obtained by developing two silver bromide grains in a "distortionless" hydroquinone developer. This developer contained only 0.02 M sodium sulfite and it developed grains which under the optical microscope appeared to be perfect pseudomorphs of the original grains. The higher resolving power of the electron microscope, however, clearly reveals the filamentary structure around the edges. It is probable that such a structure exists throughout the mass of the grain, but the restricting action of the gelatin layer prevents the filaments from spreading out in the loosely knit structure obtained when the single grain is developed on the microscope slide. The electron microscope can no longer resolve the individual filaments except around the edges.

Apart from the work of Rabinovich and some earlier observations of Meidinger (56) no quantitative studies of the reduction of individual grains have been reported. Some studies of the development of the photographic emulsion as a whole, however, have been carried out under conditions such that the kinetics of development of the individual grain can be inferred from those of the ensemble.

The kinetics of development by hydroquinone, catechol, and hydroxylamine of a motion picture positive type film of simple composition have been investigated under relatively simple conditions (James, 35, 57–61). At sufficiently high exposure (not, however, greatly beyond the normal photographic range) the rates of development of the grain ensemble closely parallel those of the majority of the individual grains and the data can be interpreted in terms of the individual grain.

The rate of development by hydroquinone over the pH range 8.0–8.9 varied approximately as the square root of the concentration of divalent hydroquinonate ion. The latter is the actual developing agent under these conditions. The rate relation is in contrast to that found for the reaction of hydroquinone with silver ions from solution, where the rate was proportional to the first power (or, in alkaline solution, to a somewhat higher power) of the concentration of reducing agent. The only explanation which has been proposed for the square root dependence is that the divalent ion becomes adsorbed prior to reaction, as originally suggested by Sheppard and Meyer (3a) on the basis of other evidence. Adsorption to silver is improbable, since no evidence for it appears in the silver-catalyzed reduction of silver ions from solution. This leaves adsorption by the silver bromide or by the silver/silver bromide interface.

Rabinovich and his coworkers (Rabinovich *et al.*, 13) have made measurements on the adsorption of hydroquinone on silver bromide from alkaline solution. They report that the measured adsorption is proportional to the area of the silver bromide surface and give a value of 0.0051 g. hydroquinone adsorbed by 35 sq. meters of surface at pH 10. This corresponds to approximately 1 molecule per 100 sq. A. surface, or about one-third of monolayer coverage. Their data are not conclusive, however, since the possibility of some oxidation of the hydroquinone by reaction with silver ions was not ruled out (see also p. 113). Indirect evidence that adsorption of hydroquinone occurs is given by the fact that the isomer, resorcinol, is adsorbed from strongly alkaline solution by silver bromide (Wulff and Seidl, 62). The reaction of resorcinol with silver bromide is much slower than the corresponding hydroquinone reaction.

The rate of development by catechol at pH 8.78 varies approximately as the square root of the catechol concentration. The rate of reduction of silver ions from solution, on the other hand, varies directly as the catechol concentration. Thus, the results obtained with catechol parallel those obtained with hydroquinone.

The pH dependence of the rate of development by hydroxylamine indicates that the monovalent ion is the active species. The rate varies as about the 0.65 power of the hydroxylamine concentration at pH 12.7 and the 0.75 power at pH 10.8. These results suggest adsorption of the hydroxylamine ion, and are in complete agreement with previous findings for the catalyzed reduction of silver chloride precipitates.

The relative rate of fog formation compared to image development increases with increasing pH of the hydroxylamine solution. This is to be expected from analogy with the studies of the reduction of silver chloride and silver bromide precipitates, where the change in nitrogen yield shows that the uncatalyzed reaction becomes more and more prominent as the pH is increased.

V. Mechanism of Direct Development

1. *Triple Interface Mechanism*

A generalized mechanism of direct development can be formulated on the basis of the preceding discussion. It is essentially an extension of Sheppard's mechanism. The developing agent, adsorbed by the silver halide, forms a complex with silver ion of the silver halide. This complex can break up in two ways: it can dissociate into silver ion and developing agent, or it can decompose into silver and oxidized developing agent. The first course presumably is completely reversible; the latter is irreversible under most practical conditions.

The decomposition of the adsorption complex into oxidized developer and silver is unselective in the absence of catalytic nuclei, and leads to the formation of "developer fog." In the presence of a nucleus of latent image silver, the complex becomes adsorbed and distorted by this silver. When hydroquinone is the developing agent, adsorption occurs through attraction between silver and the silver ion. When the developing agent is an amine, adsorption through an attraction between silver and the amino group may also play a part in the deformation of the complex. The activation energy for the decomposition into silver and oxidized developer is then smaller than that for the decomposition of the complex between silver ion and developing agent in the absence of a catalyst, and the specific rate of reaction is higher. Silver formed in the initial reaction catalyzed by the latent image can, in turn, act as catalyst for further reaction until the entire grain is reduced.

In the absence of initial catalyst, reaction occurs slowly. This reaction yields silver as a product, however, and catalytic silver centers eventually form. Further reaction then proceeds as a catalytic process. Some unselective reduction of silver salt always occurs under practical developing conditions, but often the major portion of it is instigated by catalytic silver or silver sulfide nuclei which are present in a small fraction of the grains before exposure. Such nuclei are formed during the manufacture of the photographic material, or during keeping of the material between manufacture and use. Most of the silver or silver sulfide nuclei formed under practical manufacturing conditions, however, either are too small or have not the proper configuration to serve in themselves as fast acting catalytic centers. They serve as sensitivity centers during exposure and become catalytic centers for development only on addition of some photolytic silver.

On the basis of the mechanism of development just outlined, the reaction rate should increase with increasing number of catalyst adsorption sites, and the relative rate of the catalytic reaction compared with the uncatalytic reaction should increase with decreasing temperature, i.e., the selectivity of the developing agent should increase with decreasing temperature. The observations of Rabinovich and his co-workers, already discussed, confirm in a general way the increase of rate with increasing extent of the interface. The experiments already cited on the reduction of silver chloride precipitates by hydroxylamine show that the selectivity of that agent increases with decreasing temperature, since the temperature coefficient of the initial reaction in the absence of light exposure is considerably greater than that of the reaction with the exposed precipitate.

Data obtained by Shiberstoff (63) on the development of photographic film in solutions of conventional developers offer further evidence in confirmation of the latter point. Shiberstoff obtained a general increase in selectivity with decreasing temperature of development for all agents tested. He calculated selectivity in terms of the ratio of the rate of image development to fog formation, using as the rate of the former the reciprocal of the time required to attain a density* of 1.5 for a fixed

* Optical, or photographic density is defined as the common logarithm of the opacity, i.e., the logarithm of the reciprocal of the transmittance. For a constant average size of developed grains, the density is proportional to the number of grains per unit area.

exposure and as the rate of the latter the reciprocal of the time required to obtain a fog density of 0.3. His data are given in Table V.

TABLE V

Selectivity of Developing Agents as a Function of Temperature

Developing agent	Selectivity at 15°	Selectivity at 20°	Selectivity at 25°
p-Aminophenol	500	500	270
Catechol	267	227	191
p-Hydroxyphenylglycine	259	210	150
Methyl-*p*-aminophenol	240	180	200
Metoquinone	182	172	136
Pyrogallol	200	136	100
Chlorohydroquinone	186	123	106
1-Hydroxy-2-aminonaphthalene sulfonic acid	150	113	106
Hydroquinone	93	75	50

Conditions existing at the triple interface, Ag/AgX/Solution, will influence the rate at which the catalyzed reaction occurs. These conditions will determine the activity of the silver ions, the concentration of adsorbed developer, and the state of the catalyst. The halide ion will be an important factor in determining the activity of the silver ions, and Sheppard has suggested that "the hydration and diffusion away of the halide ion" is the dominant factor in determining the specific rate of reaction at the interface (Sheppard, 15). The known order of reactivity of the halides, AgCl > AgBr > AgI, follows as a natural consequence from this point of view, whereas it would not be predicted on the basis of the electrode mechanisms.

The concentration of adsorbed developer ions or molecules at and near the interface should vary according to the conditions existing there. No direct data on this point are available. Some predictions can be made, however, that are in general agreement with rate data. Excess halide ion adsorbed by the grain surface should diminish adsorption of the developing agent. When a molecular developing agent, such as dimethyl-*p*-phenylenediamine, which is active in the form of the neutral molecule, is involved, the halide ion is competing with the agent for the adsorption site. When a negatively charged ionic developing agent such as hydroquinone is involved, electrostatic repulsion plays an important role. The silver halide grain containing excess adsorbed halide ions is protected by a negative electrostatic sheath, and only those negative ions with kinetic energy sufficient to overcome the repulsion can reach the grain surface. Under normal conditions of development, only a fraction of the developer ions will possess the required energy. The larger the charge at the grain surface and the larger the charge of the developer ion, the smaller will be the fraction of the ions which can reach the grain surface (James, 57, 60, 61).

The influence of charge on the kinetics of development becomes very important for the developing agents with two or more negative charges on the active form. Under simplified conditions, certain properties of the agents, such as the relative magnitude of the induction period and the photographic speed obtained in the early stages of development, are determined primarily by the charge of the active developing agent. The chemical nature of the agent often is of secondary importance. Thus, hydroquinone, ascorbic acid, ferro-oxalate ion, and *p*-hydroxyphenylglycine all show about the same relative induction period under simplified conditions of development. All are active as divalent ions. The divalent ion of methyl-*p*-aminophenol sulfonic acid, however, shows exceptional behavior in some respects, and its properties lie between those exhibited by a typical divalent ion and a typical monovalent ion. A likely explanation of this is that the initial adsorption of the methyl-*p*-aminophenol sulfonic acid alters the surface conditions of the silver halide in the direction of decreasing the effective charge, thus increasing the tendency towards further adsorption of the developing agent.

As we have seen, adsorption of gelatin and certain dyes by silver chloride markedly decreases the rate of reduction of the silver salt by hydroxylamine. Gelatin is always present as a modifying influence in the reduction of a photographic grain. The gelatin is strongly adsorbed by the silver halide, and probably interferes with adsorption of the developing agent. Dyes such as the cyanines are held to the surface primarily by the excess halide ion layer, but the adsorption is very strong and the silver ions below are less accessible to the developing agent. However, this steric effect may be partially or completely offset under certain conditions by the tendency of the cationic dyes to depress the surface charge. Actual acceleration of development by divalent developer ions has been achieved with some dyes. Acceleration can be obtained easily by bathing the sensitive layer in a solution of certain compounds which likewise contain quaternary heterocyclic nuclei, e.g., lauryl pyridinium bromide (Lottermoser and Steudel, 64).

Dankov (65) has suggested that the activity of the silver/silver bromide interface may be enhanced as a result of a unique circumstance which exists if cube faces of the silver and silver bromide are joined. The positions of silver atoms in the cube face of the silver structure almost exactly match the positions of the silver and bromide ions in the cube face of the silver bromide structure when one of these faces is rotated 45 degrees with respect to the other in its plane. The distance between silver atoms along a cube edge in the silver crystal is 4.07 A. whereas the distance from a silver to a bromide ion along a cube face diagonal in the silver bromide crystal is 4.08 A. Thus, it would be possible for silver ions of the silver bromide crystal to be transformed into silver atoms of the silver crystal without movement or with very slight movement of the atomic nuclei, so long as the change was confined to two dimensions. The distance between silver and bromide ions in the plane normal to this, however, is only 2.88 A., 29% less than the corresponding Ag spacings.

A similar coincidence does not exist for silver chloride or silver iodide. The diagonal of the cube face of the silver chloride is 3.92 A., or 3.8% smaller than the edge of the silver lattice. Dankov suggests that this accounts for the lesser usefulness of silver chloride as a photographic material. However, as Sheppard points out (Sheppard, 15) the defects of silver chloride relative to the bromide are in the matters of spectral sensitivity and of greater solubility and reducibility. A latent image in silver chloride grains actually develops at a rate as great as or greater than one in silver bromide, other things being equal.

2. *Electrode Mechanisms*

Several writers have expressed the view that the silver nucleus acts as an electrode which is charged by the developing agent. A cardinal postulate in such mechanisms is that the electrons can be transferred from the developing agent to the silver nucleus at any point where its surface is in contact with the developer solution. Silver ions are reduced primarily at the silver/silver halide interface.

The suggested mechanisms differ in detail (Mott, 66, 67; Berg, 68; Anastasevich, 69; Frank-Kamenetskii, 70; Bagdasar'yan, 17, 71) but all involve the idea that electrons can be transferred to silver much more readily than to a silver halide crystal. Each mechanism can be criticized on some detail (cf. Sheppard, 15; James, 72). As a general criticism, however, none of the mechanisms has explained the fact that the rate of development under simplified conditions varies with the square root of the hydroquinone and catechol concentrations, whereas the rate of reduction of silver ions from solution by the same agents varies as the first or somewhat higher power of the concentration.

The mechanism suggested by Bagdasar'yan has been formulated in more concrete terms than the others. It is in good agreement with some experimental data, but not with all. As in his treatment of the reduction of silver ions from solution, Bagdasar'yan treats the transfer of electrons from the developer to the silver nucleus as the rate-controlling process, and derives the rate equation for the reduction of a silver bromide grain:

$$d(\mathrm{Ag})/dt = k_1[\mathrm{Red}][\mathrm{Br}^-]^{-\alpha}\cdot S \tag{11}$$

where [Red] is the concentration of the reducing agent, S is the area of the silver/solution interface, and α is a fraction.

Equation (11) is in good agreement with the established dependence of development rate upon bromide ion concentration for hydroquinone and ferro-oxalate development. It does not account for the dependence of rate on the concentration of the developing agent. It is in adequate

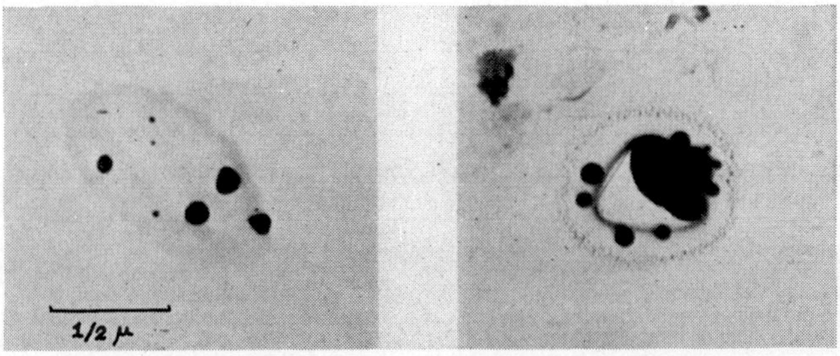

FIG. 6. (a) Gold particles formed on a single silver halide grain by bathing the exposed film for 64 hours in the aurous thiocyanate solution. The silver halide was dissolved out before the electron micrograph was made. (b) Gold particles in contact with the original silver halide grain. The dark area on the right is silver printed out by the electron beam.

agreement with Rabinovich's observations on the apparent dependence of rate upon values of S calculated on the assumption that a smooth silver surface is formed, but as already pointed out, the validity of this assumption is questionable. More recent experiments suggest that the area of the metal/solution interface is not the important factor in determining the development rate (James, 32). In these experiments, the latent image nuclei on the grain surface were built up by physical development in a gold solution to a point where they were readily visible under the electron microscope. The nuclei grew in roughly spherical form (see Fig. 6) and appeared to be in contact with the grain surface at only a relatively small area. Kinetic studies were made of the initial rate of development of grains containing such enlarged latent image nuclei.

With hydroquinone and *p*-hydroxyphenylglycine as developers, the induction period decreased as the size of the nuclei increased, but the induction period was not eliminated even when the nuclei attained a size of 20 to 80 mμ. The initial rate of development had increased by only three to four fold over that obtained with film which had been given a brief gold treatment that could do little more than replace silver by gold. Since the area of the original latent image nuclei is not known, the actual increase in area when the nuclei are enlarged to 20 or 80 mμ cannot be calculated, but the increase in area obviously is much greater than the increase in development rate. This result suggests that the rate-determining step in the development reaction occurs at the metal/silver halide interface rather than along the metal/solution interface.

3. Size of the Latent Image Nuclei

The minimum size of a nucleus which is able to initiate direct development of a silver halide grain is not known. Presumably it is smaller than the minimum size for physical development, since the threshold exposure to light is smaller for direct than for physical development. However, mere size may be of secondary importance to the properties of the silver/silver halide interface (Sheppard, 15, 73). We may be justified in speaking of a critical size only in connection with otherwise equal interface conditions. A critical size in this limited sense apparently does exist for development times of the order of those used in practical work. At least some grains of the most sensitive emulsions are made developable by the absorption of one or two quanta, corresponding to the formation of one or two silver atoms. The atom or atoms are added on to the silver or silver sulfide already present in the sensitivity speck, and presumably are not capable by themselves of initiating development. However, the addition of the one or two silver atoms to a suitable sensitivity speck has transformed it from a speck incapable of initiating development into a nucleus for development. Moreover, the removal of one atom apiece from certain latent image nuclei destroy their ability to initiate development (Kornfeld and James, 74).

4. Latensification

The effective speed of many photographic materials can be increased by suitable treatment following exposure. Some grains which had absorbed light during exposure, but not enough to form developable nuclei, are made developable by the auxiliary treatment. This phenomenon is termed "latensification." Its existence demonstrates the presence in the exposed but not developable grain (prior to treatment) of a latent "sub-image" consisting of nuclei which are not capable of initiating development under the test conditions.

Latensification can be brought about in several ways. One in which the effect must be ascribed to an increase in size of the sub-image nuclei makes use of a secondary exposure to light of low intensity. A second method in which the effect most probably is brought about by an increase in size of the sub-image nuclei involves treatment of the exposed material in an atmosphere of mercury vapor (Dersch and Dürr, 75). In this method, which is reminescent of Daguerre's development process, the increase in size is brought about by condensation of mercury atoms on the nuclei rather than by the addition of silver atoms to them.

The mechanisms of the other methods of latensification are more in doubt (cf. review by Sheppard *et al.*, 76). These methods include bathing the photographic material in a solution of silver salt, in a solution of hydrogen peroxide or sodium perborate (Vanselow *et al.*, 77), in a solution of aurous thiocyanate (James *et al.*, 31) or by fuming the material in the vapor of certain organic acids (Mueller and Bates, 78) or of ammonia. Such treatment may result in an increase in the effective size of the sub-nuclei, or simply in bringing about more favorable conditions for development at the silver/silver halide interface.

A distinction should be made between a true latensification and a simple acceleration of development. For example, bathing the sensitive layer in a solution of thallous ion or of lauryl pyridinium bromide after exposure will accelerate the early stages of development in a hydroquinone solution (Lottermoser and Steudel, 64) and the acceleration will be greater in the low than in the high exposure areas. If a short time of development is employed, the bathing will apparently result in marked latensification. If development is extended to longer times, however, the apparent latensification will disappear. The test for true latensification is to determine whether the treatment causes some grains to develop as image which, without the treatment, could not be distinguished from fog grains on prolonged development. Results obtained by James and Vanselow (77a) indicate that there is no true latensification when a powerful developing solution is used and development is carried to the maximum. An increase in effective speed results from the latensification operation only if a weak developer is used or if development in a powerful developer is not carried to completion. In the usual commercial processing, however, development is terminated before the maximum differentiation between exposed and unexposed areas is achieved.

Most latensification treatments accelerate development as well as increase the effective speed of the photographic material. The gold latensification is especially notable in this respect, and is of interest in relation to the mechanism of development. A simple motion picture positive film responds well to the gold treatment, which consists simply in bathing the exposed film for a few minutes in a solution containing the aurous thiocyanate complex in the presence of excess thiocyanate and bromide ions. Figure 7 shows the developed densities for several different exposures of treated and untreated film. All densities have been corrected for fog. The densities of the treated film increase rapidly with time of development at first, then level off and become essentially independent of development time. Density already is at the maximum in 4 minutes, which is about the normal time of development under the conditions used. Density of the untreated film increases much more slowly. At the higher exposures, the densities of the untreated film eventually equal those of the gold-treated film. The lower exposure

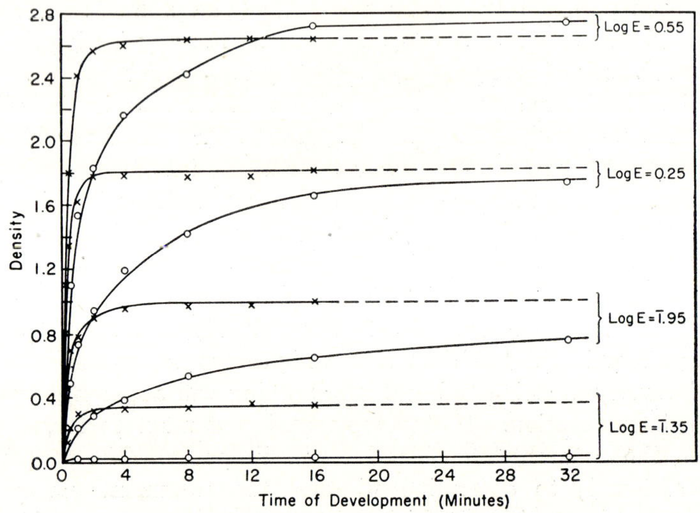

FIG. 7. Effect of gold latensification on the rate of development of a motion picture positive film. —x—x—, latensified film; —o—o—, untreated film. Curves are plotted for several values of log E, where E is the energy of the light exposure.

steps show the genuine latensification for prolonged development in the rather weak developer used. Even this latensification disappears, however, if a sufficiently powerful developer is used.

The gold treatment undoubtedly results in a replacement of at least some latent image silver by gold. It is possible, as already mentioned, that the nuclei have been increased in size by a physical development effect, but this seems unlikely for the short times involved. The maximum effect is obtained within 5 minutes under the experimental conditions used. Accordingly, it appears that the increased rate of development is the result of the changed properties of the metal/silver halide interface, which now involves metallic gold instead of silver.

5. *Filament Formation*

The silver atoms formed in direct development obviously cannot simply replace the silver ions in space along all crystal directions, since the space requirements are not the same in all directions. Moreover, electron micrographs of the silver obtained by developing the photographic grain on the microscope slide show an irregular and often decidedly filamentary structure which has no counterpart in the silver halide crystal (Ardenne, 54; Hall and Schoen, 55). In some instances,

FIG. 8. Electron micrograph of a partially developed grain, showing triangular enlargements in the filaments. The developer was a hydroquinone solution of pH 8.7 which contained no sulfite.

thin filaments project well beyond the boundaries of the original silver bromide grain. Filaments also are formed in pre-fixation physical development (Küster, 55a) but not in post-fixation physical development.

The filaments often show angular enlargements in the plane of the ribbon and these frequently take the form of flat, triangular crystals. "This is a strong indication that the sides of the filaments correspond to the octahedral (111) faces of the silver crystals, so that the filaments may be considered as silver crystals which have grown chiefly in one dimension" (Jelley, 79). Figure 8 shows typical triangular enlargements obtained by partial development in a slow-acting hydroquinone solution of a silver bromide grain in a typical motion picture positive material. The developer was of low pH (8.7) and contained no sulfite.

The process of filament formation in development probably is simply a matter of crystal growth (James, 80). The newly formed silver atoms should be quite mobile, just as they are in thin films of silver formed by condensation from an atomic beam (cf. Cockcroft, 44). With favorable initial orientation, silver crystals can grow primarily in one direction (Howey, 81). As development of the grain proceeds, microscopic pits may form around the growing nuclei, owing to elimination of bromide ions into the solution, to the solvent action of concentrated bromide ion solutions, and to migration of reduced silver. Such ulceration of the crystal lays bare silver bromide surfaces which are then no longer protected by adsorbed gelatin and which are as a consequence more susceptible to direct attack by the developer. New silver nuclei could form on such surfaces and grow catalytically in the same fashion as the original latent image nuclei. Furthermore, filaments growing from one part of the grain may make contact with another part and start development at that point. Thus, although development may start at only a few points on the crystal surface, a much larger number of nuclei may form during development. This would account for the "seaweed" cluster of filaments often observed in electron micrographs of developed grains.

VI. Simultaneous Direct and Physical Development

Under the usual conditions of commercial practice, the development reaction does not occur entirely at the silver/silver halide interface. Some reduction of silver ions from solution takes place. Such reduction presumably can occur at any point on the silver/solution interface, and the mechanism should be the same as that for post-fixation physical development. The relative extent of the physical development in comparison with that at the silver/silver halide interface will depend upon the silver halide solvent action of the developing solution and upon the rate of the direct development.

The silver halide solvent, sulfite ion, is always present in practical developing solutions which employ organic agents. The conventional solutions contain up to 100 g. sodium sulfite per liter. The action of the sulfite is manifold. It is added primarily to decrease the rate of loss of developing agent by aerial oxidation and to prevent the accumulation of quinone or quinonelike oxidation products of the developing agents. A third phase of its activity, the solvent action, is well-known but the extent to which it can alter the nature of the development process under proper conditions is often overlooked.

A rather extreme example of the alteration of development by solvent action is given by some experiments with the hydroxylamine developer (James, 58). Sulfite does not react with the oxidation products of this agent; hence no complications arise from such side reactions. Small amounts of sulfite had no observed effect upon the rate of development or the appearance of the developed image. When larger amounts of sulfite were added, the overall rate of development decreased (Table VI) and, as the sulfite concentration was further increased, the appearance of the developed image shifted over to that typical of physical development.

TABLE VI

Effect of Sulfite upon Hydroxylamine Development

NH_2OH, 0.04 M; pH, 10.8; excess Br^-, 0.00067 M

Salt concentration	R (Na_2SO_3)	R (Na_2SO_4)
M		
0	0.062	0.062
0.0025	0.063	
0.010	0.062	0.061
0.050	0.055	0.060
0.100	0.027	0.059
0.200	0.003	0.057

To account for the sharp drop in rate at the higher sulfite concentrations, it was suggested that the solvent action of the sulfite can isolate the latent image nuclei from the main body of the grains, so that development can proceed only by the slower physical development process.

TABLE VII

Rate of Solution of Silver Halide of a Motion Picture Positive Film in 2% Sodium Sulfite Solution at 23°

Duration of washing (minutes)	Total Ag/100 sq. cm. remaining in film	Ag/100 sq. cm. removed from film
0	0.048	0.000
15	.037	.011
30	.030	.018
45	024	.024
60	.016	.032
75	.012	.036

This suggestion is supported by recent experiments by the writer. Ten feet of exposed motion picture positive film, 35 mm. wide, was bathed in 4 liters of a 2% sodium sulfite solution. Vigorous agitation of the solution was maintained, and the solution was replaced every 3 minutes. Two feet of film was removed at suitable intervals. One was used for determination of total silver halide content; the other was developed in a metol-hydroquinone solution of standard composition. Table VII shows the amounts of silver halide dissolved out by the sulfite solution within various periods of time. Three-fourths of the silver halide had been dissolved out within 75 minutes.

Figure 9 shows the family of curves representing the developed strips. The sulfite washing not only decreased the maximum developed density, but actually caused a reversal in the higher exposure regions. Moreover, the decrease in maximum density is considerably larger than

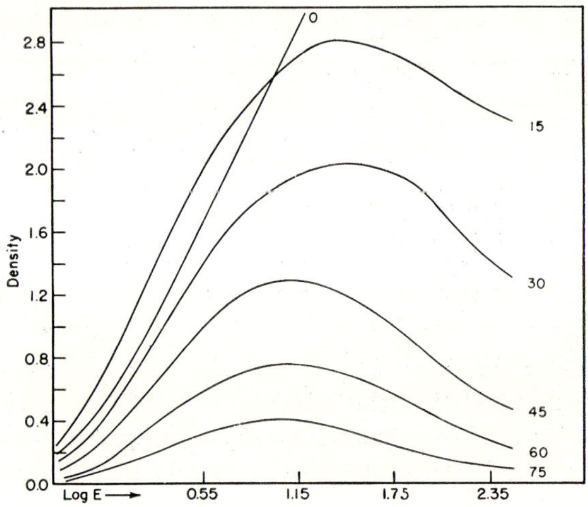

FIG. 9. Effect of washing exposed motion picture positive film in a 2% sodium sulfite solution before development. The numbers on the curves indicate the washing times in minutes at 25°.

can be accounted for solely by the decrease in the amount of available silver halide. The latent image nuclei themselves had not been destroyed, since the sulfite washing resulted in only a slight decrease in the developability of the film in a post-fixation physical developer. No reversal appeared for physical development. Evidently some of the nuclei simply had lost contact with the silver halide grains. The larger nuclei were more readily isolated from the grain by the action of the sulfite than the smaller ones.

The solvent action of sulfite tends to promote "fine-grain development," and some of the commercially used fine-grain developers contain as much as 100 grams of sodium sulfite per liter. Some of these developers contain other solvents as well, e.g., thiocyanates and amines. Any of these solvents should cause some shift to occur in the relative rates of the reactions at the silver/silver halide and at the silver/solution interfaces.

Solvent action need not always cause a decrease in the net rate of development. Latent image nuclei which were beneath the surface may be brought into action by dissolving off the outer silver halide sheath (Kogelmann, 82; Kempf, 83; Kornfeld, 84; Berg *et al.*, 85). Such nuclei could then initiate development at the silver/silver halide/solution triple interface and could also serve as nuclei for physical development. An actual increase in development rate can thus occur under proper conditions. Some developing agents, e.g., *p*-phenylenediamine, have enough solvent action in themselves to uncover internal latent image. The importance of the internal image relative to the external for development will vary widely with the nature of the photographic material.

The solvent action of sulfite on silver bromide and the resultant tendency to isolate latent image nuclei from the grain accounts for the failure of sulfite itself to act as a direct developer in spite of the autocatalytic character of its reduction of silver ion. The active nuclei simply are isolated from the grain before development gets under way. The same phenomenon enters to prevent sulfite-containing hydroquinone solutions of low pH (e.g., 8.5) from developing readily even though the thermodynamic conditions are suitable for reaction and the hydroquinone develops readily at the same pH when sulfite is absent.

References

1. Sheppard, S. E., *Phot. J.* **65,** 380 (1925).
1a. Carroll, B. H., and Hubbard, D., *J. Research Natl. Bur. Standards* **1,** 565 (1928).
1b. Mitchell, J. W., *Phil. Mag.* **40,** 249 (1949).
2. Piper, C. W., *Brit. J. Phot.* **55,** 196 (1908).
3. Sheppard, S. E., *Phot. J.* **59,** 135 (1919).
3a. Sheppard, S. E., and Meyer, G., *J. Am. Chem. Soc.* **42,** 689 (1920).
4. Volmer, M., *Z. wiss. Phot.* **20,** 189 (1921); *Phot. Korr.* **58,** 226 (1921).
5. Weissberger, A., and Thomas, D. S., *J. Am. Chem. Soc.* **64,** 1561 (1942).
6. James, T. H., *J. Am. Chem. Soc.* **61,** 648 (1939).
7. James, T. H., *J. Phys. Chem.* **45,** 223 (1941).
8. Euler, H. v., *Z. Elektrochem.* **28,** 446 (1922).
9. Proskurnin, M., and Frumkin, A., *Z. physik. Chem.* **155A,** 29 (1931).
10. Veselovsky, V. I., *Acta Physicochim. U.R.S.S.* **11,** 815 (1939).
11. Euler, H. v., and Zimmerlund, G., *Arkiv. Kemi, Mineral Geol.* **8,** No. 14 (1921).
12. James, T. H., *J. Am. Chem. Soc.* **62,** 3411 (1940).
13. Rabinovich, A. J., Peisakhovich, S., and Minaev, L., *Ber VIII Int. Kong. Phot. (Dresden)* p. 186 (1931).
13a. Rabinovich, A. J., and Peisakhovich, S., *Z. wiss. Phot.* **33,** 94 (1934).
14. Perry, E. S., Ballard, A., and Sheppard, S. E., *J. Am. Chem. Soc.* **63,** 2357 (1941).
15. Sheppard, S. E., in Colloid Chemistry. Edited by J. Alexander, Vol. 5, Reinhold, New York, 1944.
16. James, T. H., and Weissberger, A., *J. Am. Chem. Soc.* **60,** 98 (1938).

17. Bagdasar'yan, Kh.S., *J. Phys. Chem. U.S.S.R.* **17,** 336 (1943); *Acta Physicochim. U.R.S.S.* **19,** 421 (1944).
18. Frumkin, A., *Z. physik. Chem.* **160,** 116 (1932); *Reports II of the Conference on Corrosion* **1,** 5 (1940).
19. Laidler, K. J., Glasstone, S., and Eyring, H., *J. Chem. Phys.* **8,** 667 (1940).
20. Knorre, G. V., and Arndt, K., *Ber.* **33,** 30 (1900).
21. Bray, W. C., Simpson, M. E., and MacKenzie, A. A., *J. Am. Chem. Soc.* **41,** 1363 (1919).
22. James, T. H., *J. Am. Chem. Soc.* **64,** 731 (1942).
23. Nichols, M. L., *J. Am. Chem. Soc.* **56,** 841 (1934).
24. James, T. H., *J. Am. Chem. Soc.* **61,** 2379 (1939).
25. James, T. H., *J. Am. Chem. Soc.* **62,** 536 (1940).
26. James, T. H., *J. Am. Chem. Soc.* **63,** 1601 (1941).
27. Arens, H., *Ber. VIII Int. Kongr. Phot. (Dresden)* p. 63 (1931).
28. Arens, H., *Z. wiss. Phot.* **30,** 49 (1931); **31,** 68, 125 (1932); **32,** 65 (1933).
29. Schoen, A. L., private communication.
30. Vanselow, W., and Quirk, R. F., private communication.
31. James, T. H., Vanselow, W., and Quirk, R. F., *PSA Journal* **14,** 349 (1948).
32. James, T. H., *J. Colloid Sci.,* **3,** 447 (1948).
33. Jablczynski, K., and Kobryner, S., *Roczniki Chemji* **9,** 715 (1929).
34. James, T. H., *J. Am. Chem. Soc.* **62,** 1654 (1940).
35. James, T. H., *J. Chem. Phys.* **14,** 536 (1946).
36. Weissberger, A., Mainz, H., and Strasser, E., *Ber.* **62,** 1942 (1929).
37. Krishna, B., and Ghosh, S., *J. Phys. Colloid Chem.* **51,** 1130 (1947).
38. Volger, H. J., and Clark, W., *Brit. J. Phot.* **74,** 670 (1927).
39. Svedberg, T., *Phot. J.* **61,** 325 (1921).
39a. Clark, W., *Science Progress* **19,** 266 (1924).
40. Baker, A., and Usher, F. L., *Trans. Faraday Soc.* **36,** 549 (1940).
41. Reinders, W., and Hamburger, L., *Z. wiss. Phot.* **31,** 32 (1932).
42. Reinders, W., and deVries, R. W P., *Rec. trav. chim.* **56,** 985 (1937).
43. Berg, W. F., *Phil. Mag.* **36,** 337 (1945); *ibid.,* back cover, Sept. 1945.
44. Cockcroft, J. D., *Proc. Roy. Soc. London* **119A,** 293 (1928).
45. Frenkel, J., *Z. Physik* **26,** 117 (1924).
46. Hass, G., *Naturwissenschaften* **25,** 232 (1937).
47. Knudsen, M., *Ann. Physik* **50,** 472 (1916).
48. Luther, R., and Leubner, A., *Brit. J. Phot.* **59,** 632 *et seq.* (1912).
49. Sheppard, S. E., Lambert, R. H., and Keenan, R. L., *J. Phys. Chem.* **36,** 174 (1932).
50. James, T. H., *J. Am. Chem. Soc.* **62,** 1649 (1940).
51. Mees, C. E. K., Theory of the Photographic Process. Macmillan, New York, 1942.
52. Rabinovich, A. J., *Trans. Faraday Soc.* **34,** 920 (1938).
53. Rabinovich, A. J., Bogoyavlenski, A. N., and Zuev, Y. S., *Acta Physicochim. U.R.S.S.* **16,** 307 (1942).
54. Ardenne, M. v., *Z. angew. Phot.* **2,** 14 (1940).
55. Hall, C. E., and Schoen, A. L., *J. Optical Soc. Am.* **31,** 281 (1941).
55a. Küster, A., *Z. wiss. Phot.* **43,** 191 (1948).
56. Meidinger, W., *Physik. Z.* **36,** 312 (1935).
57. James, T. H., *J. Phys. Chem.* **44,** 42 (1940).
58. James, T. H., *J. Phys. Chem.* **47,** 597 (1943).
59. James, T. H., *J. Franklin Inst.* **240,** 229, 327 (1945).

60. James, T. H., *J. Franklin Inst.* **240,** 83 (1945).
61. James, T. H., *J. Franklin Inst.* **243,** 235 (1947).
62. Wulff, P., and Seidl, K., *Z. wiss. Phot.* **28,** 239 (1930).
63. Shiberstoff, V. J., *Proc. IX Congr. Intern. Phot. (Paris)* p. 357 (1935).
64. Lottermoser, A., and Steudel, R., *Kolloid Z.* **82,** 319 (1938).
65. Dankov, P. D., *Compt. rend. acad. sci. U.R.S.S.* **24,** 773 (1939).
66. Mott, N. F., *Repts. Progress Phys.* **6,** 186 (1939).
67. Mott, N. F., *J. Phys. radium* **7,** 249 (1946).
68. Berg, W. F., *Trans. Faraday Soc.* **39,** 126 (1943).
69. Anastasevich, V. S., *Acta Physicochim U.R.S.S.* **16,** 296 (1942); *J. Tech. Phys. U.S.S.R.* **14,** 467 (1944).
70. Frank-Kamenetskii, D. A., *Acta Physicochim. U.R.S.S.* **12,** 13 (1940).
71. Bagdasar'yan, Kh. S., *Acta Physicochim. U.R.S.S.* **20,** 441 (1945).
72. James, T. H., *J. Chem. Education* **23,** 595 (1946).
73. Sheppard, S. E., *Phot. J.* **69,** 330 (1929).
74. Kornfeld, G., and James, T. H., *J. Optical Soc. Am.* **33,** 615 (1943).
75. Dersch, F., and Dürr, H., *J. Soc. Motion Picture Engrs.* **28,** 178 (1937).
76. Sheppard, S. E., Vanselow, W., and Quirk, R. F., *J. Franklin Inst.* **240,** 439 (1945).
77. Vanselow, W., Quirk, R. F., and Leermakers, J. A., *PSA Journal* **14,** 675 (1948).
77a. James, T. H., and Vanselow, W., *PSA Journal* (in press).
78. Mueller, F. W. H., and Bates, J. E., *J. Phot. Soc. Am.* **10,** 586 (1944).
79. Jelley, E. E., *J. Phot. Soc. Am.* **8,** 283 (1942).
80. James, T. H., *J. Chem. Phys.* **11,** 338 (1943).
81. Howey, J. H., *Phys. Rev.* **55,** 578 (1939).
82. Kogelmann, F., Die Isolierung der Substanz des latenten Bilds. Dissertation, Graz (1894).
83. Kempf, A., *Z. wiss. Phot.* **36,** 235 (1937).
84. Kornfeld, G., *J. Optical Soc. Am.* **31,** 598 (1941).
85. Berg, W. F., Marriage, A., and Stevens, G. W. W., *J. Optical Soc. Am.* **31,** 385 (1941).

Catalysis and the Adsorption of Hydrogen on Metal Catalysts

OTTO BEECK

Shell Development Company, Emeryville, California

CONTENTS

I. INTRODUCTION

In any field of science as complex as that of catalysis, the only approach to its understanding is through a complete analysis of all factors involved, investigating each variable as completely separated from all other variables as possible. Heterogeneous catalysis always involves adsorption. In high temperature catalytic decompositions the adsorption complex may be of very short duration. In bimolecular reaction the reactants must often both be adsorbed to yield the product; sometimes it is sufficient that one reactant be adsorbed. The rate determining step in heterogeneous reactions may be the rate of adsorption of the reactant (or reactants) on the surface or it may be the rate of desorption of the product (or products) from the surface. Only in rare cases is the rate of surface reaction proper determining. By definition the catalyst reduces the energy of activation of a reaction, and in heterogeneous catalysis this is achieved through the formation of an activated adsorption complex with the catalyst. This does not necessarily mean that strong adsorption is equivalent to great catalytic activity. In fact, too strong an adsorption may simply mean that the surface is covered with either reactant or product which effectively poisons the surface for any further reaction. This shows that the activation energy necessary to form the activated adsorption complex must be considered in relation to the activation energy necessary for the reaction proper within the activated adsorption complex. For instance, it has been conclusively shown by the author and his coworkers (1) that hydrogen must be adsorbed in the form of atoms in order to hydrogenate ethylene in the neighborhood of room temperature and atmospheric pressure. It has further been shown in this work (2) that ethylene itself is not adsorbed on the surface but that it merely picks up the two hydrogen atoms from the surface. This does not mean, of course, that ethylene, in the act of picking up the two hydrogen atoms, does not form momentarily an additional activated adsorption complex with the two hydrogen atoms and the surface. It only means that ethylene could not have been adsorbed itself prior to hydrogenation, since the sum of the separate heats of adsorption of hydrogen and ethylene is three times as great as the heat of hydrogenation. As may be readily concluded, adsorbed ethylene is a poison for this reaction. It was furthermore shown by the same authors (1) that the spacing of the hydrogen atoms on the surface—that is, the crystal parameter of the surface—plays a very important part in the hydrogenation reaction, a 110 oriented nickel film, for instance, being five times as reactive as a randomly oriented film. The large spacing in rhodium of about 3.8 A. was found (2) to be a thousand times more effective than the respective spacing in nickel of about 3.5 A. While this spacing may be of primary importance in the reaction at low temperatures, it is evident that at high temperatures—that is, with most of the molecules having a high energy content—the importance of the spacing should become less critical. While the required spacing for reactions at low temperatures calls for complete dissociation of the hydrogen molecule, complete dissociation is probably not necessary at high temperatures. Under conditions of incomplete dissociation the heat of adsorption of hydrogen will necessarily be lower. Consequently hydrogen will be adsorbed to a very much lesser degree on surfaces on which it is not completely dissociated, and in order to provide adequate coverage of the surface with adsorbed hydrogen molecules, very high pressures must be applied. But even in the regime of complete dissociation of the hydrogen molecules into adsorbed atoms, the heat of adsorption—that is, the binding energy of the atoms to the surface—must be considered in addition to spacing of the atoms. Atoms adsorbed with a high heat of adsorption may require (although not necessarily) a higher activation energy for the reaction between ethylene and adsorbed hydrogen atoms than those which are adsorbed with a low heat of adsorption. Hence it is of great importance to investigate whether the surface is uniform or nonuniform with respect to adsorption.

It is obvious that no truly uniform surfaces will exist. Even a perfect crystallographic plane is nonuniform due to periodic energy variations inherent in the atomic arrangement of the crystallographic sites. Next, one may ask whether all regions of highest surface energy—that is, the crystallographic sites—are equivalent. The answer is yes if the surface is a perfect crystallographic plane. If the plane is not perfect either through crystal imperfection or due to the incomplete atom layers which may form ledges and ridges, the crystallographic sites may vary in energy, but the maximum difference would be expected to be not more than 30% of the absolute value for a perfect surface, depending on the type of binding and on the size of the adsorbed molecule in relation to the size of the crystallographic sites. More serious is nonuniformity of the surface due to impurities. A foreign atom adsorbed on the surface with a large binding energy will make the surface nonuniform and will also decrease the surface energy of the neighboring sites. But the foreign atom need not necessarily be adsorbed on the surface; it may merely occupy a place in the crystallographic layer next to the surface. In this case too the resulting binding energy between the surface metal atoms and the foreign atom will lower the surface energy of the sites next to the foreign atom and thereby change the heat of adsorption for hydrogen atoms in this vicinity. Adsorbed hydrogen atoms themselves will have the same effect. They also will make the surface nonuniform. Beeck *et al.* (3) found the heat of adsorption of hydrogen molecules to be 30,000 calories per mole for a sparsely covered nickel surface. This means that the hydrogen atom is bound to the surface with an energy of roughly 65,000 calories. For simplicity we may assume that this binding energy is equally distributed over the three or four neighboring metal atoms, depending on the type of the crystallographic site. It is obvious, therefore, that if a free site is surrounded by several other sites occupied by adsorbed hydrogen atoms, the energy of adsorption remaining for the adsorption of a hydrogen atom onto this free site must be much lower than the energy available for hydrogen atoms to be adsorbed on sites whose neighbors are still free. It is obvious too that under these conditions the decrease of heat of adsorption with increasingly covered surface will initially change little with surface coverage, since—assuming mobility of adatoms from site to site—the atoms can take positions far enough apart so as not to interfere with each other.

It will be shown later in the text that the heat of adsorption of hydrogen on porous evaporated metal films is often typical of this behavior, and mobility of adatoms will have to be assumed notwithstanding the observed relatively high heats of adsorption and the very low activation energy for adsorption. In this case, adsorption will take place on every site, and the redistribution of, for instance, adsorbed hydrogen will take place as adatoms through migration from site to site. If the adsorption process necessitates a high heat of activation as, for instance, in the case of the adsorption of nitrogen on iron at elevated temperatures, the assumption of mobility is not necessary, inasmuch as nitrogen molecules will initially not be adsorbed except on the sites of highest surface energy. The English school under the late J. K. Roberts of Cambridge (4) has been quite successful in explaining the decrease of heat of adsorption with surface coverage of hydrogen on tungsten by the interaction of adatoms just discussed. Recently, Halsey and Taylor (5) have rightly pointed out, however, that this interpretation must be limited to cases where the curves representing heats of adsorption vs. surface coverage intersect the ordinate with nearly right angles, the limiting value being an initial slope (change of heat of adsorption with surface coverage) not greater than about 0.25 the initial heat of adsorption. For larger angles heterogeneity of the surface will have to be assumed.

It is the purpose of this article to point out the need for caution in the interpretation of many experimental results so far published in the field of catalysis and for a critical attitude toward the problem of sorption of gases, which may include adsorption on the surface as well as absorption into the interior of the structure, and may easily lead to faulty conclusions.

II. Definitions

In order to prevent misunderstanding of the various terms used in this article, a brief definition of these terms appears to be desirable.

Sorption. The word sorption will be used for the total amount of gas taken up by a given solid. This includes adsorption on the surface as well as absorption into the interior of the structure. The latter may be true endothermic solution or may be exothermic solution, which infers compound formation such as the formation of metal hydrides when hydrogen is involved.

Adsorption. Adsorption refers to the adsorption of gas on the gas-solid interface only. It may be van der Waal's adsorption or chemisorption, the latter including activated adsorption.

Van der Waal's Adsorption. Van der Waal's adsorption involves forces which are of the order of magnitude of the forces between the molecules in a liquid. The heats of adsorption are therefore not very much larger (sometimes even smaller) than the heats of condensation of molecules onto their own liquid or on solid surfaces. The difficulties of defining the upper limit will become apparent later in the text.

Chemisorption. Chemisorption involves heats of adsorption which are large as compared to the heat of van der Waal's adsorption. The term chemisorption implies formation of semi-chemical bonds of the adsorbed gas with the solid surface. Chemisorption may be a process involving measurable activation energy—that is, a measurable rate of adsorption and a measurable temperature coefficient of rate of adsorption. As in the case of hydrogen adsorption on metals, chemisorption may have no measurable rate of adsorption, the adsorption being essentially instantaneous.

Activated Adsorption. Activated adsorption—that is, adsorption with a measurable rate of adsorption and a measurable temperature coefficient of rate of adsorption—is a type of chemisorption which is, for instance, found in the adsorption of nitrogen on certain metals at elevated temperatures. The difficulties of deciding whether or not true van der Waal's adsorption exists in cases where the heats of adsorption exceed considerably the heats of condensation will become apparent later in the text.

Absorption. Absorption is that part of the total sorption process which does not involve the surface. Absorption may be true solution or it may be compound formation (for instance hydrides).

Adatoms. Following the custom of Langmuir and others, the term adatoms will be used for adsorbed atoms.

III. The Extent of Surface of Metal Catalysts

It is shown in this section that the low-pressure Langmuir type isotherm of hydrogen at room temperature and those of nitrogen at liquid air or liquid nitrogen temperatures represent, in their relatively flat portion between 10^{-4} *and* 10^{-1} *mm., pressure a reliable measure of the surface of evaporated metal films. The surface values obtained are in agreement with those derived from low temperature isotherms by the Brunauer-Emmett-Teller (B.E.T.) method using krypton, methane, and butane. Nitrogen is unsuitable for evaluation by the B.E.T. method because of its high (10,000 to 5,000 Cal., depending on fraction of surface covered) heat of adsorption at liquid nitrogen temperature leading to step-wise adsorption, not observable by the Brunauer-Emmett technique.*

In order to be able to evaluate and measure adsorption relative to total sorption and to measure the heat of adsorption in its dependence on surface coverage, and in order to be able to discuss heterogeneity of the surface, it is of paramount importance to have a reliable method for measuring the total available surface ("true surface") as represented by the gas-solid interface. For more than ten years the Emmett-Brunauer method has been widely used for this purpose. The method has been invaluable from the academic standpoint as well as from the practical industrial standpoint. Notwithstanding the fact that not all the premises on which the theoretical treatment by Brunauer, Emmett, and Teller (6) (B.E.T. method) is based appear to be tenable (for instance, the forces between the adsorbed molecules are not taken into account), there has been very little reason to be disturbed by this inadequacy of the theory because it has led to a method which through one or two simple measurements allows a very rapid determination of the surface giving the same result which Emmett and Brunauer (7) had previously deducted from the shape of the isotherm itself. Because of the convincing conclusions drawn from the shape of the isotherm itself, we can be reasonably certain that any modified and improved theory of the van der Waal's adsorption must arrive at approximately the same value for the monolayer as is obtained from the B.E.T. method for truly van der Waal's adsorption. It is almost needless to say that in most practical applications relative surface values are of greater importance than the absolute values themselves.

However, there is another consideration which has worried this author as early as 1937 when he and his coworkers measured the adsorption isotherm of nitrogen on nickel and iron at liquid oxygen temperature and found that the isotherm was essentially flat between 10^{-4} and 10^{-1} mm. Hg pressure. It was about this time that the Emmet-Brunauer method was first made public, and off hand the two results appeared to be irreconcilable. This uncertainty was aggravated by the fact that the nitrogen adsorption isotherm as published by the author and his coworkers could at that time not be extended to pressures higher than 0.1 mm. due to the smallness of the surface of the evaporated metal films used, the maximum surfaces obtained being of the order of 10,000 square cm. Beeck and coworkers (1) took the view that because of the flatness of the adsorption isotherm of nitrogen at liquid air temperature in the region of 10^{-4} to 10^{-1} mm. the adsorption of the monolayer was essentially complete at 10^{-1} mm., each nitrogen molecule occupying two crystallographic sites in agreement with the hydrogen adsorption, where each hydrogen atom occupies one crystallographic site, and also in agreement with the carbon monoxide adsorption in which each carbon monoxide molecule occupies a single crystallographic site. Not until several years later in the author's laboratory did Dr. Wright (8) succeed in measuring the surface of a relatively heavy evaporated iron film using nitrogen and the B.E.T. method at liquid nitrogen temperature. The surface as obtained by the B.E.T. method was found to be 1.55 times larger than was indicated by the low pressure isotherm of nitrogen, the values obtained by the latter method being in agreement with values from hydrogen and CO chemisorption at room temperature. One other experimental fact available at that time was taken as an indication that chemisorption of nitrogen at liquid nitrogen temperature may be involved in making the B.E.T. value for nitrogen adsorption on iron and nickel and other metal surfaces appear too large: It was found that at approximately room temperature nitrogen shows a strong activated adsorption on iron films covering about half the surface with a measurable rate at 100°C. When such iron film, after having been half covered with activatedly adsorbed nitrogen, was cooled to liquid nitrogen or liquid oxygen temperatures, the subsequently occurring adsorption of nitrogen was found to be appreciably lower than the initial low temperature adsorption on a clean iron film for the same gas pressure. This result indicated that the low temperature adsorption of nitrogen was measurably reduced by an underlying chemisorbed nitrogen layer, a conclusion which is supported by additional evidence as will be described later.

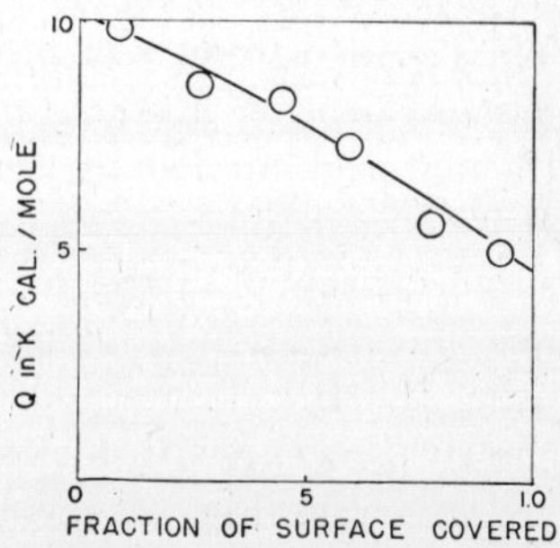

FIG. 1. The heat of adsorption on iron at −196°C. as a function of surface covered.

Later, a calorimeter was constructed with which the heats of adsorption on evaporated porous metal films could be measured directly (3). The calorimetric method and the results obtained with it will be discussed in detail in Section V. It is sufficient to state here that the heats of adsorption of nitrogen on nickel and iron at −183° were found to be 10,000 calories per mole for the sparsely covered surface, decreasing slowly to 5,000 calories for the completely covered surface. The results on iron, which are essentially identical with those on nickel, are shown in Fig. 1. The high heat of adsorption explains at once the shape of the low pressure isotherm of nitrogen on iron and nickel with their flat portion between 10^{-4} and 10^{-1} mm. pressure. Evidently the flat portion measures the monolayer in relation to the crystallographic character of the surface. Nitrogen is probably lying flat on the surface, each molecule occupying two crystallographic sites. This hypothesis is supported by the fact that the number of hydrogen atoms adsorbed is twice as large as the number of nitrogen molecules adsorbed. The forces exerted on the nitrogen molecule under these conditions may be assumed as not to be sufficient to dissociate the nitrogen molecule in contrast to that obtained at higher temperature through activated adsorption of the nitrogen molecule, which latter process leads to dissociation of the nitrogen molecule into atoms as has been verified through the fact that ammonia is formed for each adsorbed atom when the iron film carrying the activatedly adsorbed nitrogen is dissolved in hydrochloric acid (9). It is quite apparent that the Emmett-Brunauer method does not take cognizance of a previously chemisorbed adsorption layer and thus measures the monolayer too large by a factor depending on the heat of adsorption and its dependence on coverage in the first adsorption layer. This shows that any complete theory of van der Waal's adsorption will have to take into account the specificity of the first adsorption layer as influenced not only by the chemical nature of the material but also by the crystallographic characteristics of the surface. It remained to be shown experimentally that the flat part of the low pressure isotherm of nitrogen at liquid oxygen or liquid air temperature would rise sharply again at higher pressures as demanded by the results obtained using the B.E.T. method on the same surfaces. This rise was recently shown to occur with nickel, iron, and tungsten surfaces, and the results are given in Fig. 2, including the hydrogen isotherms on nickel and tungsten at 23°C. for comparison. It is this low pressure adsorption step which represents adsorption of the monolayer, the observation of which by Emmett and Brunauer was impossible under their experimental conditions. Also shown in the figure are the nitrogen isotherms on a hydrogen covered tungsten surface and on a tungsten surface on which nitrogen was chemisorbed at higher temperature* prior to the low temperature measurements. It is seen that prechemisorbed hydrogen and nitrogen influence the low temperature adsorption of nitrogen markedly, eliminating entirely the low-pressure, flat portion of the nitrogen isotherm in the case of hydrogen and lowering the step drastically in the case of preadsorbed nitrogen. It is unfortunate that these isotherms could not be extended to higher pressures for evaluation by the B.E.T. method.

It is apparent from the foregoing that the B.E.T. method will give values for the monolayer which are too large when applied to clean metal surfaces using a gas whose heat of adsorption is large relative to that in ordinary van der Waal's adsorption. With a heat of adsorption of 10,000 calories on iron, nitrogen is unsuitable for surface measurements of such metals by the B.E.T. method when absolute best values are of interest.

* The tungsten film was left in contact with nitrogen over night during which time the liquid nitrogen in the cooling Dewar had evaporated, warming the film to an undetermined temperature.

At present, the occurrence of stepwise adsorption in the low pressure region is the best criterion for caution in the application of the B.E.T. method. Since it is highly probable that less distinct steps or even several steps (heterogeneous systems) will occur in other systems from which it may be difficult to discern the monolayer, one is forced to the conclusion that final judgment of the accuracy of the B.E.T. method

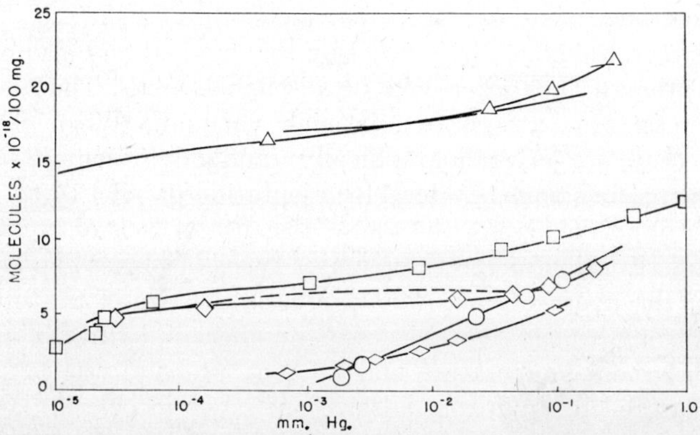

FIG. 2. Nitrogen adsorption isotherms at −196°C. and hydrogen adsorption isotherms at 23°C.

N_2 at −196°C.

— △ — △ — Tungsten
— □ — □ — Iron
— ◊ — ◊ — Nickel
— ○ — ○ — Tungsten (hydrogen preadsorbed)
— <> — <> — Tungsten (nitrogen preadsorbed with high temperature chemisorption)

H_2 at 23°C.

———————— Tungsten
— — — — — — Nickel

will depend on the agreement between the B.E.T. method and more adequate treatment from the theoretical standpoint which is at present still lacking. It is of interest to note that while the use of nitrogen with its initial heat of adsorption of 10,000 calories for iron or nickel gives a value about 50% too high, the use of ethylene at liquid oxygen temperature with an initial heat of adsorption of about 20,000 calories at this temperature gives a surface area by the B.E.T. method which is 100% too high. This clearly shows that in the case of ethylene a complete monolayer is adsorbed in the low pressure region and that this monolayer, plus the first van der Waal's adsorbed monolayer above it, is actually measured by the B.E.T. method when applied as usual in the medium pressure range.

It may be estimated that a gas suitable for the B.E.T. method should have a heat of adsorption at the temperature at which the measurements are made of preferably less than 5,000 calories and at best not more than 3,000 calories per mole. Noble gases and the lower saturated hydrocarbons appear to be suitable in this respect. In the author's laboratory, B.E.T. values have been obtained for krypton, methane, and butane on evaporated porous nickel films weighing up to a gram and with surface areas of 7 to 8 sq. meters. The surface of a great number of metal films was measured by this method and also by hydrogen adsorption. The latter will be discussed first. The data given refer to metal films evaporated in high vacuum. When these films are thin, they are crystallographically randomly oriented, but they show an increasing tendency for 110 orientation parallel to the backing with increasing thickness of the film. Because of lack of knowledge of the exact distribution of the three major types of planes exposed to the gas phase, the average plane was chosen as the size of the sites. The size of site multiplied by twice the number of hydrogen molecules adsorbed at room temperature and 0.1 mm. pressure was taken as the surface measured by hydrogen adsorption. As will be seen later on, it would have been perhaps more accurate to use the hydrogen adsorption at liquid nitrogen temperature. However, the two values are so closely the same for the type of film used that in view of the uncertainty of the size of the crystallographic sites no serious error can arise from using the room temperature adsorption value.

The B.E.T. surface areas were obtained by constructing master isotherms from several individual isotherms obtained from each of the evaporated nickel films. From these master isotherms B.E.T. plots were obtained and are shown in Fig. 3. Both the krypton and the methane values were obtained from isotherms at liquid nitrogen temperature, while the butane isotherms were obtained at $-78°$. The B.E.T. areas obtained with krypton, methane, and butane, respectively, are shown in Table I, where they are compared with surface areas obtained from hydrogen adsorption as described above. The same values were obtained for the adsorption of these gases when hydrogen was preadsorbed on the nickel films, indicating that the adsorption of krypton is not affected by the condition of the substrate. Also given in the table are the areas per molecule used in the B.E.T. method and the area per hydrogen atom site as calculated from the average plane in nickel. It was already pointed out that heavy films as had to be used in these experiments have a tendency to orient with their 110 plane parallel to the backing while being deposited. If one were to assume, therefore, that more 110 planes are exposed to the gas than of the other types, the agreement would be even better. Thus, it appears to be well established that the B.E.T.

method using krypton, methane, or butane gives values which can serve as reliable basis for surface measurements of the available surface of

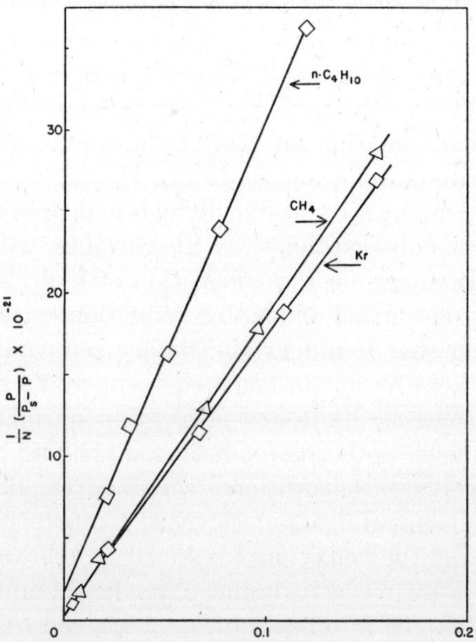

FIG. 3. B.E.T. plots from Kr and CH_4 isotherms at $-196°C$. and from C_4H_{10} isotherms at $-78°C$.

clean metal catalysts such as iron and nickel on which the heat of adsorption of nitrogen at liquid nitrogen temperature is too high to give reliable values.

TABLE I

B.E.T. Areas of Nickel Films in Comparison with Areas Measured by Hydrogen Adsorption

Gas	Uni-layer molecules $\times 10^{-18}$ per 100 m g.	Surface area sq. meters per g.	Area per H atom site* A^2	Area per H atom site A^2 (crystallographic)
Kr	5.87	8.38	6.07	5.87
CH_4	5.40	8.46	6.14	5.87
C_4H_{10}	3.48	8.53	6.18	5.87

* The area per hydrogen atom site is calculated from the surface area in column three by dividing this number by the number of hydrogen atoms adsorbed per gram of adsorbent.

IV. Hydrogen Adsorption Isobars and the Effect of Sintering

It is shown in this section with the help of hydrogen sorption isobars on evaporated metal films sintered at various temperatures that the sorption process consists of absorption into the interior of the structure as well as of adsorption on the surface. When properly evaluated, perfect correlation is obtained between hydrogen adsorption, catalytic activity, surface measurements by the B.E.T. method, and CO adsorption. Hydrogen adsorption is always fast even at the lowest temperatures. The absorption into the interior of the structure is slow and of the "activated" type.

Beeck *et al.* (1), in their work on the influence of crystal structure on the rate of hydrogenation on evaporated metal films, observed little difference in the amount of adsorption of hydrogen on nickel at liquid air and at room temperature on films which had been deposited at room temperature. This observation was at variance with all other work reported in the literature on the adsorption of hydrogen on nickel catalysts prepared from nickel oxide by reduction with hydrogen, where greater adsorption was found at the higher temperatures. Moreover, Beeck and coworkers found, in contrast to previous studies on reduced oxide catalysts, that the hydrogen adsorption on nickel film was practically instantaneous both at room and at liquid air temperatures, whereas slow "activated adsorption" was reported in the literature for typical hydrogenation catalysts at elevated temperatures. The striking discrepancy between the behavior of reduced nickel oxide and evaporated nickel films has more recently found a ready explanation through the work of Beeck *et al.* (10). These authors obtained hydrogen adsorption isobars at 0.1 mm. pressure on evaporated nickel films over the temperature range of $-196°$ to 400°C., with the general result that the shape of the isobar is dependent on the temperature at which the film has previously been sintered. As shown in Fig. 4, films sintered at 23°C. gave relatively flat isobars, whereas films sintered at 200 and 400°C. gave isobars with pronounced sorption maxima at intermediate temperatures. No maxima were obtained with decreasing temperature. Equilibrium was reached slowly when proceeding toward the maximum on the ascending branch (rising temperatures) but was reached very fast on the descending branch. The peculiar shape of the isobar obtained for the film sintered at 400° calls for further explanations. This film was sintered at 400° previous to the adsorption experiment and obviously did sinter still further when again heated to 400° in the hydrogen atmosphere. This additional sintering in hydrogen is probably linked to the formation of a new phase of hexagonally close packed nickel. Formation of this phase of nickel has previously been reported in the metallurgical literature as taking place only in the presence of hydrogen, probably through the intermediate formation of hexagonal nickel hydride. It is likely that the change from cubic face-centered nickel to hexagonal close-packed nickel is accompanied by additional sintering—that is, the formation of larger crystallites.

It is obvious that the isobars shown in Fig. 4 cannot be explained by adsorption. In the first place, the difference between the adsorption at $-196°$ as obtained initially and the total sorption found at $-196°$, after obtaining the isobars, decreases little with increased temperature of sintering. Films sintered at 23° have about ten times the initial adsorption at $-196°$C. of films sintered at 200° and about 175 times the

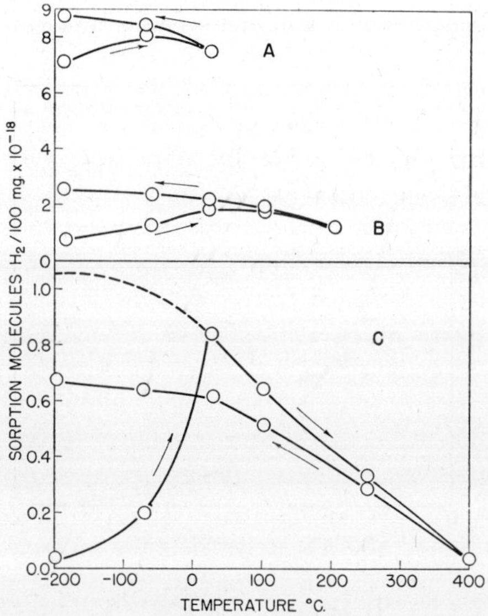

FIG. 4. Sorption isobars of hydrogen ($-196°$C. to temperature of presintering, up and down) at 0.1 mm. Hg pressure on evaporated nickel films presintered at various temperatures.

A: Presintered at 23°C.

B: Presintered at 200°C.

C: Presintered at 400°C. (Although this film was presintered in high vacuum at 400°C. for ½ hour, additional sintering apparently occurred on reheating in the presence of hydrogen. The dotted line indicates the curve to be expected with decreasing temperature if no additional sintering had occurred.)

initial adsorption of films sintered at 400°. As can be seen from the figure, the slow absorption is dependent on the weight of the film rather than on the surface available for fast initial adsorption at $-196°$C. This was also borne out by comparing isobars for metal films evaporated in vacuum and in 1 mm. of nitrogen. Beeck *et al.* (1) have reported that films produced in 1 mm. nitrogen were oriented and had a surface for fast adsorption twice as large as nonoriented high vacuum films of the same film weight. Both films also show a small amount of slow sorption. More recently, however, the small slow sorption was found to be equal for both films, again indicating a weight effect rather than a relation to the surface. In the second place, it was shown by the same authors that for oriented and nonoriented films sintered at higher temperatures the CO adsorption at room temperature decreased proportionally to the activity of both types of films for hydrogenation of ethylene. Experiments similar to those discussed in the previous section, have shown that the surface of films sintered at various temperatures and measured by the B.E.T. method using van der Waal's adsorption of krypton at

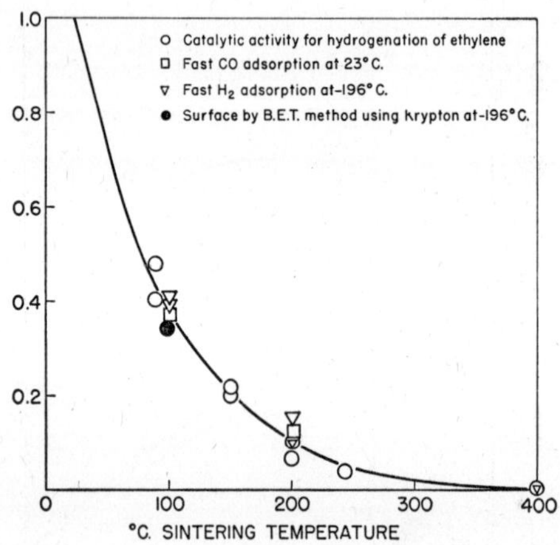

FIG. 5. Correlation of catalytic activity with adsorption and the surface area. (The ordinate represents fraction of respective values obtained for films sintered at 23°C.)

−196°C. also decreases proportionally to the decrease in activity for hydrogenation of ethylene at room temperature. Thus both the relative chemisorption of CO at room temperature and the van der Waal's adsorption of krypton are reliable measurements for the surface available for hydrogenation of ethylene. The same is true for the fast adsorption of hydrogen at −196°C., so that the latter presents a third criterion by which it is possible to determine the catalytically active surface of sintered films.

Figure 5 shows data for catalytic activity, CO adsorption at 23°, surface area by the B.E.T. method using krypton at −196°, and the fast hydrogen adsorption at −196° plotted against the temperature at which the various films were sintered. All quantities were taken as unity for films sintered at 23°C. These experiments clearly indicate that the previously observed "slow adsorption" of hydrogen on nickel catalysts is not *ad*sorption but is *sorption* consisting of adsorption and of *ab*sorption of hydrogen, the latter into positions inaccessible to CO, krypton, and the fast hydrogen adsorption at −196°C. and also inaccessible to ethylene.

In order to study the effect of absorption in the nickel-hydrogen system in more detail, Beeck *et al.* (11) have investigated the hydrogen sorption isobars between 20°K. and room temperature. As shown in Fig. 6, the solid curves represent the isobars for increasing and decreasing temperature. With increasing temperature (the part between 20 and 80°K. will be discussed later), sorption increases fast between 80 and

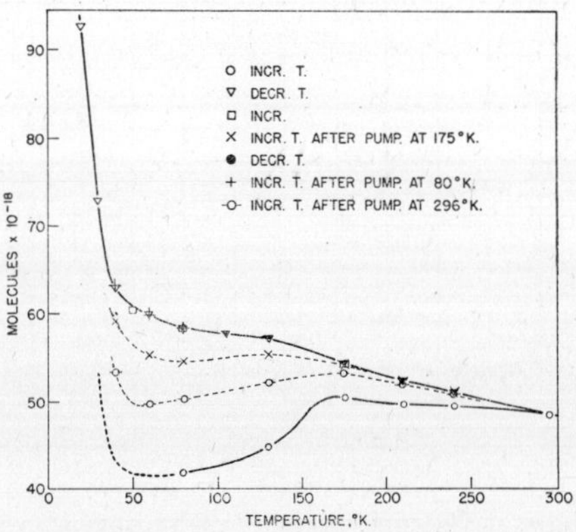

FIG. 6. Sorption isobars at 0.1 mm. pressure of hydrogen on an evaporated nickel film sintered at 23°C. (To obtain the ordinate in molecules $\times 10^{-18}$/100 mg., divide by 9.07.)

170°K. While initial sorption at each point, after raising the temperature, was fairly rapid, it was followed by a very slow sorption, and true equilibrium values were not obtained. It is possible, therefore, that the maximum at about 170°K. would have shifted to lower temperatures if long enough waiting periods had been used. In the descending branch of the isobar true equilibrium was obtained very rapidly and, as indicated by the squares, this isobar could be traced exactly when again increasing the temperature, without any intermediate outgassing of the system. When the system was pumped to high vacuum for 15 minutes at 80°K., the subsequently measured isobar starting at 20°K. with increasing temperature exactly retraced the isobar obtained initially on the branch of decreasing temperatures. After the system was pumped to high vacuum for 15 minutes at 175°K., the isobar for increasing temperature shows a pronounced minimum and maximum, the amount taken up at the 80°K. minimum agreeing well with the amount which was taken up in the first isobar when 175°K. had been reached on the branch of ascending temperatures. After pumping the system to high vacuum for 15 minutes at 296°K., the subsequently measured isobar for increasing temperature (starting again at 20°K.) shows at its minimum value an uptake practically identical with the uptake found at 296°K. on the ascending branch of the very first isobar. In the three experiments just mentioned, identity of the respective isobars at the temperature of pumping is assumed. These facts are of interest because they show that the amount of hydrogen which can be removed

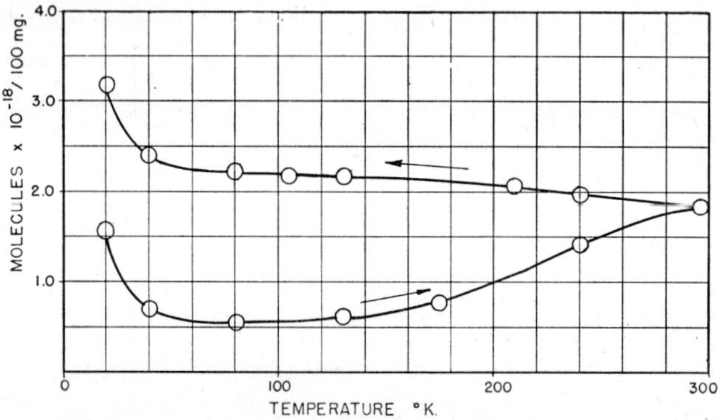

FIG. 7. Sorption isobars at 0.1 mm. pressure of hydrogen on an evaporated nickel film sintered at 200°C.

from the system by pumping to high vacuum for 15 minutes at 175°K. and 296°K. is being resorbed almost completely in the lower temperature region between 20 and 80°K. More important, however, is the fact that the rise of the initial isobar with decreasing temperature down to about 80°K. must also partly be due to *ab*sorption. Further proof of this will be given in the next paragraph.

Figure 7 shows the ascending and descending branch of a hydrogen isobar at 0.1 mm. pressure on a nickel film previously sintered at 200°C. If the 80°K. point of the ascending isobar in this figure is compared with the 80°K. point of the ascending isobar in Fig. 6 (after converting the ordinate in Fig. 6 to molecules $\times$ 10^{-18}/100 mg. of nickel film as indicated below the figure), it is noted that the initial hydrogen adsorption at 80°K. is nine times larger per weight unit for the film sintered at 23° than for the film sintered at 200°C. At the same time it is seen that the difference of hydrogen sorption between the descending and ascending branch of the isobar at 80°K. is the same for both types of films and, thus evidently a weight effect. Furthermore, the ration of total sorption at 80°K. to that at room temperature on the decreasing branch of the two isobars is the same for both types of film, indicating that the relative decrease of total sorption with temperature is independent of surface. This shows clearly that the major portion of the hydrogen sorption of a film sintered at 200° is absorption into the interior of the structure. It is interesting to note that the total van der Waal's adsorption as given by the differences of sorption between 20° and 80°K. is practically the same for the ascending and the descending part of the

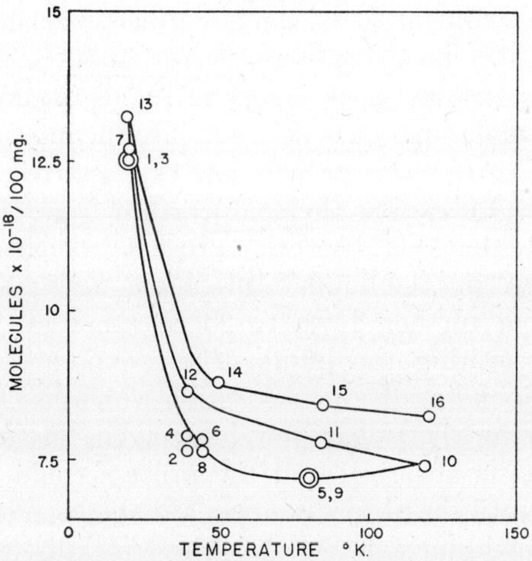

FIG. 8. Sorption isobars at 0.1 mm. pressure of hydrogen on an evaporated nickel film sintered at 23°C.

isobar. Furthermore, the numerical value for the van der Waal's adsorption for the film sintered at 200°C. is seen to be approximately 4 times lower than the van der Waal's adsorption observed on a film sintered at 23°C. as shown by Figs. 6 and 7. The reason that the ratio of the van der Waal's adsorption in the two types of film is not equal to the ratio of the surfaces as measured by the adsorption at 80°K. in the ascending part of the isobars of the two types of film will become clear after the following discussion of Fig. 8, which represents an isobar on the same type of film in the region between 20° and 120°K. All points of this isobar with increasing or decreasing temperature are numbered in the sequence in which they were measured. It is seen that, starting with point No. 1 at 20 and proceeding to 50°K., the isobar can be traced back to point No. 3 at 20°K., which coincides with point No. 1. Ascending again with the temperature, point No. 4 falls on the same curve and, after obtaining point No. 5, the isobar can again be traced back and forth to point No. 9. Since the isobar flattens out completely at 80° with the value of 7.2×10^{18} molecules per 100 mg., this means that the total adsorption of hydrogen at 20°K. consists of 7.2×10^{18} molecules per 100 mg. Ni of chemisorption and of 6.0×10^{18} molecules per 100 mg. Ni of van der Waal's adsorption. As will be shown in Section V, the heat of adsorption of hydrogen at liquid air temperature was found to be 30,000 calories per mole for the sparsely covered surface and decreases to 18,000 calories for the completely covered surface, in complete agreement with the values obtained for the heat of adsorption at room temperature. Since the isobar starting at 20°K. can be retraced back and forth to 80°K., it is evident that the chemisorption observed at 80°K. must already have occurred at 20°K. Tracing the isobar in Fig. 8 further to point No. 10, it is seen that a small amount of sorption into the interior of the structure has taken place. After this process has been started by supplying the activation energy by raising the temperature from 80 to 120°K., it is interesting to note that this absorption process will continue, even increasingly so, upon lowering the temperature again as shown by point No. 11. Tracing the isobar further down to 20°K. the total sorption value of point No. 13 is (within the limits of the experimental error) higher by a value corresponding to the absorption which has taken place in the meantime. Tracing the isobar now back again through points 14, 15, and 16, more gas is absorbed and the process could presumably have been continued until the absorption would have been identical with that of the descending branch in Fig. 6. These measurements show clearly that after the absorption process has once been started by raising the temperature from 80 to 120°K.* it will continue even at lower temperature. There is no doubt, therefore, that this process is an exothermic type of absorption or solution, possibly the formation of a hydride. In comparing Figures 8 and 7 it is seen that the van der Waal's adsorption at 20°K. of an unsintered film as measured by the difference of the ordinates of points 1 and 5 in Figure 8 is five times as large as that of the sintered film in Figure 7 (difference between adsorption at 20°K. and 80°K.), while the chemisorption as measured by the fast hydrogen chemisorption at 80°K. is thirteen times larger for the unsintered film than for the sintered film. This discrepancy may possibly find its

* Several repetitions of this experiment have shown that absorption was sometimes initiated at temperatures as low as 70 and 80°K. if more time is allowed. The increase observed with lower temperatures seems to be due to the higher surface concentration of molecules in the van der Waal's adsorption layer. Once the formation of the "hydride phase" was initiated at higher temperatures the growth of this new phase, even at temperatures below the initiation temperature, is probably due to the assistance of nuclei of the new phase which have opened up the passage into the interior.

explanation in increased capillary condensation of the sintered film. Further studies at very low temperatures are necessary to clarify this point.

It is of interest to note that in the measurement of the hydrogen absorption on the ascending isobar of Fig. 6, a sudden desorption of hydrogen took place, after raising the temperature, followed by a slow

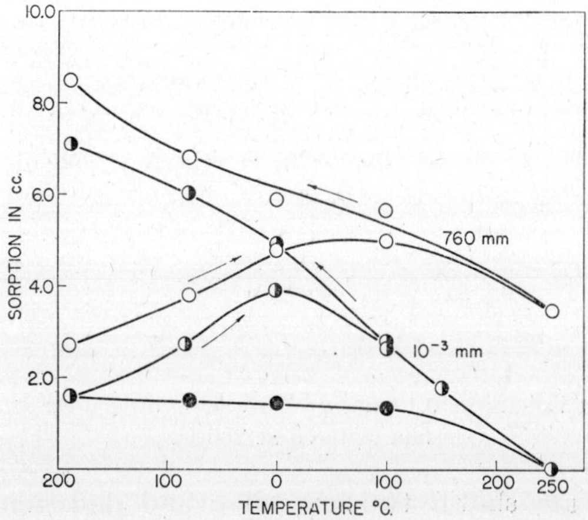

FIG. 9. Sorption isotherms on reduced nickel catalyst plotted from the data of Maxted and Hassid. (Catalyst was degassed at 250°C. before making measurements.) ● H_2 adsorption at 10^{-3} mm. when catalyst was contacted with H_2 at −190°C. and then evacuated prior to raising the temperature.

sorption that exceeded the initial desorption. This phenomenon is identical to that reported by Taylor and Chou-Shou Liang (12) and earlier by Frankenburger and Messner (12a) and suggests that their experiments may deserve re-examination in the light of the findings on nickel, inasmuch as it is possible that Taylor and Chou-Shou Liang may have interpreted an activated *ab*sorption process as activated *ad*sorption in the absence of further proof. The experiments of Taylor and Chou-Shou Liang will be discussed further on p. 192 of this article.

Before leaving the nickel experiments, it may be well to refer to the experiments on hydrogen adsorption variously reported in the literature. As an example, the work of Maxted and Hassid (13) had as its main objective the measurement of the slow "activated adsorption" of hydrogen on reduced nickel oxide catalysts. It has been proved by the foregoing that the slow *ad*sorption is actually *ab*sorption. When plotting their data as isobars, as was done in Fig. 9, the similarity between these isobars and those obtained with sintered nickel films is evident.

A few measurements have been carried out on iron, and Fig. 10 shows an ascending and descending isobar at 0.1 mm. pressure (analogous to that shown for nickel in Fig. 4) for an iron film sintered at 200°. The behavior is clearly very similar to that of nickel.

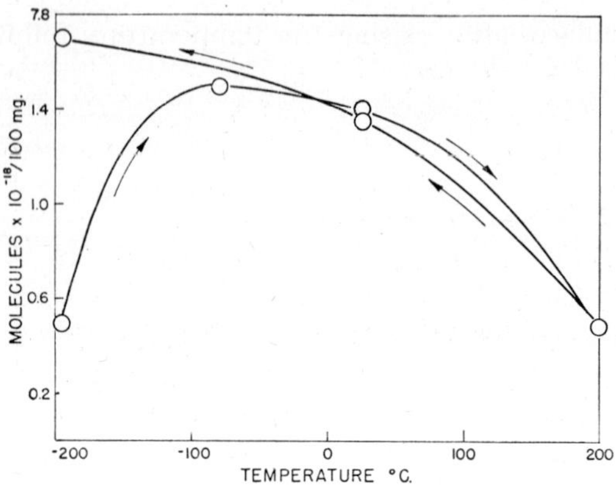

FIG. 10. Sorption isobars at 0.1 mm. pressure of hydrogen on an evaporated iron film sintered at 200°C.

In Fig. 11 are shown two ascending and descending isobars, at 0.1 mm. pressure for evaporated tungsten films sintered at 23° and at 500°. While the shape of these isobars is different from those on nickel and iron, evidence is obtained that for the film sintered at 23°, absorption

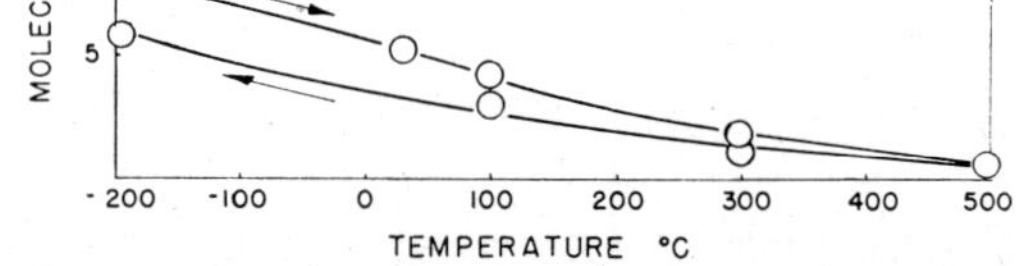

FIG. 11. Sorption isobars at 0.1 mm. pressure of hydrogen on an evaporated tungsten film sintered at 23°C. (upper curves) and at 500°C. (lower curves).

into the interior of the structure has taken place, while the film sintered at 500° shows an effect analogous to that shown for nickel in Fig. 4, in which the presence of hydrogen causes additional sintering after the temperature of 400°C. had been reached so that the descending isobar falls below the ascending part of the isobar. It is also seen that the effect of sintering and consequently the absorption is relatively small as compared to that found for nickel and iron.*

From the nature of the isobars, especially of nickel and iron, it is clear that the *ab*sorption process must be exothermic and that, like adsorption, it is reversible at higher temperature. The heat of absorption must be considerably lower than the heat of adsorption, inasmuch as the absorption process, which is definitely of the activated type with a measurable temperature dependent rate, is more readily reversible at high temperatures than the adsorption process. It will be shown in Section V that this is also verified by heats of sorption measurements on sintered films. An interesting side light on the nature of the *ab*sorption process was obtained through the measurement of sorption isobars on palladium. The adsorption on palladium at liquid nitrogen temperature was shown (through rates of hydrogenation measurements and by CO adsorption) to be comparable to the adsorption on the other metal films investigated, although the initial fast hydrogen adsorption is always followed by a slow sorption which seems to continue indefinitely. When the temperature is raised, the uptake of hydrogen increases in the neighborhood of −100°C. to an amount which is approximately equal to the absorption of one hydrogen atom per atom of palladium. This absorption decreases again rapidly when increasing the temperature to room temperature and more slowly when increasing the temperature further to 100°C., at which point the sorption again is low (about four times larger than that obtained at −196°C.). The 1/1 ratio of hydrogen atoms to palladium leaves no doubt that hydride formation is involved, and reports in the literature (14) have indeed shown that low temperature solution of hydrogen in palladium leads first to a hydride of cubic structure and that raising of the temperature causes this hydride to lose hydrogen with the formation of a hexagonal hydride with a palladium-hydrogen ratio of 3/1. These reports show also that upon raising the temperature still higher the hydride decomposes again.

V. The Heat of Adsorption of Hydrogen

In this section a method for the direct calorimetric determination of heats of adsorption on evaporated metal films is described and results for the heats of adsorption of hydrogen on nickel, iron, and tungsten are reported. In all cases the heats of adsorption decrease with the fraction of surface covered in a mode that can satisfactorily be explained by interaction of adsorbed atoms. A criterion for mobility of the adsorbed atoms is developed

* According to the literature, the solubility of hydrogen in tungsten is very low as compared to the solubility in nickel or iron.

which together with experiments on the adsorption of hydrogen on surfaces partially covered with other adsorbed gases gives evidence for the uniform character of these surfaces.

It has been shown in Section III that krypton isotherms at liquid nitrogen temperature give correct values for the surface of evaporated metal films when evaluated by the B.E.T. method. It has also been shown that these values are in excellent agreement with those obtained from the flat portion of low pressure isotherms of nitrogen at −196°C. and from hydrogen or carbon monoxide adsorption isotherms at room temperature, although in the case of hydrogen the hydrogen adsorption at liquid nitrogen temperature with certain corrections would have been preferable, as has been pointed out in Section III. Since the last-mentioned fact was not known until recently, the heats of adsorptions reported in this section have been referred to the hydrogen adsorption at 23° with the assumption that the adsorption at 0.1 mm. pressure approximately represented the monolayer. The low temperature measurements of the heat of adsorption were made at −183° and also for these measurements coverage of the surface at 0.1 mm. at room temperature was taken as the reference point for 100% coverage.

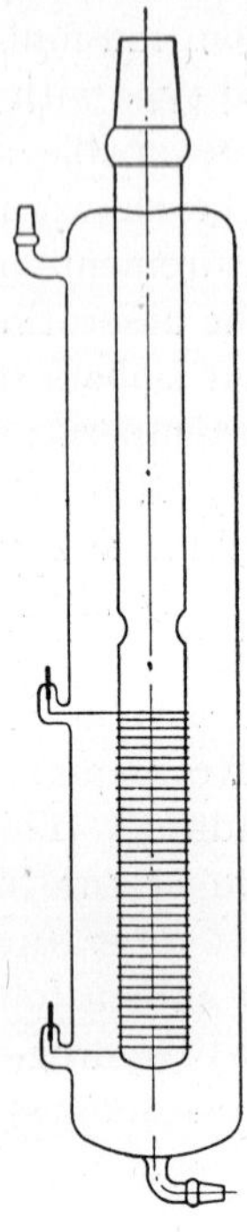

FIG. 12. Schematic sketch of heats of adsorption calorimeter.

The method of making the heats of adsorption measurements has recently been described by Beeck *et al.* (3). The calorimeter employed is shown in Fig. 12. The jacket around the inner glass tube was used for cooling water during evaporation of the metal film onto the inner surface of the inner glass tube. The jacket was evacuated before the calorimetric measurements were started, thus turning the device into a vacuum calorimeter. The inner glass tube was etched down to a very small wall thickness, thereby reducing the heat capacity to a minimum. In the upper part of this tube the glass walls were thinned by blowing and drawing so as to keep conduction through the end of the tube to a minimum. Temperature measurements were made by means of a platinum resistance thermometer. The resistance wire was wound onto the outer side of the calorimeter tube proper, in a single layer, thus measuring the average temperature change over the total part of the tube which is covered by the evaporated metal film on the inside. Typical cooling curves are shown in Fig. 13, the upper curve being an example of fast hydrogen adsorption on nickel, which offers no difficulties for accurate evaluation, and the lower one being typical of a slow activated adsorption of nitrogen on iron for the proper evaluation of which the reader must be referred to the original article. With the heat capacity carefully predetermined, the properly extrapolated deflection of the ballistic galvanometer at the instance of admitting the gas will give the heat input directly. This can then be related to the number of molecules adsorbed. The experimental procedure was to admit the gas in

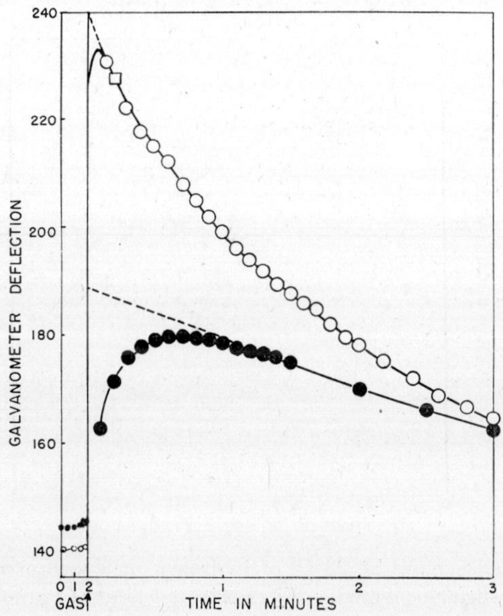

FIG. 13. Typical calorimeter cooling curves.

small increments and to obtain average values of the heat of adsorption for these increments. If the increments are not large, the curve through the heats of adsorption obtained from the various increments will then closely represent the differential heat of adsorption on the surface in terms of molecules or atoms adsorbed or in terms of fraction of surface covered in case total coverage has been defined. The heats of adsorption presented were usually obtained from measurements on several separate films.

Some of the results of these investigations as far as they seem to appear pertinent from the standpoint of the purpose of this article will now be presented.

1. Nickel

In Fig. 14 are presented the heats of adsorption of hydrogen on nickel for oriented as well as nonoriented evaporated films. It may be recalled that the oriented films were five times as active catalytically for the hydrogenation of ethylene, leading to the conclusion that the 110 plane is preferentially exposed to the gas phase. It is of particular interest, therefore, to note that the heats of adsorption of hydrogen of oriented and of nonoriented films are the same. Mention may be made at this point that recent surface measurements using the B.E.T. method with krypton at liquid nitrogen temperature have shown that this modified B.E.T. method gives about 20% larger values for oriented films as

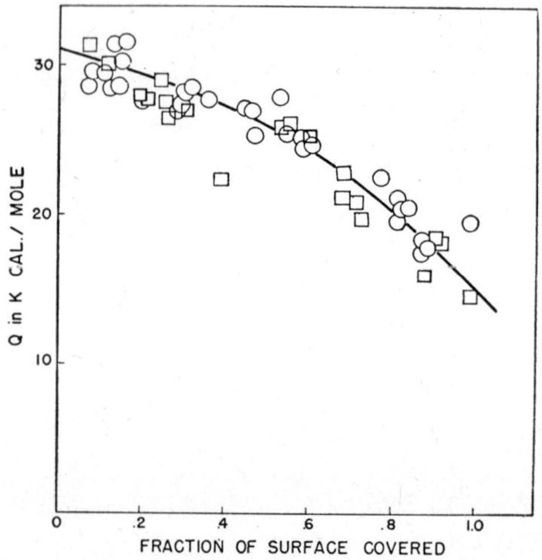

FIG. 14. Heat of adsorption at 23°C. of hydrogen on evaporated nickel films as a function of surface covered. (□ unoriented films; ○ oriented films.)

compared to hydrogen adsorption than for unoriented films. This would be expected since the individual sites of the 110 plane are the largest of the three major planes and the 100 and 111 planes would therefore adsorb less hydrogen per square centimeter than the other planes. It is hoped to make more precise measurements of this type using the hydrogen adsorption at liquid nitrogen temperature instead of at room temperature. The fact that the differential heats of adsorption of hydrogen are the same for both oriented and nonoriented films shows that heterogeneity with regard to crystallographic sites is without effect on the heat of adsorption of hydrogen, although the hydrogenation velocity of an oriented film is fivefold greater than that of an unoriented film. Equal heats of adsorption for hydrogen on oriented and nonoriented films would imply that the activation energy for the hydrogenation should be the same. This, in fact, is the case, and differences in rate of reaction will have to be ascribed solely to a difference in entropy of activation (for a more detailed discussion of entropy of activation and the dependence of rate of activation on crystal parameter, see (2)).

Mention has been made variously that the rate of adsorption of hydrogen on nickel and iron is very fast at room temperature. In fact, the rate is so fast that the limiting factor appears to be the rate at which gases enter the reaction chamber through the stopcock from the reservoir. This essentially instantaneous adsorption has been observed for many other metals, including platinum, rhodium, palladium, tungsten, tanta-

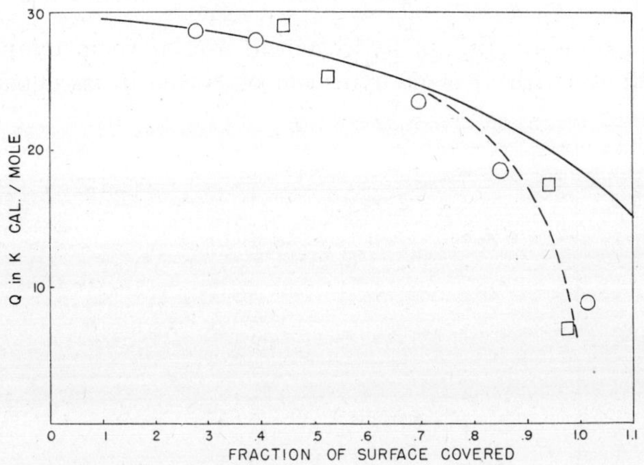

FIG. 15. Heats of adsorption of hydrogen at $-183°C.$ on evaporated nickel films (points) in comparison with the heat of adsorption at 23°C. (curve) as a function of surface covered.

lum, chromium, and others. As has been shown in Section IV, for nickel, iron, and tungsten slow adsorption of hydrogen is probably not taking place on any of these metals, and if a slow rate of disappearance of hydrogen is observed, it is *ab*sorption into the structure itself and not *ad*sorption. Since the rate of adsorption of hydrogen is essentially instantaneous, indicating a very low activation energy for adsorption, it must be expected that the heat of adsorption at lower temperature would have the same high value as at room temperature. That this is the case is shown in Fig. 15, in which the solid curve represents the heat of adsorption of hydrogen at room temperature and its dependence on surface coverage, and the points represent the heat of adsorption of hydrogen at liquid air temperature for both oriented and nonoriented films. It is of particular interest to note that in the case of hydrogen adsorption on nickel at $-183°C.$ the heat of adsorption falls to very low values for the last fractions of surface covered. This is undoubtedly due to the fact that hydrogen is not sorbed into the interior of the structure at the low temperature, whereas at room temperature increasingly more hydrogen is taken up with increased pressure. When, after the last heat of adsorption measurement at 0.1 mm. at room temperature, the pressure is, for instance, suddenly raised to 2 mm., an additional amount of hydrogen approximating 20% of that adsorbed a 0.1 mm. pressure is sorbed relatively fast with a heat of adsorption of about 10,000 calories per mole.

2. Iron

The heat of adsorption of hydrogen on iron at room temperature is almost identical with that of hydrogen on nickel in its dependence of

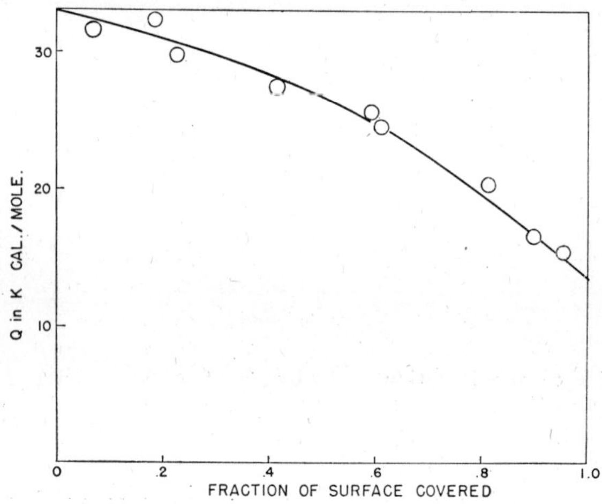

FIG. 16. Heats of adsorption of hydrogen on evaporated iron films at 23°C. as a function of surface covered.

surface coverage as seen in Fig. 16. It should also be mentioned that the heats of adsorption of deuterium on both nickel and iron are the same as for hydrogen within the limits of experimental error.

The heat of adsorption of hydrogen on iron at −183° is shown in Fig. 17, where it is compared with the heat of adsorption of hydrogen on iron at room temperature as represented by the solid curve. It is seen that the heat of adsorption of hydrogen on iron at −183° stays essentially constant until the surface is completely covered, at which point it drops to very low values. This gives a very important clue with regard to the mobility of the adatoms on the surface—that is, their ability to migrate from site to site.

3. *Criterion for Mobility*

As was already mentioned in the introduction, we have to assume that adatoms of hydrogen adsorbed on a metal surface with a high heat of adsorption must be mobile so that they either can find the sites of highest surface energy or can move as far apart from each other as possible if the heterogeneity of surface energy of the sites is due to interaction of adsorbed atoms on neighboring sites. A rapid redistribution of adsorbed atoms by evaporation and readsorption is out of the question

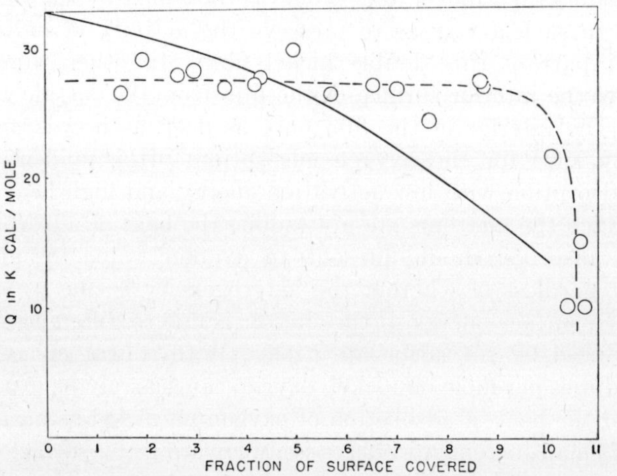

FIG. 17. Heats of adsorption of hydrogen on evaporated iron film at −183°C. (points) in comparison with the heats obtained at 23°C. (curve) as a function of surface covered.

because of the high heats of adsorption even for those atoms which are adsorbed on a nearly completely covered surface. This is shown by the fact that not even at room temperature will desorption take place to a measurable extent within 15 minutes or half an hour under high vacuum except for about 20% of coverage. This slow rate is, of course, expected since the activation energy for desorption must at least be as large as the heat of adsorption. It cannot be expected, therefore, that without mobility of adatoms of hydrogen, positions of highest surface energy will be preferentially occupied. If the atoms were not mobile, the adsorption process with its immeasurably low heat of activation would demand that every hydrogen molecule would stay on the two sites on which the two hydrogen atoms were originally adsorbed irrespective to variations in the surface energy of the neighboring sites. In terms of our porous metal films this would mean that the molecules entering the interior of the film through pores from the outside would gradually fill up the internal surface from the outside; that is, the first small increment of molecules would cover a thin outermost section of the film, the next increment would cover an additional section, and so on until the total internal surface is completely covered with adsorbed molecules throughout the thickness of the film. This type of adsorption was already reported for oxygen on evaporated nickel films (1), where it was shown that a so-called double film—that is, a film consisting of two layers of different catalytic activity, such as an oriented nickel film on top of a nonoriented nickel film—could be poisoned by the right amount of oxygen in such a way as to preserve the activity of only the lower unoriented part of this double nickel film. In other words, oxygen admitted to the interior surface of the film from the outside would penetrate into the interior of the film only as deep as it was necessary to find empty sites for the oxygen adsorption. It is evident, therefore, that for adsorption with low activation energy and high heat of adsorption for which the adatoms are not mobile the heat of adsorption for an increment adsorbed should present an average value for all the sites occupied regardless of whether the heterogeneity is due to an intrinsic difference in surface energy of the sites or is due to differences in surface energy induced by the adsorbate itself. With a heat of adsorption of 120,000 calories per mole on nickel, oxygen can scarcely be expected to be mobile and the heat of adsorption of oxygen on nickel as measured by a number of small increments successively covering the surface was found to be constant until the film was completely covered, after which it was found to drop to very low values. Since the same behavior is found for hydrogen on iron at $-183°$, one is led to the conclusion that hydrogen atoms are not mobile on iron at this temperature and that the film is successively covered from the outside, each increment of gas adding a layer until the film is completely covered. As is seen from Fig. 17, the value for the heat of adsorption of hydrogen on iron at $-183°$ is indeed an average value of the heats of adsorption found at room temperature. The low temperature measurements are also eminently suited to give credence for the value chosen for 100% coverage, insamuch as the values stay essentially constant over the whole surface, finally dropping very abruptly to very low values near 100% coverage of the surface.

4. Tungsten

Figure 18 shows the heat of adsorption of hydrogen on tungsten at 23° in its dependence on surface coverage. The data plotted are those from five different experiments using small increments in each case. Figure 19 shows the same values as represented by the solid curve in comparison with the values obtained by Roberts (4) on a single tungsten wire and with the values obtained by Frankenburg (17) as represented by the dotted line. The remarkable thing about these three measurements is that the surface used by Beeck and coworkers is

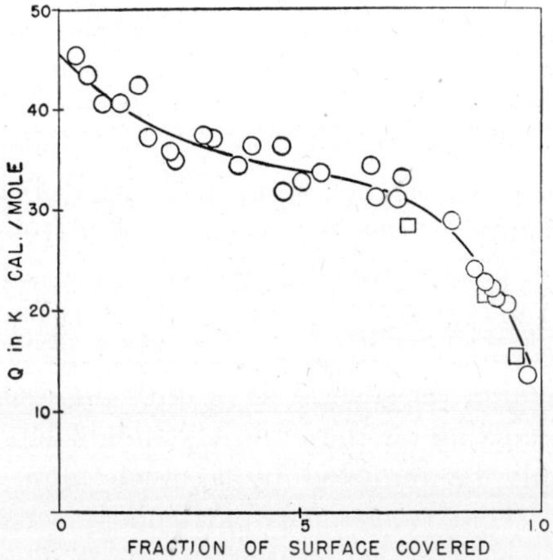

FIG. 18. Heats of adsorption of hydrogen on evaporated tungsten films at 23°C. as a function of surface covered.

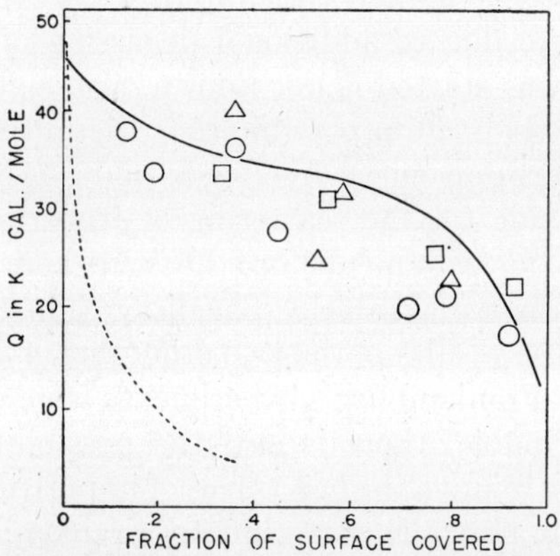

FIG. 19. Heats of adsorption of hydrogen on evaporated tungsten films at 23°C. in comparison with the results of Roberts (points) and of Frankenburg (broken line) as a function of surface covered.

about a thousand times as large as the surface of the single wire used by Roberts and that the surface of the tungsten powder used by Frankenburg is again about a thousand times larger than the surface of the evaporated tungsten film used by Beeck and coworkers. In Roberts' case the heat of adsorption was measured by the rise of temperature of the tungsten wire itself when small increments of hydrogen were adsorbed upon it. The temperature changes of the wire were deduced from changes of its resistance. The values of Frankenburg were obtained from isotherms through application of the Clausius-Clapeyron equation and represent, therefore, indirect measurements. Roberts had by far the cleanest conditions as far as the wire is concerned, since he was able to degas the wire at temperatures close to the melting point prior to the adsorption measurements. The evaporated tungsten films of Beeck and coworkers are probably nearly as clean as Roberts' tungsten wire, and the measurements on the films do not suffer from the objectionable minuteness of the effects to be observed. The surface of tungsten powder used by Frankenburg suffers from the drawback that the powder had to be reduced with hydrogen, and although it was degassed at 750°C. in high vacuum for several days until no evolution of gas could be observed upon cutting off the pumps, it might have retained impurities which can only be removed by a much more drastic treatment (see Section VI). That tungsten is particularly sensitive to impurities is shown by the fact that thin tungsten layers evaporated on a well degassed glass surface did not show any ability whatsoever to adsorb hydrogen. This had not been the case for any of the other evaporated metals investigated. Whether the impurities were derived from the glass surface or were due to additional degassing of the platinum wire when evaporation was started is not known, although the latter appears to be more likely. It is quite possible that the initial convex curvature of the heat of adsorption curve by Beeck and coworkers is due to the same cause. Allowing for the scattering of Roberts' points, the agreement between the measurements on the wire and on the evaporated tungsten films is quite remarkable. The disagreement between Frankenburg's values and the results of the two other sets of data is more difficult to explain. While Frankenburg's experiments were carried out with the greatest care imaginable, there is still the possibility that he has not been able to entirely remove the last traces of impurities from the tungsten surface. It is also possible that due to Frankenburg's technique of approaching equilibrium from higher to lower temperatures his isotherms do not actually represent adsorption isotherms but mixtures of adsorption and sorption into the interior of the structure. As has been shown in Section III, evaporated tungsten films do show a marked effect of sintering on the relative amount of sorption to adsorption even when sintered at only 23°C. From the discussions in Section III it appears, furthermore, likely that the surface measurements by the B.E.T. method on Frankenburg's tungsten powders may have given values possibly too high by a factor of two if the surface were really clean. This in itself would raise the heats of adsorption values somewhat, when plotted against the fraction of surface covered, but would, of course, not change the convex character of the heat of adsorption versus surface coverage. The heat of adsorption of nitrogen on evaporated tungsten films at room temperature was found by the author and his coworkers to be very high, of the order of 95,000 calories, and was found to be constant over the whole surface as in the case of hydrogen on iron at −183°. Frankenburg's results will be discussed further in Section VI. Whatever the cause of the rapid decrease of heat of adsorption with surface coverage in Frankenburg's case may be, the remarkable fact remains that the initial heats of adsorption—that is, for small coverage—agree for all three investigations.

5. The Adsorption of Hydrogen on Metal Films Partially Covered with Other Adsorbed Gases

It has been reported by Beeck *et al.* (1) that evaporated nickel films will adsorb carbon monoxide as molecules (one per crystallographic site), hydrogen as atoms (one atom per site), and oxygen as atoms (four atoms per site). It has also been shown that if a film is partially covered with carbon monoxide or oxygen, the remaining sites can be covered with hydrogen atoms, the sum of both being exactly equal to the number of sites which each gas alone would have occupied. Hydrogenation experiments have shown that hydrogen adsorbed on a surface partially poisoned with CO or oxygen is capable of supporting hydrogenation to the exact extent of that fraction of the surface which is capable of adsorbing hydrogen. This means that the nickel surface is poisoned nonselectively and proportionally to the extent of fraction of the surface covered by the poison.* This was taken as a powerful argument for the homogeneity of the surface of evaporated nickel films with respect to the hydrogenation of ethylene. Later the heats of adsorption of oxygen and CO on nickel were measured. It was found that the heat of adsorption of oxygen was about 130,000 calories per mole and stayed constant until the surface was completely covered, after which the heat fell to very low values. It is interesting that the heat of sorption of oxygen

* In the older literature one of the strongest supports for "active points" or active centers has been the finding that poisons such as CO destroy the catalytic activity completely, even if they are in such small amounts as to cover only a fraction of the surface. This would indicate that the catalyst surfaces referred to in the older literature were either very impure or very heterogeneous, or that since the surfaces were often measured by hydrogen adsorption, surfaces very much too high were obtained because the *ab*sorption of hydrogen into the interior of the structure as discussed earlier in this article was not realized.

on nickel corresponds, within the limit of experimental error, to the heat of oxidation to nickel oxide. Flatness of the heat of adsorption curve would indicate that the adsorbed film is not mobile, as is to be expected for such high heats of adsorption and because of the fact that the film is actually four atom layers deep. Electron diffraction shows it to be nickel oxide. The heat of adsorption for carbon monoxide was found to be 35,000 calories per mole and constant over about 85% of the surface covered, after which it fell to very low values. It may be rightfully argued that because of the flatness of the heat of adsorption curve for both oxygen and carbon monoxide, which indicates nonmobility of the adsorbate, the poisoning by these two gases of the activity for hydrogenation must be nonselective for reasons pointed out previously in this section.

Nickel films completely covered with oxygen will not adsorb hydrogen when exposed to this gas immediately after oxidation, but will regain their ability to adsorb hydrogen after several hours. Hydrogen thus adsorbed is not able to hydrogenate ethylene. Heats of adsorption measurements of this type of hydrogen adsorption have not been made.

Iron films behave in every way similar to nickel films except that upon admitting oxygen, about seven oxygen atoms are sorbed instantaneously for every crystallographic site, forming an iron oxide film seven atom layers deep, and except that the heat of adsorption of hydrogen on such an oxide covered film is almost identical with that on the clean iron surface even immediately after the oxidation has taken place.

Iron is particularly interesting from the standpoint that it will also chemisorb nitrogen. We have already seen that such chemisorption takes place instantaneously at liquid oxygen and liquid nitrogen temperatures with a heat of adsorption of 10,000 to 5,000 calories per mole with increasing surface coverage. At room temperature chemisorption of nitrogen was found to be of the activated type, and the heat of adsorption was found to decrease from 40,000 to 16,000 calories over the fraction of the surface which nitrogen will cover at that temperature—that is, 20% of hydrogen adsorption. At room temperature the iron surface will therefore be covered to the extent of one-fifth of the total surface. As has been stated earlier, this nitrogen adsorption is of the atomic type (yields NH_3 on acid treatment) and on the basis of each nitrogen atom occupying one crystallographic site, the remaining four-fifths of the surface can be filled up with hydrogen, each hydrogen atom occupying one crystallographic site, the sum being exactly equal to the total adsorption of hydrogen alone. The heat of adsorption of hydrogen under these circumstances is normal in the sense that the values are within the experimental limits of error just as if the first 20% of surface were covered with hydrogen and not with nitrogen. If the surface is completely covered with hydrogen, no nitrogen adsorption at room temperature will take place. If the surface is partially covered with hydrogen, nitrogen adsorption will take place to the extent of one-fifth of the remaining empty surface. The amount of nitrogen adsorbed as a function of fraction of surface covered with hydrogen is shown in Fig. 20. It is a perfect straight line relationship. This is very important from the standpoint of the heterogeneity of the surface. If the decrease

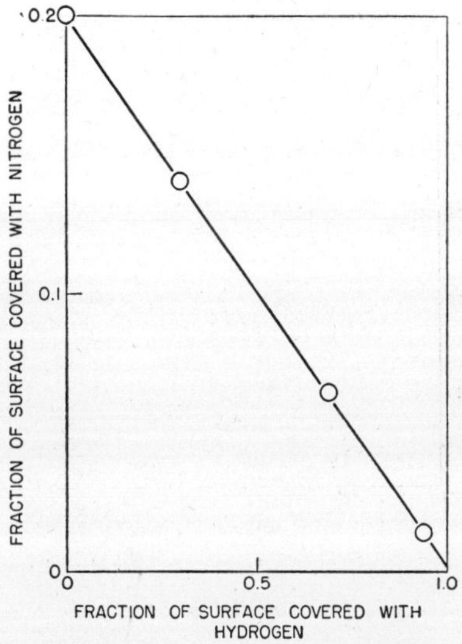

FIG. 20. The adsorption of nitrogen at 23°C. (ordinate) as a function of surface covered with hydrogen (abscissa).

of heat of adsorption of hydrogen with surface coverage were due to heterogeneity of the surface, one would have expected that all sites of high surface energy would have been initially covered with hydrogen atoms (especially since the heats of adsorption of hydrogen and nitrogen are practically the same in this case). If the surface were covered with hydrogen to an extent larger than one-fifth, one would have expected that no nitrogen would be adsorbed. The fact that this is not the case is a powerful argument for the homogeneity of the surface.

The heat of adsorption of nitrogen on tungsten at room temperature was found to be 95,000 calories per mole and constant over the whole fraction of the surface which it will cover—that is, about 60%. The adsorption is instantaneous. In Fig. 21 is shown the heat of adsorption of hydrogen on a tungsten surface covered 60% with nitrogen.

The solid curve is the heat of adsorption of hydrogen alone in its dependence of coverage, and the points represent measurements of the heat of adsorption on the 40% of surface not covered by nitrogen when the latter is preadsorbed. It is seen that these values fall essentially on the curve as if the total surface were covered with hydrogen. Since the hydrogen measurements can be made with exceptionally great accuracy, it is quite possible that the small deviations from the curve toward lower heats of adsorption values are real and indicate that the

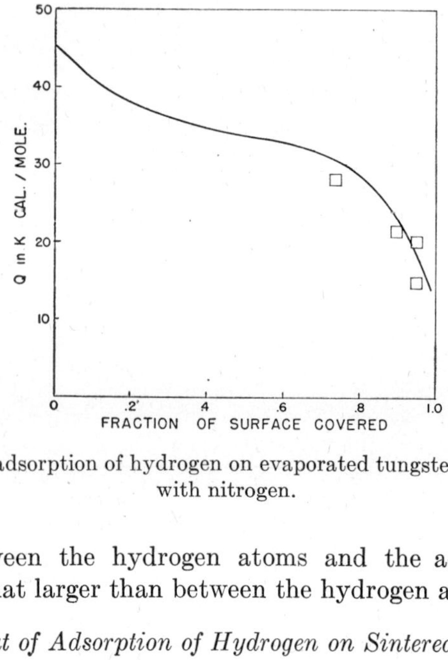

FIG. 21. Heats of adsorption of hydrogen on evaporated tungsten films 60% covered with nitrogen.

interaction between the hydrogen atoms and the adsorbed nitrogen atoms is somewhat larger than between the hydrogen atoms themselves.

6. *Heat of Adsorption of Hydrogen on Sintered Films*

It has been shown in Section IV that sintering of nickel and iron films decreases the surface very markedly, although the total hydrogen sorption decreases less, so that the ratio of total sorption of hydrogen to adsorption on the surface becomes larger the higher the temperature of sintering to which the film has been subjected. It was also shown that the heat of sorption must be very much lower than the heat of adsorption on the surface. Tungsten also shows sorption, although to a much lesser extent. When a nickel film deposited at 23°C. is sintered at about 150°, the surface decreases to 20% of the surface of a film sintered at only 23°. However, the total relatively fast hydrogen sorption decreases only by 50%, after which further slow sorption continues. It was therefore of interest to measure the heat of total sorption of hydrogen on a sintered film. In Fig. 22 is given the heat of adsorption on a nickel film sintered at 23° in the usual manner. The points for the sintered film could obviously not be related to the actual surface which was now decreased to about 20%, but the heat measurements were carried out in increments until the pressure of 0.1 mm. was reached—that is, equivalent to total surface coverage of the unsintered film. It is seen that the heat of total sorption is considerably lower than the heat of adsorption for a

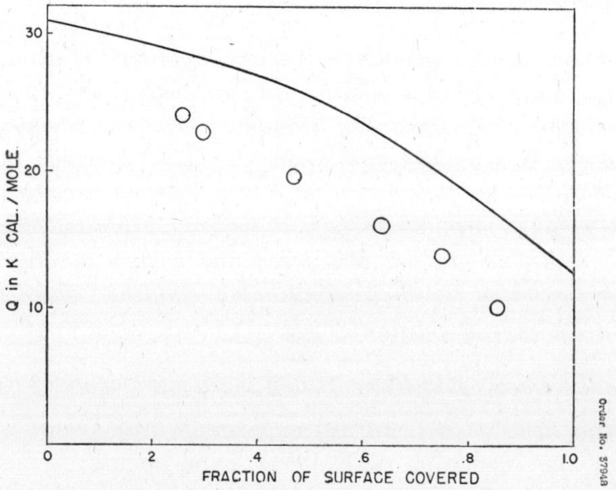

FIG. 22. Heats of sorption of hydrogen on an evaporated nickel film sintered at 150°C. (points) in comparison with the heats of adsorption of a film sintered at 23°C. (curve). (Coverage of surface at equilibrium pressure of 0.1 mm. was taken as 100% in both cases.)

film sintered at 23°. As has become clear earlier, the films sintered at 23°C. show also a small amount of absorption so that the heat of true adsorption curve on such film would probably be slightly flatter than actually measured, especially in the region of high coverage. Similar results can undoubtedly be expected for iron, and perhaps to a lesser degree for tungsten.

It must be expected that the effect on the heat of absorption will be particularly pronounced at higher pressures and in the middle temperature range, let us say between −170° and 200°. The reason for this is that absorption, which is a slow activated process, will affect the fast adsorption at low temperatures very little. Again at the very high temperatures absorption will be less than adsorption because of its lower heats. On the other hand, at the very high temperatures another absorption process, the endothermic solution, can possibly enter the picture, although this effect is undoubtedly small at low pressures, let us say up to several centimeters. As already stated in Section III, the presence of absorption will have to be particularly carefully considered in equilibrium isobars, isotherms, and isosteres, and in the heats of adsorption derived from isosteres. For all these systems, the effect of additional adsorption can be expected to be particularly marked in the middle temperature region, especially at higher pressures.

VI. Discussion

Starting with a comparison of the heats of adsorption of hydrogen obtained by Roberts on single tungsten wires, by Beeck et al. on evaporated tungsten films, and by Frankenburg on tungsten powders, the writer is led to the conclusion that the surface of tungsten powders is chemically impure, thus making for a heterogeneous surface. These findings, together with those reported earlier in this article, are then used in a critical discussion of the nature of pure metal surfaces with the conclusion that they are homogeneous with regard to heat of adsorption and surface coverage but heterogeneous with regard to the hydrogenation of ethylene, which is markedly dependent on the crystal parameter.

1. *Frankenburg's Experiments and the Experiments of Roberts*

Those who delight in beautiful experimentation cannot have failed to marvel at the ingenuity of Roberts' experiments of measuring the heats of adsorption of hydrogen and the fraction of surface covered on a single tungsten wire and at the painstaking, thorough experiments of Frankenburg on tungsten powders. It is a feat in itself that both investigators should have obtained the same value for the heats of adsorption of the sparsely covered tungsten surface, if one remembers that the surfaces used by Frankenburg were about 10^6 times larger than the surface of the single tungsten wire used by Roberts. In the following an attempt will be made at reconciliation of the discrepancy of the heats of adsorption at higher surface coverage as they are pictured in Fig. 19, using the experience gained on evaporated metal films by the author and his coworkers.

For the sake of entering into the discussion, let us assume that Frankenburg's tungsten powder surface is not clean—that is, that it has not a truly bare surface. It appears to be very probable that all oxygen was effectively removed from the surface and even from below the surface by reduction with hydrogen at 750°C. It also appears very probable that thorough degassing at 750° for many hours and sometimes days had effectively removed the last traces of hydrogen from the surface. Even the 0.01% of alkali has probably been effectively removed from the surface by this treatment. That leaves us with the main impurity of 0.1% of silica as given in Frankenburg's paper. Although not specifically stated, this figure undoubtedly represents weight per cent. This amount of silica is more than enough to cover the whole surface of the tungsten powder with silica if all the silica molecules were on the surface. In fact, it is not only possible but likely that a given heat treatment will produce an equilibrium surface condition at which a certain fraction is covered with silica. We now assume that the fraction covered with silica is approximately three-fourths of the surface as measured by the B.E.T. method. This value will remain approximately correct even if we consider the fact that the bare part of the tungsten surface may have been measured by a factor 1.5 too high in analogy to the experiments described in Section III. Since Frankenburg finds that his surface is 25% covered at the equilibrium pressure of 5×10^{-3} mm., and since this value is numerically in good agreement with the value of Beeck and coworkers on evaporated tungsten films for the totally covered metallic surface and also in good agreement with the value of Roberts for what he has concluded is complete coverage, the assumption that Frankenburg's surface is only 25% bare can be rationalized. Frankenburg himself rationalized the discrepancy between his values and the values of Roberts by assuming that Roberts' surface was 4.5 larger than the geometrical surface of the wire. This is not borne out by experiments on evaporated tungsten films, which agree well with Roberts' measurements and conclusions. In addition, sorption of hydrogen into the interior of the structure will also have a lowering effect on the heat of adsorption, and this may have to be taken into account, especially when using Frankenburg's technique for obtaining the isotherms—that is, by lowering the temperature after a given amount of gas has been admitted at a very high temperature. However, sorption is obviously a minor factor and cannot explain the high surface values obtained by the B.E.T. method.

Furthermore, if one acknowledges that the heat of adsorption values as a function of fraction of surface covered can be satisfactorily explained by interaction of the adsorbed atoms, then it becomes evident that there must be a limiting number of atoms on the surface which are far enough apart so that no interaction may occur. Assuming square sites and assuming that all neighboring sites will be subject to interaction, we see that coverages of one-ninth of the surface and less should show independence of heat of adsorption with coverage. Frankenburg's value of 0.8% for the coverage below which the heat of adsorption becomes constant, becomes 3.2% in the new alignment for only one-quarter of his surface being actually bare. While the value of 3.2% is not in agreement with the expected value of about 11%, we must not forget that three-fourths of the surface is covered with silica, which will have the effect of cutting down the 11% figure in an unknown manner (distribution of silica unknown) if one assumes that the interaction of a hydrogen atom adsorbed next to silica is of approximately the same order as the interaction between the two hydrogen atoms adsorbed next to each other.

Not only the hydrogen adsorption but also the nitrogen adsorption as observed by Davis (17a), as well as by Beeck and coworkers fits well into this picture. It was observed by Beeck *et al.* that 60% of the surface of evaporated tungsten films could be covered by nitrogen at room temperature, 50% being adsorbed instantaneously with a constant heat

TABLE II

*Hydrogen Isobars on Evaporated Tungsten Films and on Tungsten Powder**

Type of surface	Equilibrium pressure in mm.	Temperature in °C. 0	100	200	300
Evaporated tungsten film	10^{-4}	85	65	48	36
	10^{-1}	100	91	81	73
Tungsten powder	10^{-4}	21.0	15.5	7.5	3.3
	10^{-1}	29.5	23.5	16.0	10.0

* Values listed are fractions of surface covered in per cent at the specified temperatures and equilibrium pressures.

of adsorption of 95 ± 5 kcal., the remaining 10% being adsorbed much more slowly with a rapidly decreasing heat. Since the heat of adsorption for 50% coverage is instantaneous at room temperature, it must be expected to be constant over a wide temperature range, and since the heat is twice as large as that of hydrogen for which Frankenburg found constancy up to 500°C. the heat of adsorption for nitrogen should be constant to still higher temperatures. Davis finds this indeed to be the case up to the highest temperature investigated of 750°C. Furthermore, he finds the initial heat of adsorption (about 78 kcal.) constant over 14.5% of the total surface. Relating this figure to 25% of his surface as actually being bare, one obtains approximately the same ratio $14.5/25 \simeq 60\%$ as was obtained with evaporated tungsten films. It would appear from the foregoing rather convincing qualitative consideration that the discrepancies between the results of Roberts and of Beeck on the one hand and of Frankenburg and Davis on the other hand can indeed be explained on the basis of three-fourths of the surface of Frankenburg's tungsten powders being covered with impurities. Additional evidence for the inherent difference of the surfaces of Frankenburg's tungsten powder and the evaporated tungsten films is given in Table II, where the adsorbed amounts in per cent of surface covered are given for evaporated tungsten films and for Frankenburg's tungsten powder for four different temperatures and two different pressures. Of particular interest is the rapid decrease of surface covered with rising temperature of the tungsten powder in contrast to the evaporated films. This shows that the powder has regions of low adsorptive power which are not present in the tungsten films. Since higher heats of adsorption speak for cleaner surfaces, and since higher heats of adsorption at considerable surface coverings were obtained by both Roberts and the author, and since cleaner surfaces are more difficult to obtain, it would appear very worth while to investigate further the actual state of the surface of the tungsten powder. The chemisorption of CO at room temperature and low pressures suggests itself at once for the measurement of the bare part of the surface of tungsten powders; the low pressure-low temperature isotherm of nitrogen could also be used (see Section II).

2. The Nature of the Surface

The most uniform surface that nature can provide is obviously a crystal surface. As has already been stressed in the introduction, such a surface is in itself heterogeneous by virtue of its atomic or molecular character. Aside from the periodic energy changes of any crystallographic plane, the heterogeneity of the surface may be increased by crystal imperfections such as ridges and ledges caused by incomplete layers or islands of layers. The surface of a metal powder or of a porous evaporated metal film will also be heterogeneous by virtue of the fact that the single crystallites can expose different crystallographic planes to the gas phase, according to crystal habits or when forced to do so as in the case of metal films that are evaporated in an atmosphere of an inert gas causing certain crystal planes to be preferentially exposed.

If a crystal is composed of two or several atomic species, it is possible that under the most perfect conditions only one atomic species is exposed on the surface, such as the oxygen atoms on certain planes of a metal oxide crystallite. In reality, this probably will happen very seldom, since the stoichiometric ratio of oxygen atoms to metal atoms of a given oxide is not likely to be perfect for a given crystallite. An oxide crystallite will therefore expose metal as well as oxygen atoms, and thus the surface will be heterogeneous for chemisorption. Nevertheless, such surface will behave like a homogeneous surface toward van der Waal's adsorption of, for instance, the noble gases.

Heterogeneity of the surface may also be produced by impurities. In case of metals, such impurities—however small—if expressed as percentages of total weight may cover a large part of the surface, especially if heat treatment has caused them to accumulate on the surface. Any attempt to establish whether a surface is homogeneous or heterogeneous must necessarily be futile because no truly homogeneous surfaces exist. As has been shown, there are various degrees and types of homogeneity or heterogeneity, and the question of heterogeneity or homogeneity should be raised only in conjunction with a well defined property.

Van der Waal's adsorption of large atoms such as krypton will be influenced neither by chemical forces nor by the size and shape of the crystallographic sites of the surface because the krypton atom is in most cases considerably larger than the dimensions of a single crystallographic site. If the surface is to be measured in square centimeters, the use of krypton is ideal, provided reasonable assumptions as to the packing of the atoms are made. In spite of its shortcomings from the theoretical standpoint, the B.E.T. method appears to give excellent results in this respect and is at present without doubt the safest method to measure surfaces with otherwise unknown characteristics. Ample evidence has been presented in this article that the B.E.T. method will not give correct results for metal surfaces if nitrogen is used as the absorbate. Nitrogen is chemically too active with most metals and will form a chemisorbed monolayer (or part of a monolayer) previous to van der Waal's adsorption on top of this layer. This does not mean that one cannot measure the surface of metals with the aid of nitrogen adsorption except that certain other criteria have to be known and kept in mind when such measurements are made. Furthermore, as soon as chemical forces are involved the effect of the size of the crystallographic sites becomes more important and the uncertainty of surface measurement increases if it is not known which crystallographic sites are exposed. As a general rule, crystallographic sites will, at most, not vary more than 50% in size for simple crystalline materials. Ignoring the factor of the size of the site, chemisorption of many gases can be used successfully and very accurately to measure the number of sites available. On most metal surfaces produced by the film technique the chemisorption of CO molecules has given the most reproducible and reliable results. One molecule of CO is adsorbed on each crystallographic site. The adsorption has been found to be instantaneous, both at low temperatures and at room temperature, and the amounts adsorbed at liquid nitrogen temperature and at room temperature differ very little. Unlike hydrogen, carbon monoxide has the further advantage that its absorption into the interior of the crystal structure is negligible. The heat of adsorption of carbon monoxide is of the order of that of hydrogen. Unlike oxygen, it will not replace hydrogen from the surface and, unlike nitrogen with its very high heat of adsorption, the induced heterogeneity as a function of surface coverage is not large enough to prevent complete coverage of the surface with CO molecules.

There seems to be little doubt that the heterogeneity with respect to the heat of adsorption of clean surfaces of nickel, iron, and tungsten and of many other metals which have been studied is an induced heterogeneity caused by lowering of the surface energy of sites adjacent to sites already occupied by hydrogen atoms. The surfaces appear to be homogeneous with regard to the absolute value of the heat of adsorption on different crystallographic planes. At the same time, it has been shown that oriented and nonoriented films show a fivefold difference in rate of hydrogenation of ethylene. In other words, mixed crystallographic planes make the surface very heterogeneous with regard to the hydrogenation reaction.

There is some evidence that the surface of tungsten powders studied by Frankenburg is heterogenous by virtue of having impurities on the surface. Such heterogeneity can be of various types depending on whether the impurities are clustered together on the surface in islands or whether they are evenly or statistically distributed over the surface. This distribution of impurities coupled with the induced heterogeneity by the adsorbate itself will decide the shape of isotherms, isobars, isosteres, and the trend of the differential heats of adsorption with increasing surface coverage.

If absorption or solution into the interior of the crystal structure is involved, the measurement of adsorption is still more complicated, as has been shown in Section IV. It is often very difficult to discern absorption from adsorption. The literature on adsorption of hydrogen is abundant with examples in which absorption has vitiated the measurements and conclusions in one or another way. Chemisorption on clean metal surfaces is very fast. If slow activated sorption of hydrogen is observed, absorption or solution may be involved.

It would be beyond the scope of this article to attempt to discuss the whole literature on hydrogen adsorption in the light of the many difficulties which have been pointed out. When metal surfaces such as nickel, iron, and tungsten have been measured by the B.E.T. method using nitrogen, they either have been measured too large if the metal surfaces were absolutely clean, or if measured correctly, the surfaces have not been clean. Reports by Brunauer and Emmett (15), for instance, that preadsorption of hydrogen had no influence on the surface measurement of synthetic ammonia catalysts leads to the conclusion that their surfaces must have been covered with hydrogen in the first place or have been impure otherwise. Their surfaces were also reported to chemisorb nitrogen with a measurable rate at 200°C. and above. As has been shown in this article, pure evaporated iron films instantaneously chemisorb nitrogen at liquid nitrogen temperature (probably initially without dissociation) and with a slower rate at room temperature (with dissociation). The observed slow sorption of hydrogen on nickel catalysts by Benton and White (16) and by Maxted and Hassid (13) has now definitely been shown to be absorption into the interior of the structure. The interesting work by Taylor and Chou-Shou Liang (12) on the adsorption of hydrogen on zinc oxide, which has already been mentioned, lacks confirmation that the slow observed hydrogen sorption is not actually absorption into the interior of the structure. Oxides, and especially mixed or promoted oxides, would be expected to have heterogeneous surfaces, and it would therefore be of utmost importance to verify that absorption or solution is not involved. If this can be shown, the method of Taylor and Chou-Shou Liang (12) could indeed become a powerful technique in the investigation of the degree of heterogeneity of surfaces.

In the face of the great complexity of the sorption process, especially in the case of hydrogen, restraint should be shown in basing theoretical considerations on experimental results reported in the literature. The results of Frankenburg (17) and of Taylor and Chou-Shou Liang (12) have recently been variously quoted as proof of the heterogeneity of catalytic surfaces. While there seems to be no doubt in the authors' opinion that these surfaces are heterogeneous and while there are good reasons for their being heterogeneous, the work does not prove that all surfaces are heterogeneous in the sense that has been implied by these authors. From the practical standpoint, catalytic surfaces with a high degree of heterogeneity are probably more common than relatively homogeneous surfaces. A theoretical treatment of heterogeneous surfaces is therefore most desirable, especially as such treatment will also embrace uniformity or apparent uniformity of surface as special cases. The work by Halsey and Taylor (5) and Sips (18) is a step forward in that direction.

At least a few short paragraphs should be devoted to the mention of poisoning of catalysts. It has been shown by the author and his coworkers that metal film catalysts are poisoned for the hydrogenation of ethylene by oxygen, carbon monoxide, and nitrogen in exact proportion to the fraction of surface covered by these gases. Before it was known that the metals commonly used as hydrogenation catalysts absorb large quantities of hydrogen, the ratio of total hydrogen sorbed to the amount of hydrogen sulfide, for instance, which would poison the catalyst has been taken as proof of the existence of active centers which are selectively poisoned by a very small amount of hydrogen sulfide. Actually, as has become clear in Section IV, most of the hydrogen taken up by these systems was not on the surface, leaving only a small fraction on the surface as contributing to hydrogenation. It is this small fraction, therefore, which is poisoned by a correspondingly small amount of hydrogen sulfide. Since the rate of hydrogenation of ethylene was found to be the same whether involving hydrogen molecules adsorbed with a high heat of adsorption or those adsorbed with a low heat of adsorption, no other active centers except the crystallographic sites with more or less favorable spacing exist for the hydrogenation reaction on a clean surface. Additional proof of this was obtained by the use of preadsorbed tritium.

It is obvious that heterogeneous surfaces can be poisoned selectively, and this phase of selectivity needs no further discussion in the framework of this article. There is, however, the interesting case where an otherwise uniform hydrogenation catalyst was found to be selective for the hydrogenation of acetylene in the presence of ethylene. The reason for this is that a metal surface completely covered with acetylene will still adsorb about one-third of the amount of hydrogen which will be adsorbed on a clean surface. Hydrogen is apparently able to reach the metal surface even if it is covered with acetylene, whereas ethylene is not able to reach the adsorbed hydrogen due to the presence of acetylene on the surface. The result is a slow hydrogenation of the acetylene from the surface until the acetylene concentration in the gas phase becomes so low that complete coverage of the surface by acetylene becomes impossible and adsorbed hydrogen is made accessible for the hydrogenation of ethylene.

3. General Conclusions

1. The adsorption (chemisorption) of hydrogen on clean metal surfaces is almost always accompanied by absorption of hydrogen into the interior of the structure. This *ab*sorption is a slow activated process and has in the past been mistaken for activated *ad*sorption of hydrogen on the surface.

2. The heat of *ab*sorption of hydrogen is considerably lower than the heat of chemisorption.

3. The heats of chemisorption of hydrogen on nickel and iron are nearly identical and decrease from about 30,000 calories for the sparsely covered surface to about 18,000 calories for the completely covered surface. The heat of chemisorption on tungsten decreases from 45,000 calories to about 13,000 calories as a function of surface coverage (Roberts, also Beeck and coworkers). The lower values of Frankenburg for higher

surface coverage can be satisfactorily explained on the basis of surface impurities on Frankenburg's tungsten powders. The heat of adsorption of hydrogen on metals has been found to be constant over wide temperature ranges (for nickel by Beeck and coworkers from −183 to 23°C., by Frankenburg on tungsten for temperatures up to over 500°).

4. The activation energy for the chemisorption of hydrogen on all metals studied is extremely low.

5. Chemisorbed hydrogen atoms on nickel are mobile—that is, they are able to migrate from one crystallographic site to another site even at −183°C. On iron, hydrogen atoms are mobile at room temperature but are not mobile at −183°C.

6. The decrease of heat of adsorption as a function of surface coverage can be explained satisfactorily by interaction of the adsorbed atoms with each other.

7. While the surface of clean metal films appears to be homogeneous with regard to heat of adsorption and surface coverage (the latter within the limits of size of different crystallographic sites), the rate of hydrogenation of ethylene is markedly dependent on the crystal parameter.

8. On nickel, chemisorption of hydrogen is still taking place at 20°K.

9. The true extent of metal surfaces cannot be measured by the B.E.T. method using nitrogen, since nitrogen is chemisorbed at −196°C. The hydrogen adsorption at this temperature measures the surface more accurately and is in close agreement with the chemisorption of carbon monoxide at both liquid nitrogen and room temperature and with the van der Waal's adsorption of krypton.

Acknowledgement

While during most of the discussion in this article reference is made to papers already published or in the process of publication, numerous minor items are mentioned which have not been published separately already or do not warrant separate publication. It is with regard to these items that the author wishes to acknowledge the collaboration of various associates at various times, in particular Dr. A. Wheeler and Mr. A. W. Ritchie.

References

1. Beeck, O., Smith, A. E., and Wheeler, A., *Proc. Roy. Soc. London* **A177,** 62 (1940).
2. Beeck, O., *Revs. Modern Phys.* **17,** 61 (1945); **20,** 127 (1948); *Record of Chem. Progress* (*Kresge-Hooker Sci. Libr.*) July-October (1947).
3. Beeck, O., Wheeler, A., and Cole, W. A., *Trans. Faraday Soc. Conference on Catalysis* April 1950.
4. Roberts, J. K., Some Problems in Adsorption. Cambridge Physical Tracts, Cambridge University Press, 1939; *Proc. Roy. Soc. London* **A152,** 445, 447 (1935); **A161,** 141 (1937).
5. Halsey, G., and Taylor, H. S., *J. Chem. Phys.* **15,** 624 (1947).
6. Brunauer, S., Emmett, P. H., and Teller, E., *J. Am. Chem. Soc.* **60,** 309 (1938).

7. Emmett, P. H., and Brunauer, S., *J. Am. Chem. Soc.* **59,** 1553, 2682 (1937); Brunauer, S., and Emmett, P. H., *ibid.* **57,** 1754 (1935).
8. Beeck, O., Ritchie, A. W., and Wright, K., *J. Colloid Sci.* 1950, in press.
9. Beeck, O., and Wheeler, A., *J. Chem. Phys.* **7,** 631 (1939).
10. Beeck, O., Ritchie, A. W., and Wheeler, A., *J. Colloid Sci.* **3,** 505 (1948).
11. Beeck, O., Givens, J. W., and Ritchie, A. W., *J. Colloid Sci.* 1950, in press.
12. Taylor, H. S., and Chou-Shou Liang, *J. Am. Chem. Soc.* **69,** 1306 (1947).
12a. Frankenburger, W. G., and Messner, G., *Z. physik. Chem., Bodenstein Festband* 593 (1931).
13. Maxted, E. B., and Hassid, N., *J. Chem. Soc.* 1532 (1932).
14. Michel, A., Bénard, J., and Chaudron, G., *Bull. Soc. Chim.* **12,** 336 (1945); Michel, A., and Gallisot, M., *Compt. rend.* **208,** 434 (1939); Owen, E. A., and Jones, J. I., *Proc. Phys. Soc.* **49,** 603 (1937).
15. Brunauer, S., and Emmett, P. H., *J. Am. Chem. Soc.* **62,** 1732 (1940).
16. Benton, A. F., and White, T. A., *J. Am. Chem. Soc.* **52,** 2325 (1930).
17. Frankenburg, W. G., *J. Am. Chem. Soc.* **66,** 1827 (1944).
17a. Davis, R. T., *J. Am. Chem. Soc.* **68,** 1395 (1946).
18. Sips, R., *J. Chem. Phys.* **16,** 490 (1948).

Hydrogen Fluoride Catalysis

J. H. SIMONS

Fluorine Laboratories, The Pennsylvania State College, State College, Pennsylvania

Hydrogen fluoride has been employed as a catalyst for organic chemical reactions for only about a decade. Although it has come into rather extensive industrial use in this time, the total number of publications on this subject is small. This is because of the technical difficulties of handling hydrogen fluoride in the college laboratory, the source of most scientific publications. In this chapter no attempt is made to give a complete literature survey nor to exhaustively treat the organic chemistry of the reactions or products. What is attempted is a discussion of the catalytically significant properties of hydrogen fluoride, the techniques employed in its use, the range of the reactions and their types, the principles involved, the advantages and disadvantages over other catalysts for the production of the same products, and the mechanism of this action.

I. Historical

Studies in the academic laboratories of The Pennsylvania State College on the action of hydrogen fluoride on organic chemical compounds began about 1933. Prior to that time it had been used as a reagent for the preparation of fluorine containing organic compounds and as a powerful polymerizing and dehydrating agent. It has been used in the degradation of cellulose and as a dehydrating agent in the nitration of benzene. What indications there were at that time of the effect of hydrogen fluoride on organic chemical substances would lead to an expectancy of detrimental results due to the corrosive nature of the material. In 1935 a paper (Ipatieff and Grosse, 76) on the alkylation of isoparaffin with olefins was published in which the catalyst used was a combination of boron trifluoride and nickel powder. Water was used as a promoter but it was also found that hydrogen fluoride could take the place of water as the promoter. This reaction is now known to be much more favorably catalyzed by hydrogen fluoride. Publications which showed the powerful catalytic powers of this substance to produce isolable products in organic reactions began early in 1938 beginning with condensation reactions and rapidly extending to a variety of other reactions.

The development of the subject of the reactions catalyzed by hydrogen fluoride then moved rapidly and in 1939 and 1940 a number of interesting uses were published and several industrial companies applied for patents on specific applications of industrial use. The war soon put a stop to the scientific studies but hastened certain industrial developments. The largest and most significant of these was the use of hydrogen fluoride as the catalyst for petroleum alkylation. As rapidly as the engineering problems could be solved, large hydrogen fluoride alkylation plants were built and relatively soon the major portion of the aviation alkylate was supplied by these plants. Although the need for aviation alkylate has now diminished, some of these plants are still in operation, and the industrial use of hydrogen fluoride as a catalyst for reactions of aromatic compounds has become important.

II. Nomenclature

The term hydrogen fluoride is used as the name for the substance containing hydrogen and fluorine, and the formula HF used regardless of the fact that the vapor has been shown to exist as an equilibrium of polymers. By implication the liquid and crystalline material are even more highly polymerized. When emphasis is needed to call attention to the fact that water free material is designated, the term anhydrous hydrogen fluoride is used. The term anhydrous hydrofluoric acid is self-contradictory as hydrofluoric acid is the name for the aqueous solution. In addition, completely anhydrous material is not always necessary or even desirable for catalytic work. The commercial material labeled anhydrous has a certain water content. The formula H_2F_2 has absolutely no foundation and is contrary to both our chemical knowledge and good practice.

III. Chemical and Physical Properties of HF

In addition to the chemical properties that enable hydrogen fluoride to catalyze organic chemical reactions, its physical properties are important in its use. The fact that it is a highly mobile liquid with a low boiling and very low freezing point gives it important advantages over other agents for the same reactions. Following are some of the important physical properties of hydrogen fluoride.

For simple laboratory experiments the low boiling point is a minor disadvantage, since closed vessels are required if the temperature needed is room temperature or above. It also means that such experiments should be done in the fume hood. It is, however, a considerable advantage in commercial use as it enables the catalyst to be readily evaporated from the product. The low freezing point ensures that the catalyst will not freeze, even when low temperatures are employed. The low viscosity and surface tension assist in intimate mixing for heterogeneous reactions, and greatly hasten settling time in large installations. The low viscosity also enables smaller piping and pumps to be used. The low surface tension and viscosity, however, make leaks in the equipment much more serious so that joints and closures must be made very carefully. Welded connections are recommended. The solubility facts are quite important in the use of the catalyst. The high solubility of oxygen, nitrogen, and sulfur containing compounds and the significant solubility of even hydrocarbons, as well as the solubility of hydrogen fluoride in these substances, enables the catalyst to function in the liquid phase. This provides the advantages of speed of reaction and high specificity.

The chemical property most important for the catalytic power is, undoubtedly, the extremely high acidity of hydrogen fluoride. It is among those liquid substances, which, in the pure state, have the highest acidity, if it is not the most acidic substance known. This is despite the fact that in aqueous solution it is an apparently weak acid. This latter is not a true criterion of the acidity of the substance, as can be seen from the fact that its molar heat of neutralization in aqueous solution is higher than that of strong mineral acids. For weak acids this is uniformly lower. An explanation of this is given elsewhere (Simons, 10). All substances appreciably soluble in liquid hydrogen fluoride behave either as bases or salts. No acid relative to the solvent has yet been found. This in itself shows the strong acidity of the liquid.

Another essential chemical property catalytically important is the powerful dehydrating action of hydrogen fluoride. No chemical drying

TABLE I

Properties of HF

Freezing point	−83°C.	(Simons, 1)
Boiling point	19.6°C.	(Simons, 1)
	19.9°C.	(Claussen and Hildebrand, 2)
Critical temperature	230.2°C.	(Bond and Williams, 3)
Heat of fusion per g.	54.7 cal.	(Dahmlos and Jung, 4)
Heat of vaporization per g. at 748 mm. and 19°	97.5 cal.	(Simons and Bouknight, 5)
Heat of formation at 32°C. per g.	3220 cal.	(Wartenberg and Schutza, 6)
Density at 0°C.	1.002 g. per cc.	(Simons and Bouknight, 7)
Surface tension at 0°C.	10.2 dynes per cm.	(Simons and Bouknight, 7)
Viscosity at 0°C.	0.256 centistokes	(Simons and Dresdner, 8)
Vapor pressure at 0°C.	360 mm.	(Simons, 1)
Dielectric constant at 0°C.	83.6	(Fredenhagen, 9)

Equations for Temperature Variation of Properties

Vapor pressurs

$$\log P = 7.37 - \frac{1315}{T}$$ (Simons, 1)

$$\log P = 7.3739 - \frac{1316.79}{T}$$ (Claussen and Hildebrand, 2)

$$P = \text{mm.} \qquad T = {}^\circ\text{K.}$$

Density in g. per cc.

$$d = 1.0020 - 0.0022625t - 0.000003125t^2$$ (Simons and Bouknight, 7)

$$t = {}^\circ\text{C.}$$

Surface tension in dyne per cm.

$$\gamma = 40.7\left(1 - \frac{T}{503.2}\right) 1.78$$ (Simons and Bouknight, 7)

Variation of gaseous density and apparent molecular weight with temperature and pressure

$$\log K = \frac{40{,}000}{4.579T} - 43.145$$

$$K = \frac{m - 20}{(m - 120)^6} \frac{10^{10}}{P^5}$$

$$T = {}^\circ\text{K.}$$

$$P = \text{mm.}$$

m = apparent molecular weight in gas at T and P (Simons and Hildebrand, 10)

or

$$\log K = \frac{8910}{T} - 43.65$$ (Long *et al.*, 11)

agent has yet been found to extract water from it. It dehydrates sulfuric acid and produces water on reaction with phosphorous pentoxide. Reasonably dry material can be obtained by distillation, but the remaining water can only be removed by electrolysis. The catalytic importance of this property is in reactions in which water is a product. There is an apparently paradoxical situation in this extremely powerful dehydrating effect. The material would be expected to degradate rapidly all oxygen

TABLE II

Variation of Properties with Temperature

Dielectric Constant (Fredenhagen, 9)

t°C.	D
−73	174.8
−42	134.2
−27	110.6
0	83.6

Heat Capacity for 20 g. at Constant Pressure (Clusius *et al.*, 12; Dahmlos and Jung, 4)

Solid		Liquid	
°K.	C_p	°K.	C_p
11.02	0.11	200	13.45
15.2	0.31	210	13.72
21.2	0.71	220	14.12
54.6	3.22	230	14.60
65.8	3.81	240	15.10
77.4	4.20	250	15.65
100	5.95	260	16.20
110	6.63	270	16.75
120	7.05		
130	7.45		
140	7.88		
150	8.35		
160	8.70		
170	9.27		
180	10.75		
190.09	m.p.		

containing substances to remove the elements of water. It does this in fact much less readily than sulfuric acid or other drying agents. The reason for this probably is in its extremely high acidity. All these oxygen containing substances, even the carboxylic acids, are bases relative to hydrogen fluoride and form positive ions in solution by the addition of a proton. (Simons, 10.) These positive ions are much more resistant to dehydration. Even acetone reacts relatively slowly.

The other halogen halides, HCl, HBr, and HI, are insoluble in liquid hydrogen fluoride and are given off as gases in reactions in which they

TABLE II-A

Variation of Properties with Temperature

°C.	Liquid Solubility in Weight Per Cent (Phillips Petroleum Co., 13) Isobutane in HF	HF in propane	In isobutane	In n-butane
0	1.95	0.42	0.22	0.17
5	2.1	0.49	0.25	0.19
10	2.2	0.55	0.29	0.22
15	2.4	0.65	0.33	0.25
50	3.6	1.8	0.87	0.67
60	4.0	2.3	1.2	0.88

The liquid solubilities of propane, isobutane, and normal butane in hydrogen fluoride and the liquid solubilities of hydrogen fluoride in these three hydrocarbons in the temperature range 0 to 50°C. are given by the following equations.

$$\log_{10} W_i = \frac{-630.3}{T} + 2.69027$$

$$\log_{10} W_n = \frac{-569.0}{T} + 2.32359$$

$$\log_{10} W_p = \frac{-605.2}{T} + 2.74152$$

$$\log_{10} W_{HF} \text{ (in isobutane)} = \frac{-1010.6}{T} + 3.30518$$

$$\log_{10} W_{HF} \text{ (in } n\text{-butane)} = \frac{-1061.8}{T} + 3.30173$$

$$\log_{10} W_{HF} \text{ (in propane)} = \frac{-1225.3}{T} + 4.24192$$

where T = °K.; W_i, W_n and W_p are the solubilities of isobutane, n-butane, and propane, respectively, in weight per cent in the hydrogen fluoride liquid phase, and W_{HF} is the solubility of hydrogen fluoride in weight per cent in the hydrocarbon phase. (Butler *et al.*, 14)

Solubility of HF in Per Cent by Weight (Klatt, 15)

°C.	Benzene	Toluene	Anthracene	m-Xylene	o-Xylene	Tetralin
−20	1.56	1.02				
−15	1.67	1.12	2.77	0.95	0.85	0.15
−10	1.88	1.25	2.88	1.01	0.87	0.19
−5	2.05	1.34	2.96	1.08	0.94	0.21
0	2.25	1.54	3.11	1.17	1.01	0.23
5	2.54	1.80	3.27	1.28	1.12	0.27
10	2.82	2.05	3.43			
15	3.11	2.43				

Benzene is about 2% soluble in HF under ordinary conditions but about 20% soluble at 50°. Oxygen, nitrogen, and sulfur containing organic compounds are in general very soluble in liquid hydrogen fluoride.

TABLE II-B

Variation of Properties with Temperature

Solubility of Inorganic Substances in HF (Simons, 16)

Very soluble	Slightly soluble	Not appreciably soluble	Soluble with reaction	Reacts but product insoluble	Insoluble and unreactive
H_2O	MgF_2	AlF_3	Alkali halides and	$AlCl_3(HCl)$	$ZnCl_2$
NH_4F	CaF_2	ZnF_2	alkaline earth hal-	$FeCl_2(HCl)$	$SnCl_2$
LiF (2.6 per 100 at	SrF_2	FeF_3	ides dissolve to	$MnCl_2(HCl)$	$NiCl_2$
18°)	BaF_2	PbF_2	form hydrogen	$CeCl_2(HCl)$	$CdCl_2$
NaF	$CaSO_4$	CuF_2	halides	MgO	$CuCl_2$
KF (36 per 100 at	$KClO_4$	HgF_2	KCN(HCN)	CaO	HgI_2
0°	H_2S	HCl	$NaN_3(HN_3)$	SrO	AgCl
RbF	CO	HBr	$K_2SiF_6(SiF_4)$	BaO	AgBr
CsF	CO_2	HI	$KClO_3(ClO_2)$	PbO	AgI
TlF		SiF_4	$Ba(ClO_3)_2(ClO_2)$	BaO_2	HgO
AgF (33 per 100 at		$Cu(NO_3)_2$	Hydroxides	Al_2O_3	PbO_2
−15°)		$Bi(NO_3)_2$		CuO	MnO_2
$Hg(CN)_2$		$Pb(NO_3)_2$			SnO_2
KNO_3		$Co(NO_3)_2$			Cr_2O_3
$NaNO_3$		$ZnSO_4$			WO_3
$AgNO_3$		$CdSO_4$			Mn_2O_3
K_2SO_4		$CuSO_4$			
Na_2SO_4		Ag_2SO_4			

are produced. This, of course, assists in the reaction going in the forward direction, in reactions in which these gases are produced, by the removal of one of the products from the reacting phase.

IV. Technique

In the laboratory neither the apparatus nor the technique of performing the reactions is as difficult as might be assumed. It is true that glass is excluded from any part of the apparatus that comes in contact with hydrogen fluoride, but copper and iron serve very well for reaction vessels. Monel and nickel are excellent but stainless steel should not be used, particularly at elevated temperatures. Zinc is removed from brass leaving a copper surface, and the corrosion of brasses and bronzes depends upon the composition. Copper beakers and flasks serve well for reactions which proceed below room temperatures. Closed vessels are required at room temperature or above. However, by the use of silver solder for joining copper parts and the use of copper tubing fittings, both flare and compression, little difficulty is encountered by the ordinarily skilled chemist regardless of how complex he may desire his design of apparatus. Soft or lead based solder should not be used except for very temporary connections since lead is unsatisfactory. The brass tubing fittings last a surprisingly long time. For gasket material and valve packing the two plastics Saran and Teflon now enable closures to be made with ease and surety. Saran is satisfactory at low temperatures only. For many reactions above room temperature the only special apparatus required is a vessel made by closing the ends of a piece of copper tubing or pipe with silver soldered copper ends, one of these containing a copper tube to which a compression fitting carrying a blank end is attached for a stopper.

The most annoying feature of the laboratory technique is the corrosion of vessels, tubing, fittings, valves, etc., that occurs when apparatus contaminated with hydrogen fluoride is allowed to stand open to the air. The acid catalyzes oxidation of metals by the air. This corrosion does not occur during the experiments. The difficulty can be avoided by cleaning all equipment immediately after use.

By simple adaptation all the usual laboratory devices such as mercury seal stirrers, dropping funnels, distilling column, etc. can be employed. In fact, many of these are more satisfactorily made of metal than of glass. This enables the usual procedures of organic chemical technique to be followed when performing a reaction. The hydrogen fluoride can be added in a number of ways. The most useful way for simple experiments is to condense the required amount from the gas stream from a cylinder containing the liquid. A condenser made of a coil of ¼-inch copper tubing jacketed with circulating cold water and with the gas entering at the top is satisfactory. The liquid output from the condenser enters a cold receiver. Four to six feet of tubing should be used since hydogen fluroide is not easy to condense. When not in use, the condenser should be kept tightly closed at all times, since hydrogen fluoride adheres strongly on metal surfaces and since water is condensed from the air upon them. This adds water to the condenser and promotes corrosion as well as contaminates the next sample. If the condenser is not kept closed it should be cleaned after every use and carefully dried. A good control valve should be placed in line from the cylinder since the cylinder valve is unsatisfactory for this purpose. The additional valve also provides for safer operation. If the gas does not come with sufficient speed from the cylinder, it can be heated from the outside. If conditions are kept reasonably stationary, the amount of hydrogen fluoride condensed can be approximated by the time between opening and closing the valve. As all commercial hydrogen fluoride contains some water and other impurities, the rate of gas evolution from the cylinder decreases as the cylinder is emptied and the impurities are concentrated in the remainder.

The relative amount of hydrogen fluoride required for a reaction depends upon the type of reaction and the properties of the reagents and products. For some reactions only a trace is necessary while for others hydrogen fluoride is used as the liquid solvent. Reactions in which water or ammonia are produced require more catalyst than those that do not produce basic substances, as the accumulation of base lowers the catalytic activity of hydrogen fluoride. Oxygen and nitrogen containing organic substances, such as alcohols, ethers, carboxylic acids, amines, etc., react strongly with hydrogen fluoride with a considerable evolution of heat. When these substances are used, much more of the catalyst is required because enough must be added to satisfy these reactions of combination before there will be any available for use as a catalyst. In adding hydrogen fluoride to these basic substances considerable care must be exercised as the reactions are quite violent. Basic substances in this connection mean all organic compounds containing oxygen or nitrogen as well as water and sulfuric acid. The usual techniques of mixing reacting substances can be followed, but it is convenient to take the sample of hydrogen fluoride first so that the amount can be ascertained by weighing and to add the other material later. If monoalkylated aromatic compounds are to be prepared, the alkylating agent should be added slowly to an excess of the aromatic compound in the presence of the catalyst because the rate of reaction of the monoalkylated material to the dialkylated is more rapid than the rate to form the monoalkylated substance in most cases. A tri- or higher alkylated product has to be forced.

In reactions in which HCl, HBr, or HI are evolved there is a convenient way of detecting and following the course of the reaction. Gas will be evolved and a bead of silver nitrate solution in a small loop of nichrome wire placed in this gas stream will become opaque, if these gases are present. Hydrogen fluoride will not do this since silver fluoride is very soluble in water. If a simple test for hydrogen fluoride is desired, a similar bead of calcium chloride will serve very well.

After the reaction is completed the products can be poured into water and ice and the aqueous hydrofluoric acid disposed of down the drain. Care must again be exercised as the mixing of hydrogen fluoride and water generates considerable heat. No hazards or disposal difficulties are incurred with hydrofluoric acid in the usual drain lines as these are made of iron pipe and the hydrogen fluoride is soon absorbed as firm complexes with the iron. It is not detectable at any great distance from the source. For some reactions this simple form of disposal is not satisfactory, for example, in the preparation of an acyl halide or other product which reacts with water. Here distillation of the hydrogen fluoride from the reaction mixture is good procedure, but for high boiling products merely letting the reaction vessel stand open in the hood will exhaust most of it. The usual alkaline wash or alkaline treatment of the product will remove the last traces. In cases where water may not be used, dry sodium fluoride will absorb the last remaining hydrogen fluoride from either the liquid or vapor.

On the pilot plant scale, where pumps, valves, gages, etc., are employed, the technique is not greatly different than when using other catalysts except that leaks are more annoying. The technique is actually simpler than when solid catalysts are used. A good grade of steel pipe and fittings is recommended, particularly forged steel fittings and seamless pipe. Cast iron is to be avoided and any castings with slag pockets are apt to leak. Neither stainless steel nor brass are recommended and stainless is worthless at higher temperatures. Good welded joints are the best but screwed connections are satisfactory, if well made. Silver solder can be used but may leak after long usage. Iron valves are best if constructed with the screw exterior to the packing, as an interior screw is apt to freeze. Teflon packing is very good but copper sheathed packing can also be used. A packing made of vinyon encased in copper is satisfactory. Periodical (daily) drenching of the exterior of the packing gland with oil is good practice since acid seepage will cause corrosion on the exterior parts exposed to air. Valves with Monel trim are the best. Gages should have iron or special alloy Bourdon tubes and connections.

As the large scale commercial use of hydrogen fluoride is now well established, particularly in the petroleum industry, the techniques of the use of large size equipment is well known. Reports are available on various aspects of industrial use. A book has been published with particular reference to paraffin alkylation (Phillips Petroleum Company, 13). Corrosion, instrumentation, materials of construction, safety measures, etc., are included. The following journal articles also contain material of interest on large scale technique (Holmberg and Prange, 17, Frey, 18, Fehr, 19). There are certain features that need to be watched, such as corrosion, embrittlement, etc., but the above references deal with these subjects. Corrosion is not particularly serious in properly constructed equipment except where air enters.

V. Hazards and Safety

Hydrogen fluoride is a dangerous material. Its effects are serious if large quantities are inhaled or if it is allowed to remain in contact with the skin. If unwashed and untreated, a small drop of the aqueous acid on the skin will cause a painful wound. The effects are not felt immediately but 5 to 8 hours later a painful throbbing will be felt and in a few days a bad-looking black abscess will develop which is very slow to heal. On opening a black pus will be found. However, hydrogen fluoride is no more dangerous than many other chemical substances handled in large volume and is less dangerous than some. If skin that has been in contact with hydrogen fluoride is quickly and properly treated, there is no after effect, which cannot be said for nitric acid. If it is remembered that hydrogen fluoride absorbs readily in tissue and that both hydrogen and fluoride ions are toxic, the ways of treatment are obvious. Copious washing with water to remove any acid remaining on the surface is the first item. The second is to apply some material which will precipitate the fluoride ion and neutralize the acidity. If the exposure is not great even calcium hydroxide is very effective. Pastes made of organic calcium salts, such as the gluconate or lactate with precipitated magnesium oxide, are good and not as corrosive as slaked lime but probably not as effective for immediate and short-time use. Such pastes are available on the market. They should be kept on and moist for an extended period, sometimes as long as three days. One of the great dangers in treatment is the application of greases, greasy ointments, or the treatment given for ordinary burns. This will cause serious wounds. Another hazard is that contact with aqueous acid is not immediately painful and the individual, thinking that it is only water, will not treat it. His negligence will be paid for later. A good rule to follow when working with hydrogen fluoride is to treat any liquid on the surface of the skin as if it were hydrofluoric acid, despite the fact that it might be pure water. Of course, rubber gloves, face shields, goggles, safety clothing, etc., should be worn. If there is a considerable inhalation or surface area contaminated, the physician will probably give the patient calcium salts internally or intervenously to counteract the precipitation of calcium ion by the fluoride ion. Even a neglected wound caused by hydrofluoric acid will be helped by proper treatment but the sooner the treatment is applied the better. Safety measures in regard to technique, equipment, treatment, etc., are carefully discussed in the following references (Phillips Petroleum Company, 13; Fehr, 19; Harshaw Chemical Company, 20; Universal Oil Products Company, 21) and various manuals of the Manufacturing Chemists' Association which deal with the unloading of tank cars, cylinders, etc.

VI. Types of Reactions

Hydrogen fluoride is very versatile in its ability to catalyze the reactions of organic chemical compounds. In what follows, reactions which have been published are listed under a number of headings. As the classification of a reaction may depend on the reader's point of view, reactions may appear under headings different from the expected. Hydrogen fluoride will catalyze many reactions in which hydrocarbons are reagents. It will cause a long list of substances to react with aromatic hydrocarbons, e.g., olefins, alkyl halides, alcohols, mercaptans, ethers, carboxylic acids, acid halides, esters, sulfuric acid, nitric acid, carbon monoxide, oxygen, etc. Aliphatic hydrocarbons also serve as reagents in some reactions. Isoparaffins are more reactive but even saturated straight chain paraffins react under certain conditions. It causes polymerization of reactive substances, such as olefins and aldehydes. It acts as a dehydrating agent in the formation of ethers and esters. It catalyzes rearrangements, isomerization and racemization. It aids ring closure and catalyzes oxidation. Other reactions may be found later, but all the above have been reported in the literature.

1. Alkylation

As alkylation was the first reaction reported which demonstrated the truly catalytic power of hydrogen fluoride and it is the one now being used on the largest scale in industry, it is appropriate to begin the discussion of specific reactions with alkylation.

If we define alkylation as a reaction in which the product contains one or more alkyl or paraffinic groups in the molecule than were in the molecule of the source material, a great number of specific reactions are included under this heading. Hydrogen fluoride appears capable of catalyzing most, if not all, of these reactions. The material to be alkylated may be either aromatic or aliphatic or many of their derivatives; and the source of the alkyl groups can be olefins, alkyl halides, alcohols, esters, ethers, mercaptans, sulfides, other paraffinic hydrocarbons, etc. An important feature of the catalytic use of hydrogen fluoride for alkylation is that it functions with all types of alkylating reactants and is not poisoned by any of them. Its use in connection with sulfur compounds is a good example of this.

Elsewhere the author has published examples of the use of hydrogen fluoride to catalyze the alkylation of benzene using a variety of alkylating agents (Simons, 22, 23, 24). The breadth and scope of the catalytic power of hydrogen fluoride can be better obtained by a more complete examination of the alkylations now published using this catalyst. For reasons of organization, the subject is divided into aromatic and aliphatic reactions.

a. Aromatic Compounds with Olefins. Olefins function extremely well as alkylating agents. There is no gaseous product to be removed, such as hydrogen chloride, and so no provision needs to be made for its escape.

No products are formed, such as water, an alcohol, or a carboxylic acid, which dissolve in and reduce the activity of the hydrogen fluoride. Olefins are but slightly soluble in hydrogen fluoride and, therefore, do not reduce its activity. For these two reasons only small amounts of the catalyst are required. It might be thought since hydrogen fluoride both reacts with olefins to form fluorides (Grosse and Linn, 25) and polymerizes them (Fredenhagen, 26), that a considerable loss of reagent would result and that impurities in the alkylated product would be found. This is not the case since both the addition of hydrogen fluoride and the polymerization reaction are relatively slow compared to the alkylation reaction. The assumption that the fluoride is formed as an intermediate in the mechanism of the reaction will be shown later to be unsound.

The first reactions concerned (Simons and Archer, 27) alkylation of benzene with propylene to form isopropylbenzene, with isobutene to form *t*-butylbenzene and di-*t*-butylbenzene, and trimethylethylene to form amylbenzene. Later on (Simons and Archer, 28) studied these and other reactions in more detail and showed that high yields could be obtained and that the product was not contaminated with tars or other obnoxious impurities. It was shown that the products obtained with trimethylethylene were mono- and di-*t*-amylbenzene, that phenylpentane resulted from the use of pentene-2, and that cyclohexene produced cyclohexylbenzene. Cinnamic acid reacted with benzene (Simons and Archer, 29) to form β-phenylpropionic acid and allyl benzene reacted with benzene to form 1,2-diphenylpropane. It is interesting to note that although allyl alcohol reacted with benzene to form 1,2-diphenylpropane, the intermediate in the reaction, allylbenzene, was isolated and identified. This shows that in this case the hydroxyl reacted at a more rapid rate than the double bond. Both di- and triisobutylene reacted with phenol (Simons and Archer, 30) at 0°, when using hydrogen fluoride containing only relatively small quantities of water, to form *t*-butylbenzene, but diisobutylene with 70% hydrogen fluoride produced *p*-*t*-octylphenol. Cyclohexene reacted with toluene to form cyclohexyltoluene and octene-1 rapidly reacted with toluene to form 2-octyltoluene (Simons and Basler, 31).

The Jackson laboratory of the du Pont Company soon became interested in the catalytic power of hydrogen fluoride. The results of its work are recorded in three excellent papers. Using acrolein as the alkylating agent and hydrogen fluoride as the catalyst, peri syntheses have been performed (Calcott *et al.*, 32), both those that are catalyzed by sulfuric acid and others that are not. By appropriate condensation, dehydration, and reduction, perylene was obtained from phenanthrene and 1,10-trimethylene-9-hydroxyphenanthrene, 4,5 benzpyrene from 9,10 dihydroanthracene, perinaphthindone from α and β-naphtol. 3-Hexene (Spiegler and Tinker, 33) was condensed with benzene to form 3(phenyl)-hexane and 1,4 di-(1′ethylbutyl)benzene; with chlorobenzene to form 4 chloro-(1′ethylbutyl)-benzene; with toluene to form 1-methyl-4-(1′ethyl-butyl)-benezene; with *m*-xylene to form an unseparated hydrocarbon mixture; with naphthalene to form 3-(naphthyl)-hexane and poly-*s*-hexyl-naphthalene and with diisopropylnaphthalene to form an unseparated mixture of hydrocarbons.

In another paper from the Jackson Laboratories of the du Pont Company (Calcott *et al.*, 34) there is reported a repetition of some of the reactions of Simons and Archer, as well as additional ones. Mono-, di-, and 1,2,4,5 tetraisopropylbenzene were obtained from propylene and benzene; both 1′-chloro-*t*-butylbenzene and di-(1′-chloro)-*t*-butylbenzene were obtained from 3-chloro-2-methyl-propene-1 and benzene; *p*-*t*-butyltoluene and di-*t*-butyltoluene were obtained from diisobutylene and toluene; tetraisopropylnaphthalene was obtained from propylene and naphthalene; naphthyl-stearic acid was obtained from oleic acid and naphthalene; mixed isopropyltetrahydronaphthalene was obtained from propylene and tetrahydronaphthalene; 2,4,6-triisopropylphenol was obtained from propylene and phenol; a mixture of monoisopropylated *m*-cresols was obtained from propylene and *m*-cresol; and di-(*s*-hexyl)-diphenyl oxide was obtained from hexene-3 and diphenyl oxide.

b. Aromatic Compounds with Alkyl Halides. In the use of alkyl halides for alkylation means must be provided for the escape of the hydrogen halides formed in the reaction as these are but sparingly soluble in hydrocarbons or in liquid hydrogen fluoride. Olefins combine directly and very rapidly and produce no gaseous product. Gaseous olefins like propylene can be added as a gas to a stirred mixture in a closed vessel of the hydrocarbon and hydrogen fluoride and are rapidly absorbed. Liquid olefins can be conveniently added slowly or merely mixed, depending on the product desired. With alkyl halides a vent for the escape of the gaseous product is necessary. On a large scale or where the reaction takes place above room temperature it is best to provide a cold reflux condenser on this vent to return hydrogen fluoride and evaporated organic substances to the reaction vessel. On a small scale this can be omitted, if a sufficiently large and strong vessel is provided so that the gases can be retained without bursting the container. For tertiary halides the reaction takes place rapidly at 0°C. or below with most substances to be alkylated. Secondary halides react more slowly, but room temperature is adequate for reaction in most cases. Primary halides react still more slowly, but 100°C. is usually sufficient for most reactions. Alkyl halides alone with hydrogen fluoride do undergo a considerable amount of reforming, as has been shown in the case of *t*-amyl and *t*-butyl chlorides (Simons *et al.*, 35); but as these reactions of polymerization and rearrangement are relatively slow compared to alkylation, they do not interfere.

Isopropyl chloride was shown to react with benzene to form diisopropylbenzene (Simons and Archer, 27). *t*-Butyl chloride with benzene formed both mono- and di-*t*-butylbenzene and *t*-amyl chloride formed both mono- and di-*t*-amylbenzene. With toluene (Simons and Archer, 36) *t*-butyl chloride formed *p*-*t*-butyltoluene, and with naphthalene, a mono- and two di-*t*-butylnaphthalenes. *n*-Propyl bromide reacted with benzene to give a product which was 88% isopropylbenzene and 12% normal propylbenzene. *t*-Butyl chloride with phenol formed *p*-*t*-butylphenol (Simons *et al.*, 37), and it formed with ethylfuroate, ethyl-5-*t*-butylfuroate. Benzyl chloride reacted readily with benzene at 100° to form diphenylmethane (Simons and Archer, 38). *n*-Butyl alcohol reacted with benzene to form *s*-butylbenzene (Simons and Archer, 39). Ethyl iodide and benzene gave ethylbenzene (Simons and Passino, 12). *t*-Amyl fluoride reacted with benzene (Simons and Bassler, 31). Cyclohexyl fluoride and cyclohexyl chloride with toluene formed *p*-cyclohexyltoluene, and 2-fluorooctane with toluene formed 2-*p*-octyltoluene. The rapidity of the reaction of fluorides, chlorides, bromides, and iodides, appears to be in the order given. This is not necessarily an indication of the strength of the carbon halide bond, as these probably are in the reverse order, nor is it an indication of ease of ionization, as the reactions most probably, as will be shown later, do not proceed by an ionic mechanism. It is related to the ease of escape of the hydrogen halide. As hydrogen fluoride is the catalyst its formation does not retard the reaction. Hydrogen chloride, hydrogen bromide, and hydrogen iodide, have boiling points in this order and so their solubilities increase and ease of escape decreases in this order.

c. Aromatic Compounds with Alcohols. With alcohols as alkylating agents larger quantities of hydrogen fluoride are required than with either olefins or alkyl halides, as here it serves as a solvent in addition to its function as a catalyst. Both the alcohols and the water formed in the reaction dissolve in the liquid hydrogen fluoride and reduce its activity. The speed of alkylation follows that of the alkyl chlorides, the tertiary reacting the fastest and the primary the slowest. The alcohols react as fast or faster than the chlorides probably coming closer to the olefins and fluorides (Simons and Bassler, 31) (Simons and Archer, 39). Alcohols also react when treated alone with hydrogen fluoride, tertiary alcohols in particular. A tertiary alcohol after solution in liquid hydrogen fluoride cannot be recovered. Despite this fact good yield of alkylated products is obtained with alcohols in alkylation reactions and no impurities produced by such polymerizations are found. This is undoubtedly because the alkylation reaction is more rapid than the polymerization.

t-Butyl alcohol and benzene gave both mono- and di-*t*-butylbenzene (Simons *et al.*, 37). Allyl alcohol reacted with benzene to produce both allylbenzene and 1,2-diphenylpropane. (Simons and Archer, 38.) The activity of the hydroxyl group is indicated in the fact that 2-phenylpropanol was not separated. Benzyl alcohol reacted with benzene to form diphenylmethane (Simons and Archer, 39) despite the fact that this reaction is reported (Calcott *et al.*, 34) to form 1,2,3,4,5,6-hexaphenylcyclohexane by the polymerization of the alcohol. Isopropyl alcohol with benzene gave isopropylbenzene, 1,4-diisopropylbenzene, 1,2,4-triisopropylbenzene and 1,2,4,5-tetraisopropylbenzene. Ethyl alcohol with benzene gave high yields of ethylbenzene and diethylbenzene at 200°C. (Simons and Passino, 40.) Cyclohexanol with toluene gave *p*-cyclohexyltoluene (Simons and Bassler, 31) and octonal-2 gave 2-*p*-octyltoluene. *m*-Xylene and *t*-butyl alcohol gave *t*-butyl-*m*-xylene (Calcott *et al.*, 34), naphthalene and *t*-butyl alcohol gave di-*t*-butylnaphthalene, phenanthrene and *t*-butyl alcohol gave mixed *t*-butylphenanthrenes, *o*-nitroanisole and isopropyl alcohol gave 1-methoxy-2-nitro-4-isopropylbenzene, *o*-nitroanisole and cyclohexanol gave 1-methyl-2-nitro-4-cyclohexylbenzene, hydroquinone and isopropyl alcohol gave monoisopropylhydroquinone, β-naphthol and isopropyl alcohol gave diisopropyl-β-naphthol, 2,3-hydroxynaphthoic acid and isopropyl alcohol gave monoisopropyl-2,3-hydroxynaphthoic acid, naphthalene-2-sulfonic acid and isopropyl alcohol gave polyisopropylnaphthalene-2-sulfonic acid; and *p*-anisidine and cyclohexanol gave monocyclohexyl-*p*-anisidine. Optically active *s*-butyl alcohol produced with benzene, *s*-butylbenzene with a small but definite rotation (Burwell and Archer, 41). The condensation of chlorobenzene with chloral, chloral hydrate, and chloral alcoholate to form D.D.T. has been successfully accomplished with hydrogen fluoride as the condensing agent (Simons *et al.*, 42).

d. Aromatic Compounds with Ethers. Ethers function well as alkylating agents and in general are resistant to the action of hydrogen fluoride, in fact, ethers may be formed by reactions in liquid hydrogen fluoride. They are in general very soluble and their reactions are similar to the reactions of alcohols.

With benzene, *n*-butyl ether gave *s*-butylbenzene and benzyl ether gave diphenylmethane (Simons and Archer, 39). Isopropyl ether with benzene gave isopropylbenzene, 1,4-disiopropylbenzene, 1,2,4-triisopropylbenzene and 1,2,4,5-tetraisopropylbenzene in about the same yields as the alcohol under the same conditions. Diethyl ether gave ethylbenzene readily with benzene (Simons and Passino, 40). Anthracene reacted with isopropyl ether to give diisopropylanthracene (Calcott *et al.*, 34), α-nitronaphthalene reacted with isopropyl ether to give monoisopropyl-1-nitronaphthalene, *o*-cresol reacted with dibenzyl ether to give both monobenzyl-*o*-cresol and dibenzyl-*o*-cresol, benzoic acid reacted with isopropyl ether to give mono-*m*-isopropylbenzoic acid, *p*-aminophenol reacted with isopropyl ether to give diisopropyl-*p*-aminophenol and 4,4′-dihydroxytetraisopropyldiphenylamine, *N*-dimethyl-*p*-aminophenol reacted with isopropyl ether to give both monoisopropyl-*N*-dimethyl-*p*-aminophenol and diisopropyl-*N*-dimethyl-*p*-aminophenol, *p*-anisidine reacted with isopropyl ether to give diisopropyl-*p*-anisidine and 4,4′-dimethoxytetraisopropyldiphenylamine, 1-diethylamino-3-ethoxybenzene reacted with isopropyl ether to give monoisopropyl-1-diethylamino-3-ethoxybenzene, and 1-amino-2-methoxynaphthalene reacted with isopropyl ether to give triisopropyl-1-amino-2-methoxynaphthalene.

e. Aromatic Compounds with Esters. Esters also serve as alkylating agents. The alkyl radical of the alcohol forms the alkyl group of the product and the free acid is simultaneously formed. As the acid can serve as an acylating agent, both alkylation and acylation can take place; but as acylation is slower and requires more extreme conditions of concentration and temperature, it can be prevented by keeping the conditions sufficiently mild. In general esters function similarly to alcohols and ethers. Using benzene as the material to be alkylated *t*-butyl acetate gave *t*-butylbenzene (Simons *et al.*, 43), isopropyl acetate gave isopropylbenzene, *n*-butyl acetate gave *s*-butylbenzene, *s*-butylisobutyrate gave *s*-butylbenzene, and benzyl acetate gave diphenylmethane.

f. Aromatic Compounds with Sulfides and Mercaptans. Alkyl sulfides and mercaptans function very similarly to ethers and alcohols. Hydrogen sulfide is produced and it escapes as a gas not being significantly soluble in liquid hydrogen fluoride. In this respect the technique of procedure is similar to that used for alkyl halides.

g. Aromatic Compounds with Hydrocarbons. Hydrocarbons themselves can be used as the source of alkyl groups. Cyclopropane to form *n*-propylbenzene, di-*n*-propylbenzene and tri-*n*-propylbenzene (Simons *et al.*, 44). Alkylated aromatic compounds can also serve as the source of alkyl groups for another aromatic compound more readily alkylated. Phenol and *t*-butylbenzene react to give *t*-butylphenol and benzene at 0° (Simons *et al.*, 45). *t*-Butylbenzene heated alone with hydrogen fluoride at 50° converts to di-*t*-butylbenzene, chiefly para. Other similar conversions take place. A highly alkylated aromatic compound when treated with the unalkylated substance and hydrogen fluoride will share the alkyl groups. Paraffinic hydrocarbons which are produced by alkylation, such as isoctane, when treated with an aromatic compound like benzene and hydrogen fluoride will supply alkyl groups for alkylation of the aromatic.

Monoalkylbenzene or other aromatic compounds react more rapidly than benzene itself in alkylation with hydrogen fluoride and the dialkylbenzene react less rapidly in general to form tri and higher alkylated products. The polyalkylated products require more strenuous conditions. To form the monoalkyl product the alkylating agent should be added slowly to a large excess of the aromatic compound.

There is one alkyl group, i.e., methyl, that has not been successfully used for alkylation with hydrogen fluoride catalyst. (Simons and Passino, 40.) In addition methyl groups do not exchange using hydrogen fluoride. Toluene mixed with hydrogen fluoride and heated at 200°C. for a considerable period can be recovered without change and without the formation of either xylenes or tars. Phenyl groups also resist reaction and phenylation as well as methylation have not been accomplished using hydrogen fluoride as the catalyst.

h. Aliphatic Compounds. Although the alkylation of aliphatic compounds has become the largest commercial catalytic use of hydrogen fluoride up to this time, there is much less concerning it in the scientific literature. The rapid extension to large scale use and development of commercial processes came about because of the beneficial use of hydrogen fluoride as the catalyst in the production of aviation alkylate and the large demand for aviation fuel during the war. War time conditions not only put a stop to fundamental scientific work but blocked publication. The aliphatic alkylation reactions are not as clean cut in the formation of single isolable products as the aromatic ones and so are not as readily adapted to graduate student problems in the college laboratory and the writing of scientific papers acceptable in our current chemical publications. Very shortly, however, after the catalytic powers of hydrogen fluoride were discovered it was found in the author's laboratory that the alkylation of aliphatic compounds could readily be accomplished. This was to be expected.

The alkylation of isoparaffins with olefins is the reaction involved in the large scale processes in the petroleum industry. Any isoparaffin and almost any olefin can be used; and although the product in largest percentage is usually with the number of carbon atoms equal to the sum of the carbon atoms in the paraffin and olefin, this is not always the case. A great mixture of branch chain products is obtained (Phillips Petroleum Company, 13).

The technique commercially used for this process is a rapid agitation of two liquid phases, one essentially hydrogen fluoride and one essentially hydrocarbon. The reaction apparently takes place rapidly at the liquid-liquid interface. Reaction also takes place in either of the liquid phases but at a much slower rate. With aromatic compounds reaction takes place rapidly and homogeneously in either a hydrocarbon liquid phase or a hydrogen fluoride liquid phase.

Just as in the case of aromatic compounds isoparaffins can be alkylated with sources of alkyl groups other than olefins. Alkyl halides, alcohols, ethers, mercaptans, sulfides, etc., can be used. When olefins are used some alkyl fluorides from a combination of olefin and hydrogen fluoride are always formed. The quantity of this in the product can be greatly reduced by providing conditions under which the alkyl fluoride is used in alkylation. The apparent paradox is provided, in that the fluoride content of the product is lessened by further treatment with hydrogen fluoride. A more thorough treatment of the details of the alkylation of isoparaffins with olefins is found elsewhere in this volume.

The exchange of alkyl groups among paraffinic hydrocarbons occurs similarly to this reaction with aromatic compounds. This is one of the reasons for the great multiplicity of products obtained in the isoparaffin alkylation. A pure isoparaffin on treatment with hydrogen fluoride at low temperature will form a range of substances or conversely, if a particular compound is removed from a mixture of isoparaffins and the mixture then given a further treatment with hydrogen fluoride, a further separation will remove more of this particular compound.

At low temperature hydrogen fluoride does not isomerize normal paraffinic compounds nor does it alkylate them with olefins or other alkylating agents. At higher temperatures this situation is changed. The addition of a small amount of boron trifluoride also changes the results at higher temperatures. Propane has been alkylated with ethylene using hydrogen fluoride containing 2 to 7% BF_3 in a contact time of 1 to 100 minutes at 40 to 120°C. and 15 to 65 atm. (Frey, 46.)

In the reactions of aliphatic compounds it has been shown that *t*-butyl chloride reacted with the olefins trimethylethylene and cyclohexene (Simons *et al.*, 37). Further study of the latter reaction demonstrated the formation of 1-chloro-3-*t*-butylcyclohexane (Simons and Meunier, 47) and a reaction was shown to occur between isopropyl chloride and cyclohexene.

Tertiary alkyl halides undergo a series of reactions when treated with hydrogen fluoride. *t*-Amyl chloride by means of reactions catalyzed by hydrogen fluoride yielded a series of teritary chlorides including *t*-butyl chloride (Simons *et al.*, 35). *t*-Butyl chloride gives the same series of products including *t*-amyl chloride. This reaction is probably related to the redistribution of alkyl groups that takes place on treating isoparaffins with hydrogen fluoride.

2. Acylation

Despite the fact that certain ketones such as acetone and acetophenone (Simons and Ramler, 48) undergo extensive polymerization when treated at elevated temperatures with hydrogen fluoride, a considerable number of important acylations have been accomplished with this catalyst. It is not as superior a catalyst for acylation as it is for alkylation, and in general the yields are not as good. The conditions for favorable acylations are a very active aromatic compound and the resulting ketone relatively stable to polymerization. If there are no hydrogen atoms alpha to the carbonyl group in the final product, it will resist polymerization even at elevated temperatures. Acyl halides can be used, but carboxylic acids and also the acid anhydrides function equally as well. As in the case of alkylation, if the halides are used, means must be provided to handle the escaping hydrogen halide. Acid chlorides react very rapidly with liquid hydrogen fluoride to form hydrogen chloride and the acid fluoride. To the acid fluoride-hydrogen fluoride mixture can be added the material to be acylated; and the reaction vessel closed as no significant additional evolution of hydrogen chloride will occur. When acid anhydrides or the free acids are used, more hydrogen fluoride is required for the same reaction due to the production of water. Esters can also be used but in this case alkylation will also take place.

Acetic acid has been found to react with toluene to form *p*-methylacetophenone (Simons *et al.*, 49), to react with benzene to form acetophenone, and to react with phenol to form *p*-hydroxyacetophenone. Acetyl chloride also formed acetophenone with benzene and acetic anhydride reacted with toluene to form both *p*-methylacetophenone and 2,4-diacetyltoluene. Valeric acid reacted with toluene to form *p*-tolyl-*n*-butyl ketone. Both benzoic acid and benzoyl chloride reacted with toluene to form *p*-tolylphenyl ketone. Acenaphthene with either benzoic acid or benzoyl chloride gave 3-benzoylacenaphthene (Fieser and Hershberg, 50), acenaphthene with succinic anhydride gave both γ-(3-acenaphthoyl)-propionic acid and the 1-isomer, hydroquinone and benzoic acid gave hydroquinone monobenzoate, acenaphthene and acetic acid gave 1-acetoacenaphthene, and acenaphthene and crotonic acid gave 1′-methyl-3′-keto-2,3-cyclopentenoacenaphthene. 3-Acetoperinaphthane was prepared from perinaphthane and acetic anhydride (Fiezer and Hershberg, 51), hydrindene-5-carboxylic acid from hydrindene and acetic acid, 5-benzoylhydrindene from hydrindene and benzoyl chloride, 5-(α-

naphthoyl)-hydrindene from hydrindene and α-naphthoic acid, acetonaphthalene from naphthalene and acetic anhydride, and both 2- and 3-acetophenanthrene from phenanthrene and acetic anhydride.

3. *Ring Closure*

The successful use of hydrogen fluoride for alkylation and acylation would indicate that it would readily catalyze the closure of rings of organic compounds which involve these reactions. It has previously been pointed out that the "peri" synthesis is successfully accomplished with hydrogen fluoride. Ring closure by means of acylation using hydrogen fluoride has been studied in a series of excellent papers by Fieser and coworkers. γ-Phenylbutyric acid gave α-tetralone (Fieser and Hershberg, 50), hydrocinnamic acid gave α-hydrindone, γ-(3-acenaphthyl)-butyric acid gave ketotetrahydroacephenanthrene, γ-(4-methyl-3-diphenyl butyric acid gave 5-methoxy-8-phenyltetralone, 1-(β-1′-naphthylethyl)-cyclohexanol gave chrysene after the product of ring closure was dehydrogenated, *o*-benzylbenzoic acid gave anthrone, 2-(α-naphthylmethyl)-benzoic acid gave 1,2-benz-10-anthrone and 2-(4′-methoxy-1′-naphthylmethyl)-benzoic acid gave 3-methoxy-1,2-benz-10-anthrone. γ-(9,10-Dihydro-2-phenanthryl)-valeric acid gave 5-methyl-8-keto-3,4,5,6,7,8-hexahydrophenanthrene (Fieser and Johnson, 52), and γ-(2-phenanthryl)-butyric acid gave 8-keto-5,6,7,8-tetrahydro-1,2-benzanthracene. Methylnaphthylmethyl benzoic acid gave 1′-methyl-2,3-benz-10-anthrone (Fieser and Hershberg, 51) and *o*-(β-naphthylmethyl)-benzoic acid gave 9-methyl- and 9-allyl-1,2-benzanthracene by cyclizing to the anthrone with hydrogen fluoride and treating with Grignard reagent. 8-Keto-3,-4,5,6,7,8-hexahydro-1,2-benzanthracene was prepared from γ-(9,10-dihydro-2-phenanthryl)-butyric acid (Fieser and Johnson, 53) and 6-hydroxy-3,4-benzpyrene from 4-chryseneacetic acid. β-Benzohydrylglutaric acid became 1,2,3,4-tetrahydro-4-keto-1-phenyl-2-naphthalene acetic acid by treatment with hydrogen fluoride (Newman and Joshel, 54). 2-(*p*-Methylbenzyl)-benzoic acid reacted to form 2-methylanthrone-9 (Fieser and Heymann, 55), 2-(*o*-methylbenzyl)-benzoic acid formed 4-methylanthrone-9, and *o*-(3,5-dimethylbenzyl)-benzoic acid formed 1,3 dimethyl-10-acetoxyanthrone-9. α-Phenylhexahydrophthalide was made from 2-(α-hydroxybenzyl)-cyclohexane-1-carboxylic acid (Fieser and Novello, 56), 5,6,7,8,9,10,10a-octahydro-1,2-benz-10-anthrone from 2-(α-naphthylmethyl)-cyclohexane-1-carboxylic acid, and 4a′-keto-5,6,-7,8,8a,9,10,10a-octahydro-1,2-benzanthracene from 5,6,7,8,8a,9,10,10a-octahydro-1,2-benzanthracene-10-acetic acid.

Hydrogen fluoride is used to cyclize 4-(2-naphthylimino)-2-petanone to 2,4-dimethylbenzo (g)quinoline (Johnson and Mathews, 57). Zinc chloride was later shown (Johnson *et al.*, 58) to give 1,3-dimethylbenzo(f)-quinoline from the same reagent. Similarly abnormal cyclization using hydrogen fluoride gave 3,4-dimethylbenzo(g)quinoline from 3-(2-naphthyliminomethyl)-2-butanone whereas zinc chloride gave the 2,3 compound. Hydrogen fluoride gave 4-methylbenzo(g)quinoline from 1-(2-naphthylimino)-3-butanone, 2,4-dimethylbenzo(h)quinoline from 4-(1-naphthylimino)-2-pentanone, 9,11-dimethylnaphtho(1,2-g)quinoline from 4-(2-phenanthrylimino)-2-pentanone, and 8,10-dimethylnaphtho(2,1-g)-quinoline from 4-(3-phenanthrylimino)-pentanone-2.

4. Rearrangements

The ability of hydrogen fluoride to catalyze the redistribution of alkyl groups among molecules of organic compounds enables it to catalyze a number of rearrangements, which we can define as a reaction in which the product is an isomer of the reactant. Benzophenone oxime underwent the Beckmann rearrangement with hydrogen fluoride to form benzanilide. (Simons *et al.*, 45.) Phenyl acetate underwent the Fries rearrangement at 100° with hydrogen fluoride to form *p*-hydroxyacetophenone. A rearrangement similar to the Fries is the conversion of a phenyl sulfonate to an hydroxy sulfone. Hydrogen fluoride at 100° converted *p*-cresyl benzene-sulfonate to 2-hydroxy-5-methyl-diphenyl sulfone. Optically active 2-butanol was racemized by hydrogen fluoride (Burwell, 59).

Normal paraffinic hydrocarbons are unaffected by hydrogen fluoride up to reasonably high temperatures. Unalkylated or methylated aromatic compounds are also resistant to any action of liquid hydrogen fluoride up to the critical temperature of HF. Alkylated aromatic compounds and isoparaffins tend to distribute and rearrange alkyl groups even at room temperature. Isooctane, for example, when treated with hydrogen fluoride at room temperature forms everything from isobutane to above dodecanes. Normal paraffinic hydrocarbons at higher temperatures do isomerize to isoparaffins, redistribute alkyl groups, rearrange, and the like. Normal butane at 175°C. and 250 atm. for 3 hours changed to 3.2% propane, 4.3% isobutane, 2.5% isopentane, and the remainder 90% was the original butane (Frey, 60). At higher temperature more conversion takes place. At 270° and 230 atm. for 2 hours and 20 minutes the products were 0.6% methane, 10.1% propane, 25.9% isobutane, 8.1% isopentane, 2.0% *n*-pentane, 2.0% hexanes, and 0.5% heptanes, with 50.8% not converted. Pentane at 300°C. gave 10% conversion in 30 minutes, namely, 17.3% propane, 31.5% isobutane, 19.7% *n*-butane, 22.5% isopentane, 6.0% hexane, and 3.0% higher boiling. A longer contact time at the same temperature (4 hours) gave the following products from pentane; 2.7% propane, 4.6% isobutane, 3.0% *n*-butane, 10.5% isopentane, 76.6% pentane, 1.7% hexanes, and 0.9% higher boiling compounds. At still higher temperatures more rapid conversion takes place. Thirty minutes at 345°C. and 300 atm. converted 75% of butane to isobutane, 10% to propane, and 15 %to heavier hydrocarbons (Frey, 61). In 20 minutes at 370° and 68 atm. pentane was converted to 60% isobutane and 25% heavier hydrocarbons.

5. *Polymerization*

Effects caused by polymerization or at least by related reactions were well known to all those who had worked with hydrogen fluoride from the early days of its original preparation. The hardening of rubber and similar substances has long been common laboratory knowledge. The first listing of substances polymerized by hydrogen fluoride (Fredenhagen, 26) gives oleic acid, linseed oil, poppyseed oil, castor oil, sunflower oil, soybean oil, amylene, butadiene, dipentene, indene, isoprene, piperylene, pyrrole, thiophene, and thionaphthene. The polymerization of ethylene, propylene, and cyclohexene with liquid hydrogen fluoride has also been observed (Grosse and Linn, 25). The polymerization of cyclohexene has been further studied (McElvain and Langston, 62) in regard to the products formed. At 100°C. a 4-hour treatment with hydrogen fluoride gave 17% of a dimer, 6% of a trimer, 8% of a tetramer, 6% of a pentamer, 6% of a hexamer, 6% of a heptamer, 8% higher boiling, and the rest of the material boiling in the range of the isolated fractions but unseparated. Ketene acetal is polymerized with hydrogen fluoride to form a cyclic trimer 1,1,3,3,5,5-hexaethoxycyclohexane (McElvain and Langsten, 63).

The polymerization products of propylene have been observed to be saturated hydrocarbon polymers and terpenelike unsaturated hydrocarbons (Kuhn, 64). The condensation of formaldehyde with phenols and cyclohexanols by means of aqueous hydrogen fluoride has also been observed (Badertscher *et al.*, 65).

In the author's laboratories the polymerization of aldehydes, ketones, and alcohols by liquid hydrogen fluoride has been repeatedly noted. Acetaldehyde polymerizes and acetone forms polymeric substances on standing for a period of time in solution in hydrogen fluoride. If the solution is separated shortly after mixing, the acetone may be recovered. The same is true of tertiary alcohols. The peculiar action of tertiary chlorides (Simons *et al.*, 35) probably results at least in part from polymerization. The products obtained most likely come from destruction of the polymers in the process of distillation. Benzaldehyde forms a shellac like resin when treated with hydrogen fluoride. A rather interesting polymerization reaction occurs upon treating aralkyl ketones with hydrogen fluoride (Simons and Ramler, 48). From many of these benzoic acid and a resin are the products. Compounds containing two or three hydrogen atoms on the carbon atom adjacent to the carbonyl group (propiophenone and acetophenone) gave benzoic acid and a resin as products at 100°C. At 50° dypnone was formed from acetophenone and 1,3-diphenyl-2-methylpentene-2-one-1 from propiophenone. Dypnone at 100° gave a resin and benzoic acid. α-Methyl-styrene and 3-phenylpentene-2 were obtained, respectively, from the resins from acetophenone and propiophenone by thermal decomposition. Isobutyrophenone gave a resin but no acid. *p-t*-Butylacetophenone gave a resin and *p-t*-butylbenzoic acid. Benzophenone and α-trichloroacetophenone gave no reaction at 100°C. A postulated mechanism assumes a tertiary alcohol as the first step. Only ketones with one or more alpha hydrogen atoms can do this. The tertiary alcohol can polymerize. It can also dehydrate to form an unsaturated ketone, if there are two or three alpha hydrogen atomes in the original ketone. This unsaturated ketone can either polymerize or react with hydrogen fluoride to form benzoyl fluoride and a substituted styrene polymer. The benzoic acid comes from the reaction of benzoyl fluoride and water.

6. Formation of Esters and Ethers

The powerful dehydrating property of hydrogen fluoride would cause it to be expected to assist in reactions in which water is a product. Such dehydration reactions would not in the true sense be catalytic. However, as the addition of water in the case of the hydrolysis of esters (Simons and Meunier, 66) has been shown to be catalyzed by hydrogen fluoride, the catalytic powers of hydrogen fluoride are probably involved in the reverse reaction, as a catalyst must necessarily accelerate the reverse reaction if it does so for the forward one. For this reason some of these reactions are included here.

The charring action of hydrogen fluoride on wood and other cellulosic substances is common experience among those who have handled the material. Cellulose itself is soluble in hydrogen fluoride but degradation proceeds slowly at low temperatures (Helferick and Bottger, 67). Starch at −20°C. for 30 minutes gives crude amylan (Helferick *et al.*, 68). The esterification of cellulose to cellulose acetate is catalyzed by hydrogen fluoride (Bethelemy, 69). Because the solution of cellulose in hydrogen fluoride will yield polyglucosans, either by evaporation or precipitation which in turn can be converted to glucose, it has been postulated that hydrogen fluoride splits the oxygen linkages to form glucosyl fluoride (Fredenhagen and Cadenback, 70).

Ethyl acetate has been prepared from acetic acid and ethyl alcohol in liquid hydrogen fluoride at 0°C. (Simons and Meunier, 66). The hydrolysis of ethyl acetate to acetic acid and ethyl alcohol has been accomplished by a small amount of water in a solution of the ester in hydrogen fluoride. Esters have also been made by the addition of an olefin to a solution of a carboxylic acid in liquid hydrogen fluoride. Cyclohexyl acetate was made from cyclohexene and acetic acid, cyclohexyl *n*-butyrate from cyclohexene and *n*-butyric acid, octyl acetate from a mixture of octenes-1 and 2 and acetic acid, and octyl-*n*-butyrate from a mixture of octenes-1 and 2 and n-butyric acid.

The formation of certain ethers can also be accomplished with hydrogen fluoride. Anisole rather than methylphenol results from a reaction between phenol and methyl alcohol at elevated temperature (Simons and Passino, 40). The addition of an olefin to an alcohol to form an ether was shown to occur in the reaction between cyclohexene and cyclohexanol for form dicyclohexyl ether (Simons and Meunier, 66).

7. *Addition of Carbon Monoxide*

The addition of carbon monoxide to alkyl halides and alcohols is probably not greatly different from the addition of olefins to the same substances. Carbon monoxide at 43 atomospheres and 160° did not add to isopropyl chloride in the presence of dry hydrogen fluoride, but with the addition of a small amount of water isobutyric acid was obtained (Simons and Werner, 71). Under nearly the same conditions the addition of methanol in place of water served to form the same product from the same reagents. Formic acid which produces both water and carbon monoxide when heated with hydrogen fluoride gave rise, in the presence of hydrogen fluoride, to isobutyric acid from *n*-propyl alcohol. A six-carbon acid was produced by the same procedure from *s*-amyl bromide. At 150°C. nickel carbonyl gave isobutyric acid by reaction with isopropyl chloride.

Carbon monoxide will also add to aromatic coumpounds such as benzene and toluene. As the product of such an addition is an aldehyde and as aromatic aldehydes readily polymerize under the conditions necessary for the addition of carbon monoxide, the simple addition product is not obtained. These reactions have been performed in the author's laboratory using a technique similar to the addition to alcohols and alkyl halides. The products obtained are the same shellac-like resins that are obtained by treating the theoretically expected aldehyde with hydrogen fluoride under the same conditions.

8. *Sulfonation*

Hydrogen fluoride has been used to produce both sulfonic acids and sulfones. Sulfuric acid, fluorosulfonic acid, and aromatic sulfonic acids or their halides can be used as reagents. At 85–95°C. benzene and sulfuric acid in the presence of hydrogen fluoride gave benzene sulfonic acid and a trace of diphenyl sulfone (Simons *et al.*, 72). At 140–150°C. the same reagents gave chiefly the sulfone. At 60–70°C. benzene and fluorosulfonic acid in the presence of hydrogen fluoride gave benzene sulfonic acid but at 160° diphenyl sulfone was the chief product. At 85–90° both benzene and *p*-toluene sulfonic acid and toluene and benzenesulfonyl chloride gave *p*-tolylphenyl sulfone.

9. Nitration

Aromatic compounds can be nitrated in hydrogen fluoride in a very rapid and vigorous reaction. Nitrobenzene can be obtained by adding potassium nitrate to a suspension of benzene in hydrogen fluoride at 0°C. (Fredenhagen, 73) and phenol can be nitrated by adding nitric acid to a solution of phenol in hydrogen fluoride (Gleich, 74). Benzene at 0° reacts so rapidly in the presence of liquid hydrogen fluoride that, if the nitric acid is added to the center of the reaction vessel, it is consumed before it reaches the wall. Thus copper or iron vessels may be used despite the fact that a mixture of hydrogen fluoride and nitric acid reacts rapidly with these metals. It is interesting also that a quantitative yield of nitrobenzene can be obtained by this reaction at 0° without the formation of detectable amounts of dinitrobenzene (Simons *et al.*, 72) and nitrobenzene fails to nitrate further. At higher temperature, however, and more drastic conditions nitration of benzene proceeds to produce higher nitro compounds than the mono.

10. Oxidation

One of the more interesting of the reactions catalyzed by hydrogen fluoride is the oxidation of organic compounds at temperatures below 200°C. Molecular oxygen is the agent, hydrogen fluoride the catalyst and aromatic, alicyclic, and aliphatic compounds the reagents (Simons and McArthur, 75). Reaction takes place down as low as 0° and with atmospheric air to furnish the oxygen, but under these conditions the reaction is slow. Pressures up to 100 atmospheres of oxygen have been used. A great variety of oxygen carriers aid in the reaction.

With aromatic compounds conditions can be found so that ring oxidation predominates and phenolic compounds are formed. Benzene is oxidized quantitatively to phenol. Toluene is oxidized to *o*-cresol, *m*-xylene to 1,3-xylen-4-ol, and naphthalene to *β*-naphthol. The addition of certain additional catalyst, such as molybdenum oxide, promoted coupling reactions and biphenyl was formed from benzene, bi- or polytolyl hydrocarbons from toluene, di- and polyxylyls from *m*-xylene, and a trinaphthyl from naphthalene. Some side chain oxidation occurred, and a small amount of benzoic acid was obtained from toluene and *m*-toluic acid from *m*-xylene. It is interesting that benzoic acid was a product of the oxidation of benzene, when molybdenum oxide was present. Apparently it came from the oxidation of one ring of biphenyl.

More drastic conditions of temperature, pressure and time gave carbon as the product from both aromatic and aliphatic compounds. This carbon is an activated char with decolorizing properties. Tars, tarry substances, the oxides of carbon, and the usual products of the oxidation of aromatic compounds (dicarboxylic acids) were all conspicuous by their absence. Benzotrifluoride oxidized to benzoyl fluoride. The aliphatic compounds cyclohexane, methycyclohexane, *n*-heptane, etc., could be oxidized to carbon and water.

11. Promoters

The use of additional substances to increase the activity of a catalyst is a well known phenomenon. Hydrogen chloride or traces of water are known to promote aluminum chloride catalyzed reactions. In the same way the reaction of isoparaffins with olefins has been shown to be catalyzed by boron trifluoride in the presence of nickel powder and with water as the promoter (Ipatieff and Grosse, 76). Hydrogen fluoride can take the place of the water and thus serve as the promoter.

After the discovery by Simons and Archer that hydrogen fluoride was a powerful catalyst for condensation reactions, it became more usual to use it as the catalyst and add small amounts of other substances to it as promoters. It has been used in this way in conjuction with sulfuric acid in the alkylation of isoparaffins with olefins (Schmerling and Pines, 77) and for the same type of reaction in conjunction with boron trifluoride (Grosse, 78). The same mixture has been used to produce saturated cyclohexane hydrocarbons from methylcyclopentane and propylene (Pines and Ipatieff, 79). This mixture has also been used in the exchange of alkyl groups among aromatic hydrocarbons (Lien and Shoemaker, 80). Previous mention has been made of the reaction between propane and ethylene catalyzed with hydrogene fluoride with a small amount of boron trifluoride as the promoter. The use of fluorosulfonic acid as a catalyst for the alkylation of isoparaffins with olefins (Standard Oil Development Company, 81) is probably the catalytic action of either hydrogen fluoride or sulfuric acid or one of them promoted by the other, as fluorosulfonic acid is a most powerful dehydrating agent and reacts violently with water to form sulfuric acid and hydrogen fluoride. Dry fluorosulfonic acid reacts violently with hydrocarbons at slightly elevated temperatures. It would be interesting to find, if under absolutely anhydrous conditions, fluorosulfuric acid is a favorable catalyst for this reaction. As used commercially for the catalytic conversion of hydrocarbons or in condensation reactions, the hydrogen fluoride is never completely anhydrous. The presence of this small amount of water is not detrimental. It acts as a powerful promoter. It is only when larger amounts of water are present that the catalytic action is retarded. This promotional effect of small amounts of water was shown (Sprauer and Simons, 82) originally in the reaction between toluene and *t*-butyl chloride as catalyzed with hydrogen fluoride. Other substances such as methyl alcohol exhibit the same promotional effect.

VII. Mechanism

It would be highly inappropriate to be specific in regard to the mechanism of organic chemical reactions catalyzed by hydrogen fluoride as many different kinds of reactions are involved. In addition these reactions take place under widely different physical conditions. There are two conditions under which homogeneous reactions take place. These are in a hydrocarbon liquid phase with the catalyst dissolved in this phase and in a hydrogen fluoride liquid phase with the reactants dissolved. Reactions take place in both these media, but the ionic conditions in both are so vastly different that different mechanisms are probably responsible for the formation of the same new product. The high dielectric constant of liquid hydrogen fluoride permits the postulation of ionic intermediates, whereas the low dielectric constant of the hydrocarbon prohibits such assumptions. As condensation reactions take place rapidly in the hydrocarbon phase certainly all these reactions cannot proceed through ionic intermediates. As actually carried out in practice two liquid phases are frequently present with rapid stirring. This is particularly true of the alkylation of isoparaffins, and it is also equally true that in the paraffin alkylation the reaction takes place much more rapidly under these conditions than in either homogeneous phase. This indicates a heterogeneous reaction at the interface between the two liquid phases. About all that can be said even in regard to the mechanism under those conditions is that it requires much more study and that this ought to be quantitative in nature. High temperature so called "vapor phase reactions" catalyzed by hydrogen fluoride are reactions in a condensed film on the walls or packing of the container. These again are heterogeneous reactions and are probably highly complex. There is one generalization in regard to the reactions catalyzed by hydrogen fluoride that can probably be made. All these reactions take place in a condensed phase, i.e., liquid, or in the interface between such phases.

There are three lines of evidence all of which must be given full consideration in any discussion of the mechanism of a specific reaction. There are, first the products formed and their relative proportions and the change of products, if any, with a change of physical conditions; second, the quantitative measurements taken while the reaction is in progress, particularly as a function of time, and any changes which occur with changes of conditions of temperature, concentration, addition of other substances, electrical nature of medium, etc.; third, the energy consideration to provide that any proposed mechanism will not involve an essential step having too high an energy of activation. Such evidence can best be used for homogeneous reactions.

The first question that arises is why hydrogen fluoride is such a powerful catalyst for a large number of organic chemical reactions. It must be related to its extremely high acidity despite the fact that in aqueous solution it is an apparently weak acid (Simons, 16). This acidity cannot be limited to the formation of solvated protons in solution as the catalytic property is experienced in a liquid hydrocarbon phase of low dielectric constant where the concentration of ions is negligible. The reactions, which it catalyzes, are catalyzed in general by acidic substances. The type of mechanisms frequently postulated for reactions catalyzed by aluminum chloride or boron trifluoride, which involve the unshared electron pair of the molecule of the catalyst, cannot be used as hydrogen fluoride does not have such a structure. In fact, hydrogen fluoride does not equally catalyze all the reactions accelerated by these other catalysts. In some cases the products are not the same; and where they are, the mechanisms may be different.

Considerations of mechanism despite their difficulties are extremely valuable and productive. The discovery of the catalytic properties of hydrogen fluoride for condensation reactions came about from considering the mechanisms of certain organic reactions coupled with a knowledge of the chemical and physical properties of hydrogen fluoride. That fundamental acidity is involved in the catalytic properties of hydrogen fluoride is confirmed by the fact that hydrogen chloride under appropriate conditions can catalyze some of the same reactions (Simons and Hart, 81).

On the basis of products formed in a number of condensation reactions only confusion results for any step by step mechanism involving specific identifiable species as intermediates. Here are some of the facts. Benzene condenses with cyclopropane to form *n*-propylbenzene (Simons *et al.*, 44). Normal propyl bromide gives chiefly isopropyl benzene (Simons and Archer, 36) as does propylene (Simons and Archer, 28). Ethyl alcohol gives ethylbenzene, but methyl alcohol does not give toluene (Simons and Passino, 40), and benzyl alcohol and benzyl chloride give diphenylmethane (Simons and Archer, 39). Optically active secondary butyl alcohol gives a secondary butylbenzene with some optical activity (Burwell and Archer, 41). The alkylation of isobutane with propylene does not give triptane as the chief product but rather a spread of hydrocarbons containing many isomers and different numbers of carbon atoms. The postulation of an olefin as a general intermediate is ruled out, despite the fact that the failure of methanol to alkylate benzene might indicate it, by the reaction of benzyl alcohol or halide, which also cannot form an olefin, and by the fact that cyclopropane forms the normal, whereas, propylene forms the iso compound. The postulation of the fluoride as the intermediate is also rules out by the cyclopropane reaction as normal propyl halides give chiefly isopropylbenzene. Under similar conditions olefins, fluorides, alcohols, and chlorides, of the same carbon structure were caused to react with benzene and toluene under the came conditions and for the same length of time. (Simons and Bassler, 31.) The fluoride alone, without the addition of excess hydrogen fluoride, reacted very slowly; and only a small amount of product formed. A larger amount of product was obtained when hydrogen fluoride was added to the alkyl fluoride benzene mixture. The olefin reacted to give considerable product but with less than enough HF to convert the olefin to fluoride and leave some hydrogen fluoride left over to serve as the catalyst. Both olefin and fluoride gave the same yield of product, when the same amount of hydrogen fluoride was used. Alcohols, chlorides, and bromides gave lesser amounts of products in this order. The results of these experiments are also difficult to justify on the basis of either an olefin or fluoride as an intermediate. They apparently react at about the same rate. The consideration of the products formed also casts extremely strong doubt on assumptions of ionic or free radical intermediates. Cyclopropane would necessarily have to preserve its ring structure in the ionic or free radical state, which is difficult to conceive; and an optically active secondary butyl ion or radical exist. Any general step by step mechanism for hydrogen fluoride catalyzed reactions is, therefore, subject to valid criticism. For a particular reaction under one set of conditions it must be supported by a considerable amount of quantitative evidence.

Fortunately a hydrogen fluoride catalyzed reaction capable of being followed kinetically by precise quantitative measurements has been found. (Sprauer and Simons, 82.) The reaction of tertiary butyl chloride with toluene at 25°C. is quantitative within the precision of the measurements to yield *p-t*-butyltoluene when the toluene is in large excess. The reaction is homogeneous in the hydrocarbon phase, proceeds at a measurable rate with no hydrogen fluoride liquid phase present but only an equilibrium concentration in solution with that present in the vapor, generates hydrogen chloride which increases the gas pressure. The increase in pressure as the reaction proceeds can be readily and precisely measured by physical means. Thirty five separate experiments are reported under a variety of conditions. These facts were ascertained. The reaction is first order with respect to the concentration of *t*-butyl chloride but the rate is proportional to a high power (5.5) of the hydrogen fluoride pressure. The rate is not greatly changed with initial hydrogen chloride pressure. The effect found is a slowing of the reaction by increased hydrogen chloride concentration. This same effect is noticed in the individual rate curves. The rate is greatly increased by the presence of extremely small amounts of water or methyl alcohol. In addition two entirely different types of rate curve were found depending on conditions. The one gave a straight line when the rate, i.e., the slope of the rate curve, is plotted against the amount of reaction. Such a curve is capable of rationalization on a rather simple theory. However, most of the rate curves, when the same plot is made, gave a hyperbolic curve. Very diligent efforts were made to fit this collection of information including the rate curves with a hypothesis of mechanism with some reaction intermediate such as a carbonium ion. The more detailed the analysis the more it became evident that a fit of this nature was impossible. Finally a satisfactory hypothesis which did fit all the observed facts was found. The mechanism involves the mutual action of an acidic molecule (hydrogen fluoride) and a basic one (the promoter or the hydrocarbon itself) on an assemblage of molecules containing the reactants. A hydrogen transfer throughout this assemblage results in the formation of the products (amphoteric medium effect). This mechanism is also reasonable on the basis of energy of activation because in the assemblage a large number of degrees of freedom are involved and the distribution of energy among them to create the product does not involve a high energy of activation, whereas any mechanism which requires the breaking of individual carbon-carbon or carbon-hydrogen bonds requires too high an energy of activation to go at an observable rate at room temperature.

In a continuation of the study of the kinetics and mechanism of this reaction, twenty four additional experiments were reported (Pearlson and Simons, 84). More care was taken in the design and construction of the apparatus and in the purity and dryness of the reagents. Although the higher precision caused some minor corrections to be made in the mathematical formulations of Sprauer and Simons, the facts are confirmed and no change is required in the theory. In addition the effect of four promoters was studied, water, methanol, diethyl ether, and hexamethylacetone; and it was found that the rate of the reaction increased with increasing concentration of promoter, and all promoters gave essentially the same effect at the same molar concentration. As the mechanism must include as major participants not only the catalyst but also the promoter, and as the only common property of these promoters is their basicity or tendency to solvate the proton, the mechanism must be a common one for all the promoters and depend upon this common property. No mechanism involving intermediates based upon these substances as different species could be expected to give the same rates. Calculations were also made of the most rapid rate of reaction, under the most favorable set of assumptions and based upon accepted theory and recent experimental measurements, that could result from a mechanism having as an intermediate either an ion or a free radical. It was found that they were so very much slower than the observed rate that no available stretch of hypothesis or assumption could bring them together. The observed rate constant was 1×10^{-5} mols per mol per second and the calculated rate for the most favorable ionic mechanism allowing for a high energy of solvation of the ions was 1×10^{-16} for the reaction under the same conditions. The calculated rate constant for a free radical mechanism assuming a chain mechanism with a chain length of 100,000 is 1×10^{-18}. Thus the most favorable ionic or free radical mechanism give rates 100,000,000,000 and 10,000,000,000,000 times slower than the observed rate.

It is hazardous to generalize in regard to mechanism from one or a limited number of studied reactions. The above reaction, which took place in a medium of low dielectric constant, does not rule out ionic mechanisms for condensation reactions taking place in a medium of high dielectric constant, as for example, in the hydrogen fluoride liquid phase. It does, however, show that an ionic intermediate is not necessary in the mechanism of condensation reactions. An ionic mechanism may not take place or may not take place exclusively to other mechanisms, even when the dielectric constant of the medium would permit it. The optical activity of the *s*-butylbenzene made from optically active *s*-butyl alcohol and benzene with a sufficient amount of hydrogen fluoride to give a medium of rather high dielectric constant is an example (Burwell and Archer, 41). It is inconceivable that a *s*-butyl ion could preserve optical activity. A plausible explanation is readily available. Part of the reaction could proceed through an ionic mechanism and result in racemization, while another part proceeds through the amphoteric medium effect and results in an optically active product.

VIII. Advantages and Disadvantages

Hydrogen fluoride as a catalyst for organic chemical reactions has certain disadvantages in the small laboratory or where a small quantity of product of not too high purity is desired. It requires the use of a fume hood, metal containers and reaction vessels, and a technique different from the usual preparations that can be done in glass. There are also certain hazards, although not as great as with the use of nitric acid, still they must be understood and provided for. Also there are many reactions where other agents are equally good, or superior. However, where hydrogen fluoride is particularly adaptable it is superior to other agents in that higher yields of product of higher purity can be obtained.

Hydrogen fluoride is particularly powerless as a cracking and tarring agent of hydrocarbons so strikingly found with Friedel Crafts reaction using aluminum chloride or similar agents. A sample of toluene with hydrogen fluoride heated in a sealed vessel at 200°C. for a week will show no degradation. In fact the toluene will be purer on recovery than it was initially since certain impurities such as sulfur compounds will be removed. Hydrogen fluoride has no oxidizing power and so cannot undergo a reaction similar to sulfonation as experienced with sulfuric acid in the formation of acid sludges. The elimination of such side reactions combined with the ease of removal of the agent makes for higher yields of purer products.

As hydrogen fluoride functions with equal ease in alkylation with olefins, alkyl halides, or alcohols, and in acylation with acids, acid anhydrides as well as acyl halides, a wide choice of reagents is possible and a separate operation of the reconversion of them is often saved. With aluminum chloride the alkyl halides and acyl halides are the preferred reagents and frequently must be made from more plentiful, cheaper, and readily available substances.

High purity of reagents is frequently not required for a high purity product with hydrogen fluoride due to the high specificity of its reactions. For example, it is not necessary to remove thiophene from benzene as thiophene does not poison the catalyst but is itself removed by polymerization. A sample of pure di-*t*-butyl benzene was desired. The *t*-butyl laboratory wastes, alcohol, halide, etc., were gathered together and mixed with crude benzene and hydrogen fluoride. A very pure sample of para-di-*t*-butyl benzene resulted.

The advantages for the use of hydrogen fluoride for large scale commercial use were appreciated before any industrial processes were in operation (Simons, 22). Due to the fact that it is a liquid with a low boiling point, ordinary piping can provide transfer either as a liquid or as a gas. The fact that HF is a liquid of low viscosity makes for small lines, valves, and auxiliaries. As a pound contains over 22 gram mols, each pound or weight unit carries a relatively large amount of chemical action, making again for smaller vessels and containers. Its low viscosity coupled with low surface tension gives it an extremely high settling rate with hydrocarbons, so settling time is reduced or settling tanks made smaller. These facts make for engineering advantages and also operational advantages. Coupling this with the chemical advantages makes hydrogen fluoride potentially the preferred agent for large scale production for reactions which it beneficially catalyzes.

IX. Discussion

From the above treatment it is seen that hydrogen fluoride as a catalyst for organic chemical reactions has many and widely diverse uses. It can be used for many different reactions over a wide range of conditions and with many different kinds of reactants. Where it is useful, it is a preferred catalyst because of its advantages. The active state of the chemical profession is seen in the fact that the discovery of its catalytic powers was only disclosed ten years ago (1938) and that already it has been employed for many and widely different reactions both scientifically and industrially. Although the development of industrial processes in the petroleum industry using hydrogen fluoride advanced rapidly during the war, the discovery and scientific research were not war time activities. They were done prior to the war (Arnold, 85) and the war actually put a stop to much of the research. The development of the processes came about from the already published advantages of hydrogen fluoride, and the fact that a great saving of steel could be had by building new alkylation units for aviation alkylate employing hydrogen fluoride rather than sulfuric acid.

In this chapter reference has been made to a few issued patents but many do not appear in the bibliography. Practically all the issued patents involving the use of hydrogen fluoride as a catalyst are for details of equipment design or operation, and no chemical principles are involved. The number of these patents has now become very large.

References

1. Simons, J. H., *J. Am. Chem. Soc.* **46,** 2179 (1924).
2. Claussen, W H., and Hildebrand, J. H., *J. Am. Chem. Soc.* **56,** 1820 (1934).
3. Bond, P. A., and Williams, D. A., *J. Am. Chem. Soc.* **53,** 34 (1931).
4. Dahmlos, J., and Jung, G., *Z. physik. Chem.* **B21,** 317 (1933).
5. Simons, J. H., and Bouknight, J. W., *J. Am. Chem. Soc.* **55,** 1458 (1933).
6. Wartenbur, H. V., and Schutza, H., *Z. anorg. u. allgem. Chem.* **206,** 65 (1932).

7. Simons, J. H., and Bouknight, J. W., *J. Am. Chem. Soc.* **54,** 129 (1932).
8. Simons, J. H., and Dresdner, R. D., *J. Am. Chem. Soc.* **66,** 1070 (1944).
9. Fredenhagen, K., *Z. electrochem.* **37,** 684 (1931).
10. Simons, J. H , and Hildebrand, J. H., *J. Am. Chem. Soc.* **46,** 2183 (1924).
11. Long, R. W., Hildebrand, J. H., and Morrell, W. E., *J. Am. Chem. Soc.* **65,** 182 (1934).
12. Clusius, K., Hiller, K., and Vaughen, J. V., *Z. physik. Chem.* **B8,** 427 (1930).
13. Hydrofluoric Acid Alkylation. Phillips Petroleum Co., 1946.
14. Butler, E. B., Miles, C. B., Kuhn, C. S., Jr., *Ind. Eng. Chem.* **38,** 147 (1946).
15. Klatt, W., *Z. anorg. u. allgem. Chem.* **234,** 189 (1937).
16. Simons, J. H., *Chem. Revs.* **8,** 213 (1931).
17. Holmberg, M. E., and Prange, F. A., *Ind. Eng. Chem.* **37,** 1030 (1945).
18. Frey, F. E., *Chem. & Met. Eng.* **50,** 126 (1943).
19. Fehr, C. M., *Petroleum Refiner* **22,** 239 (1943).
20. Anhydrous Hydrofluoric Acid. Harshaw Chemical Co., 1942.
21. Safety in the Operation of Hydrogen Fluoride Alkylation Plants. Universal Oil Products Co., Booklet No. 252.
22. Simons, J. H., *Ind. Eng. Chem.* **32,** 178 (1940).
23. Simons, J. H., *Petroleum Refiner* **22,** 83 (1943).
24. Simons, J. H., *Petroleum Refiner* **22,** 189 (1943).
25. Grosse, A. V., and Linn, C. B., *J. Org. Chem.* **3,** 26 (1938).
26. Fredenhagen, K., *Z. physik. Chem.* **A164,** 190 (1933).
27. Simons, J. H., and Archer, S., *J. Am. Chem. Soc.* **60,** 986 (1938).
28. Simons, J. H., and Archer, S., *J. Am. Chem. Soc.* **60,** 2952 (1938).
29. Simons, J. H., and Archer, S., *J. Am. Chem. Soc.* **61,** 1521 (1939).
30. Simons, J. H., and Archer, S., *J. Am. Chem. Soc.* **62,** 451 (1940).
31. Simons, J. H., and Bassler, G. C., *J. Am. Chem. Soc.* **63,** 880 (1941).
32. Calcott, W. S., Tinker, J. M., and Weinmayr, V., *J. Am. Chem. Soc.* **61,** 949 (1939).
33. Spiegler, L., and Tinker, J. M., *J. Am. Chem. Soc.* **61,** 1002 (1939).
34. Calcott, W. S., Tinker, J. M., and Weinmayr, V., *J. Am. Chem. Soc.* **61,** 1010 (1939).
35. Simons, J. H., Fleming, G. H., Whitmore, F. C., and Bissinger. W. E., *J. Am. Chem. Soc.* **60,** 2267 (1938).
36. Simons, J. H., and Archer, S., *J. Am. Chem. Soc.* **60,** 2953 (1938).
37. Simons, J. H., Archer, S., and Passino, H. J., *J. Am. Chem.* **60,** 2956 (1938).
38. Simons, J. H., and Archer, S., *J. Am. Chem Soc.* **61,** 1521 (1939).
39. Simons, J. H., and Archer, S., *J. Am. Chem. Soc.* **62,** 1623 (1940).
40. Simons, J. H., and Passino, H. J., *J. Am. Chem. Soc.* **62,** 1624 (1940).
41. Burwell, R. L., Jr., and Archer, S., *J. Am. Chem. Soc.* **64,** 1032 (1942).
42. Simons, J. H., Bacon, J. C., Bradley, C. W., Cassaday, J. T., Hoegberg, E. I., and Tarrant, P., *J. Am. Chem. Soc.* **68,** 1613 (1946).
43. Simons, J. H., Archer, S., and Randall, D I., *J. Am. Chem. Soc.* **61,** 1821 (1939).
44. Simons, J. H., Archer, S., and Adams, E., *J. Am. Chem. Soc.* **60,** 2955 (1938).
45. Simons, J. H., Archer, S., and Randall, D. I., *J. Am. Chem. Soc.* **62,** 485 (1940).
46. Frey, F. E., U.S. Patent 2,391,148 (1945).
47. Simons, J. H., and Meunier, A. C., *J. Am. Chem. Soc.* **65,** 1269 (1943).
48. Simons, J. H., and Ramler, E. O., *J. Am. Chem. Soc.* **65,** 1390 (1943).
49. Simons, J. H., Randall, D. I., and Archer, S., *J. Am. Chem. Soc.* **61,** 1795 (1939).
50. Fieser, L. F., and Hershberg, E. B., *J. Am. Chem. Soc.* **61,** 1272 (1939).
51. Fieser, L. F., and Hershberg, E. B., *J. Am. Chem. Soc.* **62,** 49 (1940).

52. Fieser, L. F., and Johnson, W. S., *J. Am. Chem. Soc.* **61,** 1647 (1939).
53. Fieser, L. F., and Johnson, W. S., *J. Am. Chem. Soc.* **62,** 575 (1940).
54. Newman, M. S., and Joshel, L. M., *J. Am. Chem. Soc.* **62,** 972 (1940).
55. Fieser, L. F., and Heyman, H., *J. Am. Chem. Soc.* **64,** 376 (1942).
56. Fieser, L. F., and Novello, F. C., *J. Am. Chem. Soc.* **64,** 802 (1942).
57. Johnson, W. S., and Mathews, F. J., *J. Am. Chem. Soc.* **66,** 210 (1944).
58. Johnson, W. S., Woroch, E., and Mathews, F. J., *J. Am. Chem. Soc.* **69,** 566 (1947)
59. Burwell, R. L., Jr., *J. Am. Chem. Soc.* **64,** 1025 (1942).
60. Frey, F. E., U.S. Patent 2,403,649 (1946).
61. Frey, F. E., U.S. Patent 2,403,650 (1946).
62. McElvain, S. M., and Langston, J. W., *J. Am. Chem. Soc.* **66,** 1759 (1944).
63. McElvain, S. M., and Langston, J. W., *J. Am. Chem. Soc.* **65,** 2239 (1943).
64. Kuhn, C. S., Jr., U.S. Patent 2,400,520 (1946).
65. Badertscher, D. E , Berger, H. G., and Bishop, R. B., U.S. Patent 2,406,339 (1946).
66 Simons, J. H., and Meunier, A. C., *J. Am. Chem. Soc.* **63,** 1921 (1941).
67. Helferich, B., and Bottger, S., *Ann.* **476,** 150 (1929).
68. Helferich, B., Starker, A., and Peters, O., *Ann.* **482,** 183 (1930).
69. Barthelemy, H. L., U.S. Patent 1,839,912 (1932).
70. Fredenhagen, K., and Cadenback, G., *Angew. Chem.* **46,** 113 (1933).
71. Simons, J. H., and Werner, A. C., *J. Am. Chem. Soc.* **64,** 1356 (1942).
72. Simons, J. H., Passino, J. H., and Archer, S., *J. Am. Chem. Soc.* **63,** 608 (1941)
73. Fredenhagen, K., German Patent 529,538 (1930).
74. Gleich, H., Russian Patent 39,775 (1934).
75. Simons, J. H., and McArthur, R. E., *Ind. Eng. Chem.* **39,** 364 (1947).
76. Ipatieff, V. N , and Grosse, A. V., *J. Am. Chem. Soc.* **57,** 1616 (1935).
77. Schmerling, L., and Pines, H., U.S. Patent 2,214,481 (1941).
78. Grosse, A. V., U.S. Patent 2,216,274 (1941).
79. Pines, H., and Ipatieff, V. N., U.S. Patent 2,340,557.
80. Lien, A. P., and Shoemaker, B. H., U.S. Patent 2,397,495 (1946).
81. Standard Oil Development Co., British Patent 537,589 (1941).
82. Sprauer, J. W., and Simons, J. H., *J. Am. Chem. Soc.* **64,** 648 (1942).
83. Simons, J. H., and Hart, H., *J. Am. Chem. Soc.* **66,** 1309 (1944).
84. Pearlson, W. H., and Simons, J. H., *J Am. Chem. Soc.* **67,** 352 (1945).
85. Arnold, P. M., *Trans Am. Inst. Chem. Engr.* **39,** 812 (1943).

Entropy of Adsorption

CHARLES KEMBALL

Department of Physical Chemistry, The University, Cambridge, England

CONTENTS

I. INTRODUCTION

The measurement of the heat of adsorption as an aid to the interpretation of adsorption data has become an important feature of experimental work in the field of catalysis during the last two decades. The magnitude of the heat evolved on adsorption will usually indicate whether the attachment to the surface is of a physical nature or whether chemisorption has occurred. In the former case the values are of the order of one to three times the heat of vaporization of the adsorbate, but with chemisorption the energies involved are much greater, as is the specificity for the adsorbent. Investigations of the variation of the heat of adsorption with the coverage of the surface can lead to a knowledge of the heterogeneity of adsorbents and the interactions between adsorbed molecules. Accounts of this subject are to be found in the books of Adam (1) and Brunauer (2). Our knowledge of entropies of adsorption is not so advanced and comparatively few determinations or theoretical investigations have been made. The chief purpose of such investigations is to determine whether or not the adsorbate is freely mobile. Barrer's work (3) indicates that substances absorbed in zeolites are not capable of translation and Foster (4) following Barrer's calculations finds the same to be true for many substances adsorbed on ferric oxide and silica gels. On the other hand Damköhler and Edse (5) find that CO adsorbed on copper oxide is freely mobile at 650°K. These conclusions are contrary to expectation because one would think that the freedom of the molecules would be greater under physical adsorption such as Barrer's and Foster's cases than under chemical adsorption. Hill (6) has shown by statistical calculations that one would expect translational freedom in the majority of cases of van der Waals adsorption and in a more recent paper (7) has found how constants in the B.E.T. equation for multimolecular adsorption depend on the ability of a diatomic molecule to rotate on the surface. Kemball (8, 9) measured entropies of substances physically adsorbed on mercury and found a range of freedem from complete mobility for acetone to complete immobility with water.

In view of the interest now being taken in the development of equations of state for adsorbates, and the attempts to derive isotherms for multimolecular adsorption covering wide ranges of relative pressure, the need for a general survey of entropy of adsorption is apparent. Apart from questions of mobility, and association and dissociation of the adsorbate, which may be determined from the entropy of the adsorbed material, a knowledge of entropy changes involved on adsorption is necessary to predict the actual adsorption under given circumstances when the heat of adsorption is known or may be estimated. This method of prediction of the extent of adsorption is more fundamental than the direct estimation of free energies, although success has been obtained along these lines notably by Traube (10), since re-interpreted by Ward and Tordai (11). Unfortunately, only a small proportion of the data available is suitable for the determination of entropies of adsorption. It is necessary that the free energy and total energy changes should be known for adsorption to a defined standard state. A further difficulty about interpreting the entropy is the need to know what the standard state means in terms of surface coverage, which generally involves a knowledge of the surface area of the adsorbent.

II. Possible Standard States for the Adsorbed Material

The most useful type of standard state is one defined in terms of a small number of molecules per unit area of adsorbent surface. In an attempt to have a definition analogous to that for three-dimensional matter—one atmosphere at any temperature—Kemball and Rideal (12) defined a standard state with an area per molecule of $22.53T$ A.^{2} where T is the absolute temperature. This corresponds to the same volume per molecule as the three-dimensional state if the thickness of the surface layer is 6A. In terms of surface pressure it corresponds to 0.0608 dynes/cm. for a perfect two-dimensional gas at all temperatures, and as such the definition may be extended to cover condensed films.

However, in practice it is frequently difficult to obtain results at such low surface concentrations either by direct measurement or by extrapolation. Consequently it is convenient to have an alternative definition in terms of surface covered, e.g., Barrer (3) and Foster (4) take $\theta = \frac{1}{2}$ as the standard state. A knowledge of the amount of material required to complete a monolayer is needed to apply this definition. Where such information is not available, one has to fall back on a definition in terms of so much adsorbate per cubic centimeter or per gram of adsorbent, and this makes interpretation more difficult. The accuracy of the determination will be greater when the heat of adsorption is known for adsorption directly to the standard state rather than as an average over the isotherm as a whole.

III. The Statistical Calculation of the Entropy of the Adsorbed Material

The recent investigation by Damköhler and Edse (5) virtually consists of the calculation of the entropy of CO adsorbed on copper oxide. They calculated the theoretical value of the constant b_0 in the Langmuir equation

$$\theta = p \cdot \frac{b_0 e^{\frac{\lambda}{RT}}}{1 + p \cdot b_0 e^{\frac{\lambda}{RT}}} \tag{1}$$

where p is the pressure, θ the fraction of the surface covered and λ is independent of temperature. They postulated a number of different assumptions about the freedom of the molecules on the surface and obtained a value for b_0 in each case. This is equivalent to calculating the entropy of adsorption when the standard state is $\theta = \frac{1}{2}$, for then

$$p = \frac{1}{b_0} e^{-\frac{\lambda}{RT}}$$

and the free energy change is given by

$$\Delta G = RT \ln p/p_0 = -RT \ln b_0 p_0 - \lambda$$

p_0 being the standard three-dimensional pressure. The entropy change is given by

$$\Delta S = -\frac{\partial \Delta G}{\partial T} = R \ln p_0 b_0 - RT \frac{\partial \ln b_0}{\partial T} \tag{2}$$

They showed that b_0 may be expressed statistically in terms of the partition function of the molecule in the adsorbed state Σ_a^* less the Boltzmann factor, $c^{\lambda/RT}$, the partition function of the molecules in the gas phase, Σ_g, and the volume occupied by a molecule in the gas phase, i.e.,

$$b_0 = \frac{\Sigma_a^* \, V_g}{\Sigma_g \, kT} \tag{3}$$

The values of b_0 ranged from 8.5×10^{-7} (mm. Hg)$^{-1}$ for complete three-dimensional translational freedom and rotational freedom to 6.6×10^{-11} (mm. Hg)$^{-1}$ for the replacement of all modes by vibrations. If these values are put in (2) with the appropriate values of the term $RT \frac{\partial \ln b_0}{\partial T}$ which are $-R$ and $\frac{3}{2}R$ the entropies of adsorption in these extreme cases become -12.6 and -37.4 cals./deg. mole. They quote Schwab and Drikos (13) as using a value for b_0 of 10^{-7} in the kinetics of the oxidation of CO on copper oxide, and therefore suggest that the adsorbed molecules are relatively free on the surface. This is a somewhat surprising result because one would not expect the translation to be entirely free in cases of chemisorption. The hindrance to free movement will, of course, depend on the energy barrier to be traversed on passing from one site to the next. It would appear that the method used by Schwab and Drikos to estimate b_0 is at fault rather than the method of attack employed by Damköhler and Edse. Nevertheless, there is an illogicality in the latters' treatment, which is common to other work in this field, namely the insertion of translational partition functions into a constant in a Langmuir equation. The equation is based on the assumption that the molecules are adsorbed on fixed sites and, as may be seen from the work of Fowler and Guggenheim (14), does not involve any partition function for the translation of the adsorbed molecules. In such cases where the entropy indicates translational freedom, the Langmuir equation is not strictly applicable, although still of considerable use.

Barrer (3) makes similar calculations for the entropies of occlusion of substances by zeolites and reaches the conclusion that the adsorbed material is devoid of translational freedom. However, he uses a volume, area or length of unity when considering the partition function for translation of the adsorbed molecules in the cases where they are assumed to be capable of translation in three, two or one dimensions. His entropies are given for the standard state of $\theta = 0.5$, and the volume, area or length associated with the space available to the adsorbed molecules should be of molecular dimensions, $v = 125 \times 10^{-24}$ cc., $a = 25 \times 10^{-16}$ cm.^{2} and $l = 5 \times 10^{-8}$ cm. When these values are introduced into his calculations the entropies in column four of Table II of his paper come much closer together, as is shown in Table I. The experimental values for different substances range from zero to -7 cals./deg. mole or entropy units, and so further examination is required in each case to decide whether or not translation is possible. Some of the experimental entropies of absorption obtained by Barrer and Ibbitson (15) will be considered in appropriate sections of this survey. The results given by Foster will also be re-examined because his conclusion that the systems he investigated showed no translational freedom in the adsorbed state, was based on Barrer's calculations.

Kemball (8, 9) attempted to calculate the entropy of various substances adsorbed on mercury and was able to explain the entropy of

TABLE I

Revision of Barrer's Values for the Entropy of Occlusion

Freedom in occluded state	ΔS Barrer's values e.u.	ΔS Corrected values e.u.
Complete translation and rotation (t^3, r^3)	109.4	9.2
Two-dimensional trans. (t^2, r^3, v)	70.93	4.13
One-dimensional trans. (t^4, r^3, v^2)	32.46	−0.94
No translation (r^3, v^3)	−7.01	−7.01
Two rotations (r^2, v^4)	−10.23	−10.23
One rotation (r, v^5)	−13.45	−13.45
Only vibration (v^6)	−16.67	−16.67

adsorbed benzene in terms of two-dimensional translational freedom and rotation only in the plane of the ring (the benzene being parallel to the mercury surface). An expression was derived for the translational entropy of a perfect two-dimensional gas, i.e.,

$$_{2}S_{\text{trans}} = R \ln MTa + 65.80 \tag{4}$$

where M is the molecular weight of the gas and a the area available per molecule. The rotational entropy was calculated from an expression given by Halford (16) for the partition function of a molecule free to rotate in n independent ways, i.e.,

$$S_{\text{rot}} = R\left[\ln \frac{1}{\pi\sigma}\left\{\frac{8\pi^3(I_A{}^a I_B{}^b \ . \ . \ . \ I_G{}^g)^{1/n}\, kT}{h^2}\right\}^{n/2} + \frac{n}{2}\right] \tag{5}$$

where $a + b + c + \ldots\ldots\ldots + g = n$, I_A, I_B etc. are the moments of inertia and σ is the symmetry number.

For immobile adsorption there may still be rotational freedom which may be estimated from the appropriate form of (5) but there will always be the combinatory term in the entropy associated with the number of ways of distributing the molecules over the surface. If the fraction of the surface covered is $1/x$ the distribution of N molecules among the Nx sites gives an entropy

$$S = R[x \ln x - (x - 1) \ln (x - 1)] \tag{6}$$

It is possible to calculate the entropies in cases where association or dissociation occurs at the same time as localized adsorption. Chang (17) gave an expression

$$W = \left[\frac{\left\{2(z - 1)\frac{N_A}{z} - N_B\right\}!}{\left(2\frac{N_A}{z}\right)!\,(2N_A + N_B)!} \right]^{\frac{z}{2}} \frac{(2N_A + N_B)!}{N_B!} \left(\frac{2}{\sigma_{AA'}}\right)^{N_A} \tag{7}$$

for the number of ways of combining N_A double molecules with N_B single molecules, where z is the number of nearest neighbors and $\sigma_{AA'}$ is a symmetry number, equaling 2, if the two ends of a double molecule are indistinguishable. This may be adapted to cover both dissociation and association. For the former we put $N_A = N$ and $2N_A + N_B = 2xN$, $1/x$ again representing the fraction of surface covered and N_B being the number of unoccupied sites each capable of holding one of the dissociated parts of a molecule. Taking $\sigma_{AA'} = 2$ and $z = 4$ the entropy is given by

$$S_D = 2R[2(x - \tfrac{1}{4}) \ln (x - \tfrac{1}{4}) + \ln 2 - x \ln x - (x - 1) \ln (x - 1)] \tag{8}$$

or if $z = 6$ by

$$S_D = R[\sigma(x - \tfrac{1}{6}) \ln (x - \tfrac{1}{6}) + \ln \sigma - 4x \ln x - 2(x - 1) \ln (x - 1)] \tag{9}$$

On the other hand, for association of the molecules into dimers we put $N_A = N/2$ and $2N_A + N_B = xN$ and the entropies under the two cases of $z = 4$ and $z = 6$ are found to be half the values given by expressions (8) and (9). Some of the values obtained from these equations with different fractions of the surface covered are given in Table II.

TABLE II

Combinatory Entropies for Immobile Adsorption

Fraction of surface covered	Molecular adsorption S e.u.	Dissociative adsorption S_D e.u.		Associative adsorption S_A e.u.	
		$z = 4$	$z = 6$	$z = 4$	$z = 6$
1/100	11.2	13.9	14.7	6.9	7.3
1/10	6.5	9.1	9.9	4.6	5.0
1/8	6.0	8.6	9.5	4.3	4.7
1/6	5.4	8.0	8.8	4.0	4.4
1/4	4.5	7.0	7.8	3.5	3.9
1/2	2.9	5.0	5.8	2.5	2.9

In some cases the adsorption is so weak that the entropy associated with the vibration replacing the translational motion perpendicular to the surface may not be negligible. The entropy associated with a vibration of frequency ν is given by

$$S_{\text{vib}} = R\left[\frac{h\nu}{kT}\left(e^{\frac{h\nu}{kT}} - 1\right)^{-1} - \ln\left(1 - e^{-\frac{h\nu}{kT}}\right)\right] \tag{10}$$

and a few results calculated from this equation are quoted in Table III.

TABLE III
Vibrational Entropies

$\frac{\nu h}{kT}$	ν sec.$^{-1}$ T = 100°K.	ν sec.$^{-1}$ T = 300°K.	S_{vib} e.u.
0.01	2.08×10^{10}	6.25×10^{10}	11.13
0.05	1.04×10^{11}	3.13×10^{11}	7.95
0.10	2.08×10^{11}	6.25×10^{11}	6.57
0.25	0.52×10^{12}	1.56×10^{12}	4.74
0.50	1.04×10^{12}	3.13×10^{12}	3.38
1.00	2.08×10^{12}	6.25×10^{12}	2.06

We are now in a position to examine some of the experimental data available, which will be treated in sections according to the apparent freedom possessed by the molecules in the adsorbed state.

IV. "Supermobile" Adsorption

In this section we deal with those systems where the entropy loss on adsorption is less than would be expected if all movement perpendicular to the surface was denied to the adsorbate. This will happen in cases where the potential energy trough is shallow and where translational motion perpendicular to the surface is replaced by a vibration of low frequency.

Cassel and Neugebauer (18) investigated the adsorption of some of the rare gases on mercury over a range of temperatures by surface tension measurements. They found that the curves for surface pressure against gas pressure were almost linear and it is possible to interpolate their results to the standard state π = 0.0608 dynes/cm., obtaining the pressure p_0 in equilibrium with a film at this surface pressure. The thermodynamic quantities for the adsorption of xenon are given in Table IV:

TABLE IV

The Adsorption of Xenon on Mercury (Standard State π = 0.0608 dynes/cm.)

Temp. °K.	p_0 mm.	ΔG cals./mole	ΔH cals./mole	ΔS e.u.
237	1.82	−2842	−3500	−2.8
293	7.54	−2685		

the standard state in three dimensions being taken as 1 atmosphere. The figure for ΔH compares with the average heat of −3400 cals./mole obtained by Cassel and Neugebauer over this temperature range. Now the entropy of xenon at 265°K. calculated from the equation

$$S = R \ln M^{3/2}T^{5/2} - 2.30 \tag{11}$$

is 39.9 e.u., and the two-dimensional entropy from equation (4) is 30.6 e.u. The minimum loss of entropy on adsorption if no motion was possible perpendicular to the surface would therefore be 9.3 e.u., and the discrepancy must be due to the weakness of the vibration replacing the third degree of translational freedom. From Table III we see that a vibration of frequency of the order of 6×10^{11} sec.$^{-1}$ would provide the 6.5 e.u. possessed by the adsorbate over and above the 30.6 e.u. for two-dimensional translation.

Another example of this type is the adsorption of acetone on mercury (Kemball, 9) where the heat of adsorption to the same standard state is 7.5 kcals./mole and the entropy change 4.4 e.u. at 310.6°K. The calculated value for loss of translation normal to the surface is 8.6 e.u., and the discrepancy of 4.2 e.u. may be accounted for by a frequency of 2×10^{12} sec.$^{-1}$ These frequencies of about 10^{12} sec.$^{-1}$ compare with the frequency found by Orr (19) for argon on KCl and with the values expected by Hill (7).

V. Mobile Adsorption

The next set of examples show an entropy of adsorption roughly equal to the entropy change on losing the degree of translational freedom normal to the surface, i.e., in the adsorbed state the molecules are equivalent to a two-dimensional gas or vapor. The data for a variety of different adsorbates and adsorbents is given in Table V. The isotherms obtained by Armbruster (20) for the adsorption of CO and N_2 on silver were not S-shaped, and they could be fitted to equations of the Langmuir type. The amount of adsorbate required to saturate the surface was given for each substance at both temperatures. Armbruster calculated the heats of adsorption by the method of Brunauer, Emmett and Teller (22) and there is some doubt about the validity of such heats.

The experimental entropies of adsorption were calculated after obtaining the free energies of adsorption at $\theta = \frac{1}{2}$ from the gas pressure in equilibrium with half the amount of adsorbate required to form the monolayer. The same principles were used to obtain the figure for the entropy of adsorption of O_2 on unreduced steel. The values for carbon tetrachloride were taken directly from Foster's paper (4). The results for adsorption in chabazite were obtained from the work of Barrer and Ibbitson (15) with the slight modification needed to allow for the different standard states in the two phases used by them. The figures in the last column

TABLE V

Examples of Mobile Adsorbates

Adsorbate	Adsorbent	Ref.	Temp. °K.	ΔH kcals./mole	ΔS e.u. exptl.	ΔS e.u. theor.
(a) *With standard state given by* $\theta = \frac{1}{2}$						
CO	Silver	(20)	78	−3.27	−16.9	−16.0
			90	−3.60	−19.3	−16.5
N_2	Silver	(20)	78	−3.05	−16.5	−16.0
			90	−3.60	−17.8	−16.5
O_2	Unreduced steel	(21)	90	−3.3	−15.8	−16.7
CCl_4	Silica gel	(4)	303	−9.0	−17.4	−19.3
CCl_4	Ferric oxide	(4)	303	−9.4	−14.6	−19.3
(b) *With standard state given by* $\theta = \frac{1}{10}$						
C_2H_6	Chabazite	(15)	428.5	−8.5	−18.8	−15.6
C_3H_8	Chabazite	(15)	438	−11.2	−20.5	−16.3
(c) *With standard state given by* $\pi = 0.0608$ *dynes/cm.*						
C_2H_5OH	Mercury	(9)	310.6	−11.4	−11.2	−8.5
C_3H_7OH	Mercury	(9)	310.6	−10.8	−7.9	−8.7

of the table give the difference in entropy associated with the change from a three-dimensional gas to a two-dimensional gas calculated from relations (11) and (4). The area available for the CO, N_2 and O_2 was assumed to be 15 A.^{2} per molecule, and in the case of CCl_4, 50 A.^{2} per molecule. For the two examples with the surface only one-tenth covered, the area was taken to be 225 A.^{2} per molecule, a figure based on the assumption that the area required by one molecule is 25 A.^{2}

There seems to be a rough agreement between the experimental values for the entropy (which in most of the cases are subject to an experimental error of 2–3 e.u., due to uncertainty as to the exact value of heat of adsorption at $\theta = \frac{1}{2}$), and the calculated values for adsorption to a perfect two-dimensional gas or vapor. Better agreement would be entirely fortuitous because in all cases there will be a certain entropy for vibration, though probably not as large as the figures given in the last section, and with the concentrated nature of the film when $\theta = \frac{1}{2}$, and to a lesser extent when $\theta = \frac{1}{10}$, there must be some interaction between the molecules. There is little doubt that the comparative freedom of CCl_4 is to be associated with the symmetrical character of that molecule. Another set of examples on a different type of adsorbent surface are the adsorption of A, O_2 and N_2 on KCl and CsI crystals followed by Orr (23). Orr evaluated the variation of the heat of adsorption as the amount of material on the surface increased and identified the completion of the monolayer with a maximum in heat of adsorption. His data are particularly valuable because the heat of adsorption is known for different amounts on the surface, and an average value does not have to be used. Some figures obtained from Orr's paper are given in Table VI. The entropy of adsorption on the basis of a perfect two-dimensional gas in all these cases would be about 16 e.u., i.e., in agreement with the first three results, but not with the last three. The latter presumably mean more restriction on the adsorbate than loss of the third degree of translational freedom. It will be seen that in each case

TABLE VI

Gases on Crystals (Standard State $\theta = \frac{1}{2}$)

Adsorbate	Adsorbent	Temp. °K.	ΔH cals./mole	ΔG cals./mole	ΔS e.u.
A	KCl	79.4	−1780	−527	−15.8
	CsI	75.0	−2300	−1027	−17.0
O_2	KCl	79.3	−1900	−573	−16.7
	CsI	83.0	−2510	−946	−18.9
N_2	KCl	79.6	−2400	−911	−18.7
	CsI	79.0	−2775	−1236	−19.5

the restrictions upon the freedom are greater with CsI than with KCl. The results with A on KCl will be discussed in detail in a later section, because it is possible to show the occurrence of two-dimensional condensation when the surface is about four-fifths covered.

VI. Adsorptions Showing Intermediate Freedom

This section includes the majority of physically adsorbed substances, and it is only in the exceptional cases that the loss of entropy can be attributed to the loss of particular degrees of freedom. In Table VII are given an extract of Foster's results for adsorptions on ferric oxide and silica gels together with the loss in entropy calculated for the loss of translational movement normal to the surface. In each case the area used in equation (4) was 50 A.^{2} per molecule. Other results of this type are those of Kemball and Rideal (12) and Kemball (9) which are summarized in Table VIII. The observed entropy of adsorption of benzene on mercury can be explained on the loss of all rotation except that in the plane of the ring and the loss of translational freedom normal to the surface. The calculated loss under these circumstances is 26.0 e.u. comparing well with the experimental value of 25.2 e.u. The suggestion was made from the entropy of adsorption in the case of n-heptane and the observed co-area of 32.7 A.^{2} that the molecules were partially curled up and possessed almost spherical symmetry. The additional loss of 6.2 e.u. was compared with the difference of 5.1 e.u. between the entropies for rotation of n-hexane and n-heptane. To give some idea of the extent of restriction imposed when toluene is adsorbed parallel to the surface, one may quote the figure of 38.5 e.u., which would correspond to the loss of all rotational freedom, and the third degree of translational freedom. However, it is unlikely that all rotation will cease before the

TABLE VII

Adsorptions on Ferric Oxide and Silica Gels (Standard State $\theta = 1/2$)

Adsorbate	Adsorbent	Temp. °K.	ΔS e.u. exptl.	$_2S_{trans} - {}_3S_{trans}$
Dioxan	Fe_2O_3	303	−23.5	−18.7
Dioxan	SiO_2	303	−26.6	−18.7
Ethyl Alc.	SiO_2	290	−27.1	−18.1
n-Octane	SiO_2	303	−26.2	−19.0
Toluene	Fe_2O_3	303	−27.6	−18.7
Toluene	SiO_2	303	−26.3	−18.7

TABLE VIII

Adsorption of Vapors on Mercury (Standard State $\pi = 0.0608$ dynes/cm.)

Adsorbate	Temp. °K.	Co-area A^2	ΔS e.u.	$_2S_{trans} - {}_3S_{trans}$
Benzene	323.1	34.4	−25.2	−9.0
Toluene	310.6	37.3	−39.6	−9.1
	310.6	23.5	−15.2	−9.1
n-Heptane	310.6	32.7	−15.3	−9.1
Methyl Alc.	310.6	29.9	−27.2	−8.1
n-Butyl Alc.	310.6	29.0	−27.8	−9.0
n-Amyl Alc.	310.6	36.1	−29.0	−9.2
n-Hexyl Alc.	310.6	43.9	−33.4	−9.3

other two degrees of translational freedom are partly hindered. The anchoring effect due to the introduction of the methyl group on going from benzene to toluene is interesting.

VII. Immobile or Localized Adsorption

1. van der Waals Adsorption

In order to have localized adsorption with only physical interaction it is clear that either the interaction must be strong, or the kinetic energy of the adsorbed molecules must be small. As an example of the latter condition we have the work of Keesom and Schweers (24, 25) for low temperature adsorption of hydrogen and neon on glass. They assumed that the actual area of the glass was equal to the apparent area, and the results in Table IX were worked out for $\theta = \frac{1}{2}$ on that basis. The

TABLE IX

Low Temperature Adsorption on Glass

Substance	Temp. °K.	ΔG cals./mole	ΔH cals./mole	ΔS e.u.	S_{gas} e.u.	$S_{adsorbate}$ e.u.
H_2	20.3	−661	−876	−10.7	12.6	1.9
Ne	20.3	−456	−777	−15.8	21.6	5.8
Ne	17.5	−504	−777	−15.6	20.9	5.3

entropies in column six were calculated from equation (11) (the rotational entropy of hydrogen being negligible at 20°K.). Now Giauque (26) gives the entropy of solid hydrogen at 13.95°K. to be about 0.52 e.u. and consequently one would expect the entropy of hydrogen undergoing localized adsorption to a state with $\theta = \frac{1}{2}$ to be a little greater than the combinatory entropy value of 2.9 e.u. The entropy of neon gas at 27.2°K. is 23.1 e.u., and Moelwyn-Hughes (27) gives the entropy of vaporization as 15.2 e.u. at that temperature, and the entropy of fusion at 24.5°K. as 3.27 e.u., making the entropy of solid neon at 24.5°K. about 4.6 e.u. When this figure is combined with the combinatory entropy of 2.9 e.u. to give 7.5 e.u. we have a value that would be expected for neon frozen on to the surface at 25°K., and the values in Table IX are of the correct size to represent neon undergoing localized adsorption at these lower temperatures.

Another similar case is the absorption of hydrogen in chabazite, investigated by Barrer and Ibbitson (15). Taking their figure for the standard entropy of adsorption of −2.2 e.u. at 89°K. with the surface covering given by $\theta = 0.215$, and adjusting to allow for the different standard state they used in the gas phase, and also for converting to a standard state on the surface of $\theta = 0.215$, the figure obtained is -17.3 e.u. The translational entropy of the gas at this temperature (11) is 22.1 e.u., and assuming that the rotational entropy may be calculated by the appropriate form of equation (5) (which is not quite true because of the low moment of inertia and high quanta for rotation with hydrogen) the total entropy of 27.2 e.u. is obtained. Thus the adsorbed hydrogen has only 9.9. e.u. as a maximum, and of these some five are required for the combinatory term as may be seen from Table II. It would appear, therefore, that the hydrogen is undergoing localized adsorption, which is in agreement with the conclusion reached by Barrer (3).

As an example of strong interaction leading to localized adsorption we have the figures for the entropy of adsorption of water on different adsorbents, which are summarized in Table X. In all cases the heat of adsorption is quite large and the entropy values for the adsorbed material

TABLE X

Water on Different Adsorbents

Adsorbent	Ref.	Temp. °K.	ΔH kcals./mole	ΔS e.u.	S_{vapor} e.u.	$S_{adsorbate}$ e.u.
Standard state given by $\theta = \frac{1}{2}$						
SiO_2	(4)	303	−15.2	−34.3	45.2	10.9
$Fe2O_3$	(4)	303	−15.7	−31.7	45.2	13.5
Stainless steel	(28)	293	−12.9	−28.6	45.0	16.4
Stainless steel	(28)	293	−13.5	−29.4	45.0	15.6
Stainless steel	(28)	293	−12.8	−28.4	45.0	16.6
Standard state given by π = 0.0608 dynes/cm.						
Mercury	(9)	310.6	−17.6	−35.9	45.4	9.5

are much nearer the figure of 12.8 e.u. (obtained by adding the combinatory entropy of 2.9 e.u. to that of 9.9 e.u. for the entropy of ice at 273°K. estimated from the work of Giauque and Stout, 30) than to the value of 27 e.u., which was calculated on the assumption of the loss of only the translational motion normal to the surface. In the case of adsorption on mercury, the entropy of adsorption is so great that it cannot be explained unless the molecules are associated to a considerable extent on the surface. For the minimum entropy for distribution alone, assuming no freedom at all, is 14.6 e.u., in the standard state of π = 0.0608 dynes/cm. (taking the area occupied by a water molecule to be 12 A.^{2}). This is considerably higher than the experimental value of 9.5 e.u. However, it was shown (9) that if all the water molecules were associated into dimers, assuming that the number of nearest neighbors to a site on the mercury surface was 4, the entropy would decrease to 8.7 e.u.

2. Chemisorption

Taylor and Sickman (31) measured the adsorption of water on ZnO in the neighborhood of 634°K. Data were not available at that time to indicate the extent of the surface available, so as standard state we adopt a system containing 1.03 cc. (measured at N.T.P.) per g. of adsorbent. As far as may be ascertained from isotherms, this represents the surface about half covered. The thermodynamic quantities for this adsorption are given in Table XI.

TABLE XI

Chemisorption of Water on ZnO (Standard State Given by 1.03cc./g.)

Temp. °K.	p_0 mm.	ΔG kcals./mole	ΔH kcals./mole	ΔS e.u.
605.5	3.0	−6.659	−29.5	−37.8
661.9	24.2	−4.527		

The heat of adsorption calculated by Taylor and Sickman was 30.3 kcals./mole and clearly indicated chemisorption and the entropy of adsorption indicates that the adsorbed molecules have little freedom. Wagman and his coauthors (29) give the entropy of water vapor at 634°K. as 51.3 e.u., i.e., a value of 13.5 e.u. for the adsorbed water. This is of the size expected for complete immobility as may be seen by examining a simple model. Suppose that on adsorption the water combines with the ZnO to from Zn $(OH)_2$ and let us estimate the entropy of this substance. There will probably be little change in the vibrational entropy from that of water. This latter may be estimated by calculating the translational entropy of water from equation (11), and the rotational entropy from the appropriate form of (5), using the moments of inertia given by Gordon (32), and subtracting the sum of this value from the 51.2 e.u. for the total entropy. The vibrational contribution is found to be 1.3 e.u. Suppose that each of the OH groups is capable of rotation about the Zn—O bond, each with a moment of inertia of half the small moment of inertia for water, i.e., 0.996×10^{-40} g. cm.^{2} According to the equation (5) the entropy for each of these is 1.9 e.u. The distributive term for dissociative adsorption when $\theta = \frac{1}{2}$ will be either 5.0 e.u. or 5.8 e.u., depending on whether $z = 4$ or $z = 6$. This means that the entropy of the water when adsorbed in this manner will be either 10.1 or 10.8 e.u., which is in agreement with the value of 13.5 e.u. within the accuracy of these calculations and the experimental error.

Two further examples of chemisorption are the reactions of N_2 with iron catalysts investigated by Emmett and Brunauer (33) and with tungsten investigated by Davis (34). In the first case, taking the standard state as 3 cc. adsorbed per 10 cc. of catalyst, the figures in Table XII were obtained for adsorption on a well-known doubly promoted iron catalyst. Emmett and Brunauer gave the value of the heat of adsorption as 35.0 kcals./mole in agreement with the value calculated from the free energies at the two temperatures. They mention the difficulty of deciding whether the nitrogen is adsorbed atomically or molecularly. There is also some uncertainty about the amount of the surface covered at the standard state of 3cc./10cc. They found that at low temperatures 16 cc.

TABLE XII

Chemisorption of N_2 on Catalyst 931. (Standard State Given by 3 cc./10 cc. Catalyst)

Temp. °K.	p_0 mm.	ΔG kcals./mole	ΔH kcals./mole	ΔS e.u.
669	35	−4.092	−35.0	−46.2
722	241	−1.647		

was required to complete a monolayer. However, from an inspection of the isotherms at higher temperatures, it would appear that 3 cc./10 cc. represents the surface about half covered. The entropy of nitrogen at 696°K. is 51.8 e.u. (Johnston and Davis (35) and Wagman and his coauthors, 29) making the entropy of the adsorbed material 5.6 e.u. Although it is clear that this value indicates localized adsorption, it is not possible to decide whether the adsorption is molecular or dissociative, mainly because of the uncertainty as to the fraction of the surface covered. The experimental value would agree with either the figure of 4.5 e.u. for molecular adsorption with $\theta = 1/4$, or with the figures 5.0 or 5.8. e.u. for dissociative adsorption with $\theta = 1/2$ (Table II).

Furthermore, even if the coverage were known accurately, there might be a small vibrational or rotational entropy associated with the molecular model, which would bring its entropy up to the same value as the atomic model, making them indistinguisable. Davis (34) investigated the chemisorption of nitrogen on tungsten, finding the heat of adsorption of 75 kcals./mole at low coverage. From his result the pressure in equilibrium with a surface $1/10$ covered at 750°C. is 0.00030 mm., making the free energy of adsorption −30.0 kcals./mole, and the entropy −44 e.u. The entropy of nitrogen at 1023°K. is 54.7 e.u., leaving the adsorbed nitrogen with 10.7 e.u. in the standard state given by $\theta = 1/10$. As may be seen from Table II, this could correspond either to atomic adsorption (9.1 e.u. or 9.9 e.u.) or to molecular adsorption (6.5 e.u.) if there were some slight rotational or vibrational freedom available in the adsorbed state.

VIII. Detection of Phase Changes

The results of the accurate work of Orr (23) on the adsorption of argon on KCl and CsI crystals have been mentioned above, and it was shown that when the surface was half covered, the adsorbate had an entropy equivalent to that of a two-dimensional gas. The curves given

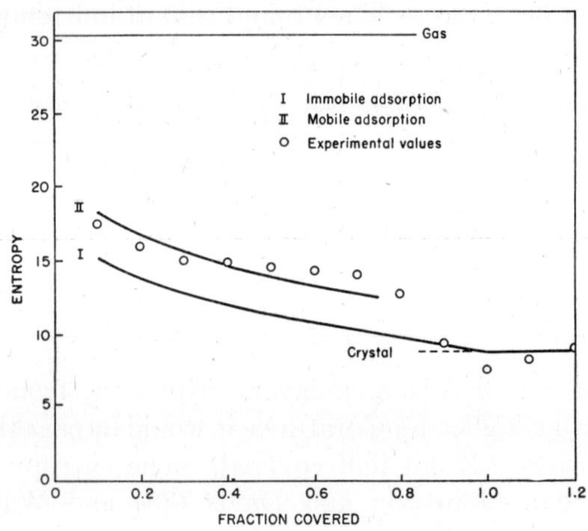

Fig. 1. Argon on KCl, 79°K.

by Orr for the heat of adsorption against coverage were interesting, showing firstly a decrease attributed to the filling of high energy sites, and then a rise attributed to interaction between the adsorbed atoms which reached a maximum at $\theta = 1$, and gradually decreased to the heat of sublimation of argon when $\theta = 2$. At this latter stage Orr showed that the adsorbate was little different from solid argon. It would therefore appear that at some stage between $\theta = \frac{1}{2}$ and $\theta = 2$, the argon must undergo a phase change. Accordingly detailed calculations of the entropy of adsorption at various points of the isotherm were made and are shown in Fig. 1. The value of the entropy of the gas was calculated from equation (11). The entropy of the solid was calculated from the heat of sublimation given by Orr and the value of the saturation pressure at 79.4°K. A distributional entropy was added to the entropy of the solid in order to obtain curve I, the distributional entropy being taken from Table II, column two.

Curve II represents the entropy of a two-dimensional gas, assuming that it occupied a volume of 15 A.^{2}, i.e., at $\theta = 1/10$ the area per molecule used in equation (4) was 135 A.^{2} and at $\theta = 1/2$ the area available was 15 A.^{2} The entropy of vibration perpendicular to the surface is neglected. The circles give the experimentally determined values calculated from Orr's curves for the heat of adsorption and the isotherm, and it will be seen that at about $\theta = 0.8$ the values shift from Curve II to Curve I. This indicates that the rise in heat of adsorption which occurs at this stage may be explained in terms of a phase change of the adsorbed molecules from a gaseous film to a condensed film. The alteration of the heat of adsorption at low values of θ does not show up in the entropy changes because in this region it is due to the variation of the adsorbent and not to the freedom or state of the adsorbate.

IX. Discussion

The examples that have been discussed show that a knowledge of the entropy of adsorption, and hence the entropy of the adsorbed material, gives some indication of the extent to which the molecules are capable of unhindered translation on the surface. Physical adsorption, although generally giving rise to mobile adsorption, may also under certain circumstances lead to localized adsorption. The smaller the heat of adsorption in relation to the heat of vaporisation, the greater is the probability of finding complete two-dimensional translational freedom. The lower the temperature at which the adsorption is carried out, the more likely is the adsorption to be of the immobile type. Symmetry of the adsorbate molecules is important to a certain extent, especially with the larger molecules, for example, CCl_4 and ethyl alcohol have considerable freedom on silica gel and mercury respectively, but butyl alcohol and toluene do not. At very low temperatures, and with relatively large heats of adsorption, physical forces may lead to immobility, although this will, to a considerable extent, depend on the nature of the force field near the adsorbent surface, i.e., upon the height of the potential energy barrier between one trough and the next.

The three cases of chemisorption examined all showed localized adsorption, but there may be other examples in which a considerable mobility is possessed by the adsorbed molecules or atoms. Again, this would be more likely to occur at high temperatures. Entropy values can give indications of dissociation or, more particularly, of association when localized adsorption takes place. This was clearly noticeable in the values for the entropy of water on mercury. Likewise a knowledge of the changes in entropy during the course of an isotherm may be of use in detecting phase changes, as was shown with the results for argon on KCl.

References

1. Adam, N. K., The Physics and Chemistry of Surfaces. Oxford Univ. Press, London, 1941.
2. Brunauer, S., The Adsorption of Gases and Vapors. Princeton Univ. Press, Princeton, New Jersey, 1943.
3. Barrer, R. M., *Trans. Faraday Soc.* **40,** 374 (1944).
4. Foster, A. G., *J. Chem. Soc.*, 360 (1945).
5. Damköhler, G., and Edse, R., *Z. physik. Chem.* **B53,** 117 (1943).
6. Hill, T. J., *J. Chem. Phys.* **14,** 441 (1946).
7. Hill, T. J., *J. Chem. Phys.* **16,** 181 (1948).
8. Kemball, C., *Proc. Roy. Soc. London* **A187,** 73 (1946).
9. Kemball, C., *Proc. Roy. Soc. London* **A190,** 117 (1947).
10. Traube, I., *Ann.* **265,** 27 (1891).
11. Ward, A. F. H., and Tordai, L., *Trans. Faraday Soc.* **42,** 408 (1946).
12. Kemball, C., and Rideal, E. K., *Proc. Roy. Soc. London* **A187,** 53 (1946).
13. Schwab, G. M., and Drikos, G., *Z. physik. Chem.* **B52,** 234, (1942).
14. Fowler, R. H., and Guggenheim, E. A., Statistical Thermodynamics. Chap. X, Cambridge Univ. Press, London, 1939.
15. Barrer, R. M., and Ibbitson, D. A., *Trans. Faraday Soc.* **40,** 195 (1944).
16. Halford, J. O., *J. Chem. Phys.* **2,** 694 (1934).
17. Chang, T. S., *Proc. Roy. Soc. London* **A169,** 512 (1939).
18. Cassel, H., and Neugebauer, W., *J. Phys. Chem.* **40,** 523 (1936).
19. Orr, W. J. C., *Trans. Faraday Soc.* **35,** 1247 (1939).
20. Armbruster, M. H., *J. Am. Chem. Soc.* **64,** 2545 (1942).
21. Armbruster, M. H., and Austin, J. B., *J. Am. Chem. Soc.* **68,** 1347 (1946).
22. Brunauer, S., Emmett, P. H., and Teller, E., *J. Am. Chem. Soc.* **60,** 309 (1938).
23. Orr, W. J. C., *Proc. Roy. Soc. London* **A173,** 349 (1939).
24. Keesom, W. H., and Schweers, J., *Physica* **8,** 1007 (1941).
25. Keesom, W. H., and Schweers, J., *Physica* **8,** 1020 (1941).
26. Giauque, W. F., *J. Am. Chem. Soc.* **52,** 4816 (1930).
27. Moelwyn-Hughes, E. A., Physical Chemistry. Cambridge Univ. Press, London, 1940.
28. Armbruster, M. H., *J. Am. Chem. Soc.* **68,** 1342 (1946).
29. Wagman, D. D., Kilpatrick, J. E., Taylor, W. J., Pitzer, K. S., and Rossini, F. D. J. Research Bureau of Standards, **34,** 143, (1945).
30. Giauque, W. F., and Stout, J. W., *J. Am. Chem. Soc.* **58,** 1144 (1936).
31. Taylor, H. S., and Sickman, D. V., *J. Am. Chem. Soc.* **54,** 602 (1932).
32. Gordon, A. R., *J. Chem. Phys.* **2,** 65 (1934).
33. Emmett, P. H., and Brunauer, S., *J. Am. Chem. Soc.* **56,** 35 (1934).
34. Davis, R. T., *J. Am. Chem. Soc.* **68,** 1395 (1946).
35. Johnston, H. L., and Davis, C. O., *J. Am. Chem. Soc.* **56,** 271 (1934).

About the Mechanism of Contact Catalysis

GEORGE-MARIA SCHWAB

Department of Inorganic, Physical and Catalytic Chemistry, Institute Nicolaos Canellopoulos, Piraeus, Greece

CONTENTS

I. INTRODUCTION

The mechanism of a chemical reaction involves the questions what sorts of molecules interact, in what order, and which atoms change their position and bonding during this interaction. This aspect of the mechanism of a reaction can be expressed by chemical constitutive formulae of the intermediate stages. The general method of obtaining such knowledge is the investigation of the reaction kinetics, i.e., of the variation of velocity with time and concentration. In its more recent development, however, chemical kinetics is directed to a more subtle purpose: To find out what particular changes of the atomic bonds occur and what reactions take place within the complexes which are formed temporarily between the reacting molecules. Studies of this kind help to elucidate the mechanism of activation. Especially in contact catalysis, insofar as it concerns constitutionally simple reactions, this may be considered as the main purpose of research on kinetics. The necessary knowledge is chiefly obtained by investigating the energy relationships, particularly the temperature dependence of the reactions. Thus the investigation of mechanisms is based on the dependence of the rates of a reaction on the concentrations and on the temperature.

II. Reaction, Flow and Diffusion

Generally, contact catalyses are carried out in *flow systems*, because these arrangements make the best use of the property of catalysts to act upon successive amounts of the unreacted substances. But in this case, the time (concentration) function of the observed overall reaction must be sharply distinguished from the time (concentration) function of the *true chemical changes within the catalyst surface.* In heterogeneous processes, transport phenomena (flow, diffusion, convection), if they are slow enough, may conceal the proper chemical processes. In practically every case a decision is possible by evaluating the temperature coefficient of the overall reactions. The temperature coefficient of transport processes is negligibly small whereas the temperature coefficient of chemical reactions is large and exponential.

Much theoretical work has been done to derive expressions relating concentration and reaction velocity at the surface of a catalyst in a flow tube from the relationships between the rate of flow and the yield. The interrelationship of flow, diffusion and reaction has been investigated especially by Damköhler (1), Förster and Geib (2) and others. Neglecting the effect by which the reaction gases are mixed by a back diffusion along the reaction zone, i.e., assuming a flow rate faster than the diffusion rate, we get the general equation (Damköhler, 1):

$$\frac{L}{w_a} = \int_{c_{ja}}^{c_{je}} \frac{\left(\nu_j - \nu \frac{c_{ja}}{c}\right)(-\,dc_j)}{\left(\nu_j - \frac{\nu c_j}{c}\right)^2 \mathbf{U}} \tag{1}$$

(L is the length of the reaction zone, w_a the flow rate of the entering gases, c_{ja} resp. c_{je} the concentrations of the jth molecule species at entrance resp. exit, ν_j the number of molecules of the jth species in the reaction equation, ν the change of molecule number, **U** the reaction velocity per unit length of catalyst). For a first order reaction without a change of the number of molecules, equation (1) gives

$$\frac{c_e}{c_a} = e^{-kt} \tag{2}$$

(k is the rate constant $\mathbf{U}/c$ and t the contact time L/w_a. This is the simple first order equation. If, on the contrary, a longitudinal diffusion effects a complete mingling of the reaction zones, we have (Förster and Geib, 2):

$$\frac{c_e}{c_a} = \frac{1}{1 + kt} \tag{3}$$

Thus, in principle, the question of mixing by diffusion can be decided.

But it should be borne in mind that this longitudinal mixing, as can be shown by the kinetic theory of gases, plays no important role at reactor lengths of at least several centimeters and at linear flow rates of at least some centimeters per minute. These conditions will certainly be fulfilled in most of the usual laboratory work, using flow rates of many cubic centimeters per minute and tube cross sections not exceeding a few square centimeters.

Another question is whether a concentration gradient may occur perpendicular to the direction of the flow of the gases, especially between such layers of the gas near the surface which are deprived of reactant by the surface reaction and the richer free gas volume. However, according to an estimate of Damköhler (1), diffusion is sufficiently fast in a normal laboratory reactor to minimize such differences: even in an empty space 1 cm. in width no larger concentration differences are to be expected than about $\pm 1\%$.

These considerations show that the actual reaction rates in the catalytic surface may be measured dynamically under properly chosen conditions. A new problem arises when the catalyst grains are *porous* and the catalytic reaction proceeds mainly within these pores. In such a case the retarded diffusion through narrow pores may eventually be inadequate to supply the reactants quickly enough, resulting in a greater concentration in the gas space than at the inner pore surface. It has been calculated (Zeldowitch, (3)) that in this case even the temperature coefficients of the overall reaction may differ considerably from the temperature coefficients of the surface reaction, e.g., the former may be only one-half as large as the latter. However, C. Wagner (4) has shown that these conditions may hardly be realized under laboratory conditions, and that the rate is always determined completely either by the speed of diffusion (zero concentration in the pore interior) or by the reaction (equal concentration inside and out). In fact, Schwab and Zorn (5) observed both cases on nickel skeleton catalysts without the occurence of an intermediate region.

Regardless of such diffusion effects, the flow rate may act as the limiting factor of the reaction velocity. Although this can easily be detected, it has sometimes led to serious errors. Cases of comparable flow and reaction rates have been treated by Schwab and Drikos (6).

A reliable dynamic determination of the kinetics requires the performance of many sets of experiments with systematic variations of flow rates and entrance concentrations. This makes the dynamic method not very well suited for thorough kinetic investigations.*

* It may be mentioned that from dynamic measurements, nevertheless, some first information about the reaction order has been obtained by Schwab and Theophilidis (7), by determining the absolute velocity.

It is preferable to establish conditions under which the concentrations of the reactants vary continuously during every single run. This is the case in *static* arrangements where an enclosed gas volume is subjected to catalytic conversion over a long period of time. Whereas it is easy to conduct a gas stream through a hot catalyst tube under uniform conditions, it is more difficult to keep a catalyst mass at a constant and uniform temperature in a static arrangement. For this reason quasi-dynamic arrangements have been developed in which the gas circulates in a closed cycle that contains the heated catalyst at one point. The circulation is effected by a pump (Dohse and Kälberer, 8) or by thermal density differences (Schwab *et al.*, 9). As long as the circulation velocity is much greater than the reaction velocity, no doubt exists that the true velocities are measured in such a system.

If the catalyst used is in form of a wire or of a foil, or if it can be attached to a wire or to a foil, it can be heated electrically in a static apparatus with the gas at room temperature. In this case again, it is questionable whether the diffusion suffices to maintain a uniform concentration ratio all over the vessel. At high pressures the thermal convection affords complete mixing, but even at lower pressures, no larger concentration gradients will be built up than differences of about 1% during a reaction that lasts a few minutes (see p. 253). The conditions prevailing in a larger vessel at pressures too low for convection have been treated numerically by Schwab and Drikos (10). For this case, up to a pressure of 10 mm. Hg, the diffusion levels the concentration differences effected by the reaction proper as well as by thermo-diffusion (Chapman-Enskog effect).

III. Reaction and Adsorption

The true surface reaction rate having been measured, the problem arises as to which partial process determines this over-all rate. Catalysis may be a series of three consecutive reactions: Adsorption of reactants, chemical change within the surface layer and desorption of the products. Which of these steps is the rate determining one? It is known that there may occur an "activated adsorption" which requires activation energy and therefore time, and that this type of adsorption occurs preferentially on catalytically active adsorbents in the temperature range of normal catalysis. Hence, the over-all velocity of the catalytic reaction may essentially be that of the activated adsorption. This has been proved in certain cases, notably for the transformations of para- and ortho-hydrogen and for exchange reactions between hydrogen and deuterium. In these reactions, involving only hydrogen modifications, the adsorption process is coupled with a dissociation of hydrogen into atoms, and no further chemical change is needed for the formation of equilibrium hydrogen on desorption. However, in other reactions, the question can be raised as to whether the activated adsorbed substrates require an additional amount of activation energy to react in the surface, or whether this surface reaction is immeasurably fast, and, accordingly, its rate determined merely by that of the activated adsorption.

This question cannot be answered by merely kinetic observations. Schwab and Pietsch (11) showed as early as 1928 that both assumptions (1) that the rate is determined by impacts of activated molecules on the surface (which nowadays we call activated adsorption) and (2) that the rate is determined by a surface reaction, lead to velocity equations of an identical form. Not even the absolute velocity gives an answer: Schwab and Drikos (12) showed that identical expressions for the absolute velocity constant result from both assumptions, and Laidler *et al.* (13) obtained the same result by means of the statistical transition state method. Luckily, it is of no essential importance for the understanding of the reaction mechanism to know whether molecules other than those actually reacting are attached to the surface.

This identity of formulation means that in normal cases correct results for the velocity equation will be obtained by assuming that adsorption equilibrium is established and the adsorbed molecules react at a rate proportional to their surface concentration when they possess the critical energy. Then, for the surface concentration σ_j of every reactant the Langmuir adsorption isotherm holds:

$$\sigma_j = \frac{b_j p_j}{1 + \Sigma_j b_j p_j} \tag{4}$$

and for the reaction velocity of a reaction of j reactants (in all practical cases $j \leqq 2$):

$$\frac{-\,dp_j}{dt} = k\Pi_j \frac{b_j p_j}{1 + \Sigma_j b_j p_j} = k \frac{\Pi_j b_j p_j}{(1 + \Sigma_j b_j p_j)^j} \tag{5}$$

Now, it is interesting to note that this last expression does not hold in most reactions with $j > 1$, but is to be replaced by:

$$\frac{-dp_j}{dt} = k \frac{\Pi_j b_j p_j}{1 + \Sigma_j b_j p_j} \tag{6}$$

This means that usually one reactant only needs to be adsorbed in equilibrium with its own gas concentration and with that of other displacing species present, while the others need not be adsorbed, but react on impact with the adsorbed species.

The above formulae lead to several special cases. Let us restrict ourselves to a single reactant, i.e., a decomposition or isomerization. Then we have

$$\frac{-dp}{dt} = k \frac{bp}{1 + bp} \tag{7}$$

the so called "reaction of broken order," which has been observed, e.g., for the decomposition of nitrous oxide on indium oxide or for the ortho-hydrogen transformation on solid oxygen. If the adsorption is very strong ($bp \gg 1$), the equation becomes:

$$\frac{-dp}{dt} = k. \tag{8}$$

This means that such a reaction is independent of the concentration, or of "zero order." Such a case is presented by the dehydrogenation of formic acid vapor on metals. If, on the contrary, $bp \ll 1$, we have

$$\frac{-dp}{dt} = k \cdot b \cdot p \tag{9}$$

a reaction of "first order," e.g., the decomposition of hydrogen iodide vapor on platinum or of methane on carbon.

We see in these formulae that the kinetically measured "velocity constant" has only in one special case, namely, in the case of zero order, the meaning of a real velocity constant. In all the other cases, it contains implicit adsorption coefficients, e.g., in the form kb = const., or (in retarded reactions) kb/b' = const. Only in the case of broken order is it possible to determine k and b separately from kinetic measurements, as e.g., Schwab and Zorn (5) did for the ethylene hydrogenation.

IV. The Adsorption Coefficient

In most cases, however, b has to be estimated by a theoretical calculation, in order that k may be extracted from a measured velocity constant. This is important because only then is it possible to compare k with the kinetic or statistical theory and so to check numerically an assumed reaction mechanism. Accordingly, we shall present here some theoretical approaches to b.

One could think about measuring b directly in a separate adsorption experiment. However, it has been found that the adsorption thus measured has no quantitative relation to the adsorption leading to a catalytic reaction. Not only the amounts adsorbed, but even the shape of the isotherm and the sensitivity to poisoning or surface alterations are totally different. The reason is that the adsorption experiment measures the total surface under favorable conditions (B.E.T. method), whereas the catalysis takes place on a quantitatively entirely different "active" fraction of this surface (see p. 260) (see Schwab and Drikos (12), and Schwab and Theophilidis (7)).

In a first order reaction k' is equal to $k \cdot b$ (see above). The experimental (apparent) activation energy (q_s) is given by

$$k' = k_0' \cdot e^{\frac{-q_s}{RT}} = k_0 \cdot e^{\frac{-q}{RT}} \cdot b_0 \cdot e^{\frac{+\lambda}{RT}} = k_0 b_0 \cdot e^{\frac{-(q-\lambda)}{RT}} \tag{10}$$

q being the true activation energy of adsorbed molecules. This yields

$$k_0' = k_0 \cdot b_0 \tag{11}$$

Hence, the temperature coefficient of k' having been measured, for an absolute calculation of k' only k_0 and b_0 must be known, and not the heat of adsorption, λ. At the moment we are concerned with b_0. A simple *statistical estimate* can be based on the assumption that in the absence of adsorption energy the adsorption space is filled at a proportion given by the ratio of the molecular adsorption volume (liquid volume V_m) to the molecular gas volume

$$\sigma_0 = b_0 p = \frac{V_m}{V_{\text{gas}}} = \frac{V_m p}{RT} \tag{12}$$

and hence:

$$b_0 = \frac{V_m}{RT} \tag{13}$$

The evaluation in pressure magnitudes (b is of the dimension of a reciprocal pressure) gives values for b_0 of the order of 10^{-7} (mm. Hg.)$^{-1}$, in agreement with the experience accrued from the studies of catalyzed reactions.

A *kinetic estimate* is also possible because at equilibrium and in absence of adsorption forces the rate of condensation per unit surface $p/(2\pi MRT)^{1/2}$ is equal to the rate of desorption $b_0 p \frac{\delta}{V_m} \cdot \nu$ (where δ is the adsorption volume per unit surface, i.e., the thickness of the adsorption layer and ν is the vibration frequency perpendicularly to the surface). This gives

$$b_0 = \frac{V_m}{\nu\delta(2\pi MRT)^{1/2}} \tag{14}$$

In its order of magnitude, this value of b_0 coincides with the statistical result. (The difference between the two derivations is that in the second one a translatory degree of freedom is replaced by a vibrational degree of freedom for the adsorbed state.)

Thirdly, the rates of adsorption and desorption could be calculated by means of the *transition state method* and be equalized. But, b_0

being an equilibrium constant, the special assumptions about the transition state vanish, and the method is bound to give results identical with the above statistical method.

These formulae have proved to be useful. Using them, Schwab and Zorn (5) were able to show that on certain Raney nickel catalysts the retardation of the ethylene hydrogenation by hydrogen and by ethylene are adsorption displacements, in contrast to the conditions for other catalysts for which a retardation by diffusion has been proved. No agreement exists in this latter case between observed and calculated "b". Using a theoretical value for b_0, Cremer (14) could estimate the heat of adsorption of ortho-hydrogen on solid oxygen. In cases of anomalous promoter effects (increased activation energy) abnormally high b_0 values are found, probably due to loose bonding and increased freedom of the adsorbate (15).

V. The Velocity Coefficient

Like b_0, the temperature independent factor of the velocity coefficient, k_0, can be calculated theoretically. For a *kinetic estimate*, we assume that k_0 equals the number of adsorption places multiplied by the probability that an adsorbed molecule reacts when in possession of the activation energy. This probability, according to usual assumptions, is a vibration frequency ν of the order of 10^{13} sec^{-1}. Thus, we arrive at:

$$k_0 = \frac{F\delta\nu}{V_m}, \tag{15}$$

F being the active catalyst surface.

This expression, giving moles per second, can be transformed into conventional units, either into pressure change per minute by multiplication with the factor $60RT_0/V$ (T_0 being room temperature and V the apparatus volume), or into the flow rate of product gas in cubic centimeters per minute by multiplication with the factor $60RT_0/p$. In the second case an order of 10^{11} per 10 cm.^{2} of catalyst surface is found. This ought to be the temperature independent factor for *zero order* (assuming that all adsorption sites are occupied). Topley (16) and Burk (17) confirmed this for the ammonia decomposition on tungsten and molybdenum and for the decomposition of hydrogen iodide on gold. In these examples the metallic surfaces of the catalysts can be considered to be equal to their geometric (smooth) surfaces. For nonuniform surfaces, we have to deal with an unknown factor F, which is increased by molecular roughness and decreased by limitation of catalytic action to active centers. Anyhow, Schwab and Theophilidis (7) showed that our expression (15) gives an upper limit for the following groups of reactions:

Dehydrogenation of formic acid on metals and alloys, decomposition of 2-propanol formiate on glass, and of ethanol on kaolin. Schwab and Karatzas (17a) found the same expression to be an upper limit for the decomposition of ethyl chloride on salt surfaces. The reasons for deviations from (15) will be given later.

For a *first order* reaction, we have:

$$k_0' = k_0 b_0 = \frac{F\delta\nu}{V_m} \cdot \frac{V_m}{RT} = \frac{F\delta\nu}{RT} \tag{16}$$

According to the authors cited above, equation (16) holds for the decomposition of hydrogen iodide on platinum and of nitrous oxide on gold, and furthermore, according to Schwab and Drikos (12), for the oxidation of carbon monoxide to carbon dioxide on copper oxide.

A comparison of zero and first order velocities is of interest. We have for zero order:

$$_0k_0 = \frac{F\delta\nu}{V_m}; \qquad _0v_0 = {}_0k_0 \tag{17}$$

and for first order:

$$_1k_0 = \frac{F\delta\nu}{RT}; \qquad _1v_0 = {}_1k_0 \cdot p = \frac{F\delta\nu}{V_{\text{gas}}} \tag{18}$$

Hence:

$$\frac{_0v_0}{_1v_0} = \frac{V_{\text{gas}}}{V_m} \sim 1000 \tag{19}$$

for atmospheric pressure, reactants of mean molecular weight, and for a temperature of 300°K. of the measured gas volume or flow. Therefore, at equal (apparent) activation energy and temperature, first order reactions must proceed about a *thousand times slower* than zero order ones (7). This is the reason why, for example, the zero order formic acid dehydrogenation may easily be measured on bulk metal catalysts at 200–300°C. whereas the approximately first order ethanol dehydrogenation requires highly activated porous metals of large specific surface in order to become measurable under the same conditions. The same has been shown for the decomposition of formaldehyde, acetic acid, and hydrazine hydrate. In these cases, the fractional surface coverage is simply 1000 times lower than in the case of a zero order reaction.

Similar estimations have been made (Laidler *et al.*,13) by means of calculations involving the transition states. This method offers the advantage of using more exact data for the available degrees of freedom in the activated state of the reactant. Beyond this, it does not differ essentially from the kinetic method. The vibration in the direction that leads from the former to the new atomic arrangement is replaced by a translation, and the frequency ν by a transition probability kT/h, giving the reciprocal time of motion over an atomic distance, again being of the order of 10^{13} sec^{-1}. In fact, the cited authors and many others could calculate a number of first and zero order reactions satisfactorily by this theory. The author wishes to express the opinion that the greater accuracy of the transition state method at present cannot yet be exploited in heterogeneous catalysis, as long as the really active surface is the main source of uncertainty.

VI. The Theta-Rule

It has been mentioned above that with nonuniform catalyst surfaces, as they mostly occur in practice, the above equations give merely an upper limit of the velocity although in such catalysts the true surface would be expected to be greater than the geometrical one. In nearly all the cases where a reaction has been measured on different catalysts, it has been found that k_0 is by no means a universal constant as would have to be expected from the simple theory. There is a regular connection between k_0 and the activation energy, of the general form:

$$k_0 = a \cdot e^{q/h} \tag{20}$$

Adadurow, Balandin, Constable, Cremer, Eckell, Schwab and other workers have found this rule to apply for very different reactions and catalysts, even for bulk metals. It means that by a certain decrease of the activation energy (by means of a suitable catalyst preparation) the reaction velocity is not, in general, increased by the expected amount because of the simultaneous decrease of k_0. With catalysts of highest activation energy, k_0 exceeds the theoretical limit just as much as may be explained by the surface roughness, and with catalysts of low activation energies, it is lower by several orders of magnitude. It is true that similar relations have been found and derived theoretically even for homogeneous reactions (Storch, 18; Christiansen, 19; Fairclough and Hinshelwood, 20) but in heterogeneous reactions a special reason seems to cause an especially wide variation of k_0. For various reasons, it appears improbable that the entire surface of a catalyst is active. Rather, the activity seems to be on "active centers" (Taylor). These centers will be of different energy contents. For statistical and thermodynamic reasons, they will be the more numerous, the lower their energy. At the same time, they will be the less efficient, the higher their q values, i.e., the lower their energy. Thus, it can be understood qualitatively that high activation energies are coordinated with high k_0 values. Constable (21) was the first to express this idea and to give it the above mathematical formulation in agreement with his experimental findings. He postulated a random distribution of the active centers. According to Schwab and Cremer (22), a thermodynamic distribution may be imagined in certain cases, this distribution being established at the temperature of formation of the catalyst, and being "frozen" by a sudden cooling or stopping this formation. In these cases, we would have

$$h = R\theta \tag{21}$$

with θ being the temperature of preparation of the catalyst. In fact, in numerous cases (not in all of them, of course), it is possible to calculate the correct preparation temperature *a posteriori* from the empirical $q - k_0$ relation. This is in full analogy to the phenomena of conductivity of semiconductors. There, Busch (23) found the same relation between the temperature independent factor, the energy of electron excitation and the temperature of preparation. He traces it to a thermodynamic distribution of perturbation terms. It may be hoped that a further refinement of these considerations and experiments will throw some light on the difficult problems of absolute catalytic activity.

VII. The Activation Energy

A study of the relationship between the rates of catalyzed reactions and the concentrations of reactants may clear up the intermolecular mechanism and such studies may help in finding out which molecules are adsorbed, whether the adsorption is weak or strong, and which fraction of the surface is active. Every theory of the intermolecular mechanism can and must be checked by a calculation of the absolute velocity.

We shall now go a step farther and consider the intramolecular mechanism by asking what changes are effected within the molecules during the reaction as a consequence of their adsorption by the catalyst. The study of temperature coefficients is useful for this purpose. This yields the energy of activation, q, according to

$$k = k_0 \cdot e^{-q/RT} \tag{22}$$

or alternatively

$$R \frac{d \ln k}{d\left(\frac{1}{T}\right)} = q \tag{23}$$

It might be asked whether q is a *heat* of activation (analogous to a thermodynamic Q-value) or a *total* energy of activation (analogous to a U-value), and likewise, whether it is the energy content of the activated molecules or only their energy excess as compared with the average molecules. All these questions are insignificant at the present state of heterogeneous catalysis, as the differences between these various values are of the order of RT ($\sim$ 1 kcal./mole) a quantity which is not greater than the usual limits of experimental errors. More importance is to be ascribed to the differences between true and apparent activation energy, mentioned above. In a zero order reaction, the measured temperature coefficient gives q directly; in a first order it is equivalent to $q - \lambda = q_s$, and in a retarded reaction $q_s = q - \lambda + \lambda_h$, the latter term being the heat of adsorption of the retarding substance. Where it is possible to measure k and b individually, these terms may be separated. An important example is that of hydrogenation, where q_s usually is small, zero or even negative, whereas the true activation energy q is of the order of 20 kcal./mole (Maxted, 24; Schwab *et al.*, 25; Herbo, 26).

To interpret the true activation energies in terms of the intramolecular mechanism, we have to ask whether there are any theoretical foundations on which to build such an interpretation. Although the theoretical understanding of activation energies has made considerable progress, their quantitative deduction, for heterogeneous reactions, is still to be achieved. Some time ago, it was hoped that activation energies were simply connected with known molecular properties of the reactants while the temperature independent factor was considered to be more or less fortuitous. Today, however, the situation is rather the inverse. For the latter term equations exist and may be quantitatively verified, but the activation energies themselves can only very roughly be estimated and still have to be determined experimentally for every system.

Polanyi (27) as early as 1921 suggested that the activated state consists of free atoms which are adsorbed or bound by large affinities (heat of adsorption or diminution of the homogeneous activation energy) to the catalyst. A more quantitative treatment became possible after the theory of wave mechanics presented equations for the mutual interaction of different covalent bonds. London (28) showed that the total interaction energy of three atoms X, Y, and Z is

$$W = A + B + C \pm \sqrt{\alpha^2 + \beta^2 + \gamma^2 - \alpha\beta - \beta\gamma - \alpha\gamma} \qquad (24)$$

A, B and C being the Coulomb energy functions, α, β, and γ the resonance energy functions over the distances XY, YZ and XZ. Hence, if these functions are known (a usual approximation is to simply subtract 10% of the total energies for the Coulomb interaction), the saddle energy may be calculated for certain distances as being the activation energy of the reaction

$$X + YZ \rightarrow XY + Z$$

The result is that for minimum activation energy, the bond YZ must be stretched and weakened before the final bond XY is established. Quantitative calculations show that the activation energy evaluated in this manner amounts, at the most, to 7% of the bond energy YZ. In accordance with this theory, recombinations of atoms of hydrogen, nitrogen, oxygen, chlorine on metal surfaces

$$H + MeH \rightarrow H_2 + Me$$

have negligible activation energies. Semiempirical potential functions, taken from spectroscopic data (Morse functions) may be introduced for α, β, and γ, and so, by calculating the total energy for all sets of distances, the saddle point in the surface of potential energies to be surmounted in the reaction may be found (Eyring and Polanyi, 29). In the same way Eyring (30) was the first to treat exchange reactions between two diatomic molecules, again with the result that only a part of the energy required to split the bonds of the reactants has to be spent, because of a previous bond stretching. Hirschfelder (31) gave useful approximations for this kind of calculations as far as our present state of knowledge permits. For the reaction $X + YZ \rightarrow XY + Z$, q amounts to about 5.5% of the bond energy YZ, and for the reaction $AB + CD \rightarrow AC + BD$ to about 28% of the sum of the bond energies of AB and of CD.

This way of theoretical approach can also be applied to heterogeneous reactions. This has been done by Ekstein and Polanyi (32) and Sherman and Eyring (33). They start from the assumption that the surface atoms of a solid catalyst can be treated as free atoms that interact with atoms of the reactants. The bonds within the reactant molecule are either split (H_2) or at least stretched and loosened (C-C). The degree of this weakening of bonds within molecules of the reactants will be significant for the activation energy which is still needed to bring an activated adsorbed molecule into chemical reaction. Rideal (34) pointed out that this picture offers a good explanation for the frequently observed relative smallness of q as compared with λ. The degree of weakening of bonds by adsorption will depend first, on the bond between catalyst and reactant, and second, on the distance of adjacent atoms of the catalyst. It may be for this reason that certain hydrogenations show a distinct velocity maximum at a certain lattice distance of the catalyst metal (Beeck, 35). Quantitative calculations along these lines can be carried out at present merely for very simple cases, because the respective Morse functions are known for only a very few types of molecules. However, the relative orders of magnitude of activation energies of related reactions can be interpreted (see also Balandin, 36) and their quantitative estimation may be assisted by introducing empirical factors of the Hirschfelder type.

VIII. The Influence of the Substrate

According to the above view, activation energies depend both on the reactant and on the catalyst. For one given catalytic reaction, the influence of the following three properties of the reactant has been studied: its zero point energy, its homogeneous activation energy and its constitution. The influence of the zero point energy may be evaluated separately from reactions of isotope molecules. The heavy isotope possesses a smaller zero point energy, and this difference is only in part compensated by the smaller zero point energy of the heavy activated molecule, so that the heavy isotope requires a higher activation energy. This has been confirmed in the decomposition of ammonia on tungsten (Jungers and Taylor, 37). The relationship of the "heterogeneous" activation energies with the homogeneous activation energies has been expressed by the gross thumb rule that the heterogeneous activation energy is about half as large as the homogeneous activation energy. Obviously, this rule can at best be a crude approximation because it does not consider the nature of the catalyst, nor an eventual change of mechanism. The constitution of the reactant may have a systematic influence. Thus, in the dehydration and dehydrogenation of different alcohols, the activation energy is lowered by constant amounts with the substitution of methyl groups in α-, β- or γ-position. (Dohse, 38; Bork, 39.)

IX. The Influence of the Catalyst

In a one-substance catalyst, its chemical nature is naturally of importance. This, and not the k_0-factor is the reason why there are good and bad catalysts. An attempt has often been made to reduce the dependence of the activation energy on the chemical nature of the catalyst to certain single parameters of the catalyst material, especially to its lattice parameters. As a rule, this has not led very far. Obviously, the chemical nature of the catalyst, with its influence on the binding forces exerted on the reactants, is another decisive factor. For example, it has been shown that the activation energy of the decomposition of nitrous oxide on oxides of the alkaline earths decreases in the order of increasing stability of the (unstable under actual reaction conditions) peroxide. In several cases, the activation energy for a certain reaction is in fact a material constant. In other cases, however, it is probably influenced by the physical state of the catalyst such as by phase transitions, by magnetic and electrical transitions, by mechanical stresses, surface activation, lattice disturbances, etc. All these factors may alter the activation energy, but need not necessarily do so.

X. Mixed Catalysts

Generally, variations of the chemical nature of the catalyst have not, so far, yielded as much information as have the systematic variations of the compositions of multicomponent catalysts. Most mixed catalysts in industry are *heterogeneous* mixtures. It would go beyond the scope of the present article to review them. But it is noticeable that, in general, two kinds of promoter action can be distinguished: (1) Structural promotion in which the activation energy of the active constituent remains unchanged, and the influence of the added substances emerges as a change of k_0 by prevention of a surface sintering of the active component (the iron contact of ammonia synthesis), and (2) synergetic promotion. In the latter type, new active centers are formed in the phase boundary of the components, which have a lower activation energy than each of the single components. Here, obviously, a specific interaction of the electronic structure of the components occurs. As for this interaction, the phenomena in homogeneous mixed catalysts may serve as a model, although they are of but little importance in industry.

Of the homogeneous multicomponent catalysts, certain alloy catalysts will be treated more thoroughly because their study has led to results of general interest. Schwab and coworkers (40 and later papers) studied the catalytic effect of Hume-Rothery alloys on the dehydrogenation of formic acid vapor. These are alloys of copper, silver, or gold with elements of the B-groups of the second to fifth column of the Periodic Table. If equal atomic amounts of these elements are dissolved, e.g., in the silver lattice, the activation energy of the formic acid dehydrogenation increases by an increment proportional to the square of the valency surplus of the solute element. This is exactly the law followed by the electric resistivity. It means that the activation energy increases with increasing electron concentration. At higher concentrations of the solute, these systems form a series of intermetallic phases, each of them being stable at a certain electron concentration, independent of the individual metals: The phases α, β, γ, ϵ, η. The activation energy on these phases invariably shows a steep maximum at the phase γ, and decreases in the phase ϵ and η, again parallel to the electric resistivity. This variation of the activation energy and of the resistance is explained by application of the wave-mechanical theory of the Hume-Rothery-alloys. This theory states that the electron concentration of a given phase can increase only up to a certain limit of stability. At this limit the wavelength coordinated to the fastest electrons is small enough to involve Bragg reflections at the lattice planes, and no further electrons can move freely in the conductivity band, or first Brillouin zone. It is remarkable that in this state not all the levels of the zone are occupied, but some levels, corresponding to electrons with anisotropic velocity distribution, are left free. In the γ-phase, because of its peculiar lattice geometry, the fraction of the Brillouin zone left free for the reason mentioned is greater than in the phases α, β, ϵ, and probably η. The important consequence is that a high activation energy indicates a high degree of completion of the Brillouin zone. This result is in direct connection with the picture given above for the mechanism of activation: in order that a covalent bond be formed between the metal catalyst and reactant atoms, electrons must be shared between them. This is clearest in the case of hydrogen where activated adsorption consists of the binding of a proton at an interlattice place, while the electrons are dissolved in the electron gas of the metal. It is plausible that the formation of this state (which would be the intermediate state of dehydrogenation) requires the larger an activation energy, the fewer and the higher are the levels which are available for electrons in the Brillouin zone of the metal.

XI. Outlook

These alloy catalysts merely represent a special example. It shows, however, that the systematic variation of catalysts of a fixed chemical nature and of a known physical structure may throw some new light on the nature of the activation process. It may be hoped that in catalysts other than metals similar variations may yield similar information, and that such knowledge may be applied in the future to phase boundary promotion and, quite generally, to the understanding of the individual values of activation energies.

Accordingly, the catalytic activity in a given catalytic reaction depends on only four factors. Two of them are specific for the system as a whole: the activation energy and the reaction order. The latter may be reduced to the heat of adsorption, as b_0 is a nearly universal constant. The other two factors are, at least in first approximation, properties of the catalyst: its surface area F and its energy distribution function. Future work will have to answer the question of which parameters control, qualitatively and quantitatively, these four factors.

References

1. Damköhler, G., Der Chemie-Ingenieur. Edited by A. Eucken and M. Jacob, Leipzig 1937, Vol. 3, part 1, p. 359 *seq.*
2. Förster, Th., and Geib, K. H., *Ann. Physik.* **20,** 250 (1934).
3. Zeldowitch, I., *Acta Physicochim. URSS* **10,** 583 (1939).
4. Wagner, C., *Z. physik. Chem.* **A193,** 1 (1943).
5. Schwab, G.-M., and Zorn, H. *Z. physik. Chem.* **B32,** 169 (1936).
6. Schwab, G.-M., and Drikos, G. *Z. physik. Chem.* **A185,** 405 (1940); **A186,** 348 (1940).
7. Schwab, G.-M., and Theophilidis, N., *J. Phys. Chem.* **50,** 427 (1946).
8. Dohse, H., and Kälberer, H., *Z. physik. Chem.* **B5,** 131 (1929).
9. Schwab, G.-M., Staeger, R., and Baumbach, H. v., *Z. physik. Chem.* **B21,** 65 (1933).
10. Schwab, G.-M., and Drikos, G., *Z. Elektrochem.* **50,** 97 (1944).
11. Schwab, G.-M., and Pietsch, E., *Z. physik. Chem.* **B1,** 385 (1928).
12. Schwab, G.-M., and Drikos, G., *Z. physik. Chem.* **B52,** 234 (1942).
13. Laidler, K. J., Glasstone, S., and Eyring, H., *J. Chem. Phys.* **8,** 659, 667 (1940).
14. Cremer, E., *Z. physik. Chem.* **B28,** 383 (1935).

15. Schwab, G.-M., and Schultes, H., *Z. physik. Chem.* **B25**, 411 (1934). Schwab, G.-M., and Staeger, R., *Z. physik. Chem.* **B25**, 418 (1934). Schwab, G.-M., and Nakamura, H., *Ber.* **71**, 1755 (1938).
16. Topley, B., *Nature* **128**, 115 (1931).
17. Burk, R., *J. Am. Chem. Soc.* **56**, 1279 (1934).
17a. Schwab, G.-M., and Karatzas, A., *J. Phys. Coll. Chem.* **52**, 1053 (1948).
18. Storch, H. H., *J. Am. Chem. Soc.* **57**, 1395 (1935).
19. Christiansen, J. A., *Z. physik. Chem.* **B37**, 374 (1937).
20. Fairclough, R. A., and Hinshelwood, C. N., *J. Chem. Soc.* **1937**, 538.
21. Constable, F. H., *Proc. Roy. Soc. London* **A108**, 355 (1925).
22. Schwab, G.-M., and Cremer, E., *Z. physik. Chem.* **A144**, 243; **B5**, 406 (1929).
23. Busch, G., *Helv. Phys. Acta* **19**, 167, 189, 462 (1946).
24. Maxted, E. B., and Moon, C. H., *J. Chem. Soc.* **1935**, 1190.
25. Schwab, G.-M., *et al.*, *Z. physik. Chem.* **A171**, 421 (1934); **B32**, 169 (1936); **B41**, 189 (1938).
26. Herbo, Cl., *Bull. soc. chim. belg.* **50**, 257 (1941); **51**, 44 (1942).
27. Polanyi, M., *Z. Elektrochem.* **27**, 143 (1921).
28. London, F., *Z. Elektrochem.* **35**, 552 (1929).
29. Eyring, H., and Polanyi, M., *Z. physik. Chem.* **B12**, 279 (1931).
30. Eyring, H., *J. Am. Chem. Soc.* **53**, 2537 (1931).
31. Hirschfelder, J. O., *J. Chem. Phys.* **9**, 645 (1941).
32. Ekstein, H., and Polanyi, M., *Z. physik. Chem.* **B15**, 334 (1932).
33. Sherman, A., and Eyring, H., *J. Am. Chem. Soc.* **54**, 2661 (1932); **54**, 3191 (1932).
34. Rideal, E. K., *Proc. Cambridge Phil. Soc.* **35**, 130 (1939).
35. Beeck, O., *Rev. Mod. Phys.* **17**, 1 (1945).
36. Balandin, A., *Z. physik. Chem.* **B3**, 167 (1929).
37. Jungers, J. C., and Taylor, H. S., *J. Am. Chem. Soc.* **57**, 679 (1935).
38. Dohse, H., *Z. physik. Chem.*, Bodenstein volume 533 (1931).
39. Bork, A., *Acta Physicochem. URSS.* **12**, 899 (1940).
40. Schwab, G.-M., *Trans. Faraday Soc.* **42**, 689 (1946).

Author Index

Numbers in parentheses are reference numbers. They are included to assist in locating references in which the authors' names are not mentioned in the text. Numbers in italics refer to the page on which the reference is listed at the end of each article.

Example: Adams, C. E., 64 (ref. 78), *80*, indicates that this author's article is reference 78 on page 64 and is listed in the bibliography on page 80.

Subject Index

B

C

T